22nd International Conference on Atomic Physics 2010

(ICAP 2010)

Journal of Physics: Conference Series Volume 264

Cairns, Australia
25 – 30 July 2010

Editors:

Hans Bachor
Peter Drummond
Peter Hannaford

ISBN: 978-1-61782-299-5
ISSN: 1742-6588

Printed from e-media with permission by:

Curran Associates, Inc.
57 Morehouse Lane
Red Hook, NY 12571

Some format issues inherent in the e-media version may also appear in this print version.

Printed by Curran Associates, Inc. (2011)

Institute of Physics
Dirac House, Temple Back
Bristol BS1 6BE UK

Phone: 44 1 17 929 7481
Fax: 44 1 17 920 0979

techtracking@iop.org

Additional copies of this publication are available from:

Curran Associates, Inc.
57 Morehouse Lane
Red Hook, NY 12571 USA
Phone: 845-758-0400
Fax: 845-758-2634
Email: curran@proceedings.com
Web: www.proceedings.com

22nd International Conference on Atomic Physics 2010

(ICAP 2010)

Journal of Physics: Conference Series Volume 264

Cairns, Australia
25 - 30 July 2010

ISBN: 978-1-61782-299-5
ISSN: 1742-6588

IOP Institute of Physics

TABLE OF CONTENTS

PREFACE

The 22nd International Conference on Atomic Physics (ICAP 2010) was held from 25 to 30 July, 2010 in Cairns, Tropical North Queensland, Australia. This conference followed on from the series of highly successful biennial ICAP conferences held in Storrs, Innsbruck, Rio, Cambridge MA, Florence, Windsor, Amsterdam, Boulder, Munich, Ann Arbor, Paris, Tokyo, Seattle, Göteborg, Cambridge MA, Riga, Berkeley, Heidelberg, Boulder, Oxford and New York. ICAP 2010 was attended by 630 participants from 37 countries.

The conference presented an outstanding program of papers covering the most recent advances in atomic physics, including atomic tests of fundamental physics and basic symmetries; precision measurements, including atomic clocks, atom interferometers and fundamental constants; ultracold gases and Bose–Einstein condensates; ultracold Fermi gases; ultracold molecules; quantum simulators with atoms and ions; few-body systems; ultrafast phenomena and free electron lasers; quantum information with atoms and ions; quantum optics and cavity QED with atoms; and hybrid and optomechanical systems. The papers in this Proceedings represent a collection of the invited talks.

The conference program consisted of 48 invited talks presented in plenary sessions, including 10 'hot topic' talks highlighting the most recent advances in the field, and about 490 poster papers presented in three afternoon sessions. The program included talks by Nobel Laureates Claude Cohen-Tannoudji, Wolfgang Ketterle and Bill Phillips, a memorium talk commemorating the scientific life of Vladilen Letokhov, and an evening lecture by Alain Aspect on 'Wave particle duality for a single photon: quantum weirdness brought to light'. The conference was preceded by a two-day workshop in Cairns on Variation of Fundamental Constants and Violation of Fundamental Symmetries P, T(EDM), CPT, Lorentz Invariance, organised by the University of New South Wales; and three-day Student Workshop at Cape Tribulation, organized by the Australian Research Council Centre of Excellence for Quantum-Atom Optics (ACQAO). A website with full details of the conference program, abstracts and other information can be found at: http://www.swin.edu.au/icap2010.

We would like to thank the participants, especially those who contributed talks, posters and manuscripts, for making ICAP2010 such an exciting and memorable conference. We thank the Program Committee for putting together an outstanding program and the ICAP International Advisory Committee for their expert advice and suggestions. We gratefully acknowledge the financial support of our sponsors: the Australian National University, the Australian Research Council Centre of Excellence for Quantum-Atom Optics, Griffith University, the Ian Potter Foundation, the International Union of Pure and Applied Physics, the National Institute of Standards and Technology, Swinburne University of

Technology, and contributors to the trade exhibition: Coherent, Coherent Scientific, the Institute of Physics Publishing, Lastek, NewSpec, Nufern, Oxford University Press, Spectra-Physics, Springer, Toptica Photonics and Warsash Scientific. Finally, we thank our Conference Secretariat, Maria Lamari, and the Local Organising Committee for their tireless and expert efforts in the organisation of ICAP2010, and the staff of the Cairns Convention Centre, whose friendly and efficient service contributed much to the success of the conference.

The next ICAP conference is planned to be held in Palaiseau, France from 23 to 27 July 2012 (http://www.ifraf.org/icap2012).

Hans Bachor
Peter Drummond
Peter Hannaford
Editors

INVITED SPEAKERS

Alain Aspect	Palaiseau
Markus Aspelmeyer	Vienna
Vanderlei Bagnato	São Paulo
Victor Balykin	Moscow
Kristian Baumann	Zürich
Jim Bergquist	NIST, Boulder
Frédéric Chevy	ENS, Paris
John Close	Canberra
Claude Cohen-Tannoudji	ENS, Paris
Jean Dalibard	ENS, Paris
Eugene Demler	Harvard
Michael Doser	CERN
Markus Drescher	Hamburg
Francesca Ferlaino	Innsbruck
Victor Flambaum	Sydney
Chiara Fort	Florence
Elisabeth Giacobino	ENS, Paris
Philippe Grangier	Palaiseau
Chris Greene	JILA, Boulder
Markus Greiner	Harvard
Eric Hessels	Toronto
Hidetoshi Katori	Tokyo
Wolfgang Ketterle	MIT
Michael Kohl	Cambridge
Wu-Ming Liu	Beijing
Francesco Minardi	Florence
Holger Müller	Berkeley
Karim Murr	Garching
Hanns-Christoph Nägerl	Innsbruck
Jeremy O'Brien	Bristol
Silke Ospelkaus	JILA, Boulder
Krzysztof Pachucki	Warsaw
Bill Phillips	Gaithersburg

Randolf Pohl	Garching
Eugene Polzik	Copenhagen
Cindy Regal	JILA, Boulder
Jakob Reichel	ENS, Paris
Helmut Ritsch	Innsbruck
Christian Roos	Innsbruck
Mark Saffman	Wisconsin
Christophe Salomon	ENS, Paris
Gora Shlyapnikov	Orsay
Richard Taieb	Paris
Masahito Ueda	Tokyo
Chris Vale	Melbourne
Andreas Wallraff	Zürich
Matthias Weidemüller	Heidelberg
Martin Weitz	Bonn
Artur Widera	Bonn
David Wineland	NIST, Boulder

IN MEMORIAM

On 21 March 2009 Professor Vladilen Letokhov passed away in Troitsk near Moscow. Letokhov was an outstanding scientist in laser physics and laser spectroscopy. He was born on 10 November 1939 in the small Siberian town of Taishet, not far from Lake Baikal. After graduating from the Moscow Institute of Physics and Technology (MIPT) in 1963, he attended the Physical Institute of the USSR Academy of Sciences. He did his postgraduate studies under the supervision of Nobel laureate Nicolay Basov. In 1969 he defended his PhD thesis on the theory of laser pulse generation and amplification, and a year later he received a second doctor of science degree in quantum radiophysics. In 1970 Vladilen went to the new Institute of Spectroscopy at the USSR Academy of Sciences in Troitsk. He became the deputy director for research and organized the department of laser spectroscopy, which he headed until his last days. Letokhov was also a faculty member at the MIPT, where he served as a professor of physics from 1972 to his death and as head of the chair of quantum optics from 1986 to 1998.

Letokhov's scientific interests included various areas of laser physics, spectroscopy, chemistry, and biomedicine. His most important contributions, however, were in the field of laser spectroscopy. He was the first to realize selective detection of atoms and molecules by multiphoton resonant ionization, which made it possible to develop methods of ultrasensitive analysis. To him belonged the discovery of nonresonance feedback in random lasers. Letokhov was also among the first to achieve laser spectroscopy with sub-wavelength spatial resolution. With his collaborators, he suggested and developed methods of laser control of atomic motion, resulting in the creation of atom traps based on gradient forces. His group carried out the first experiments on cooling, collimation, and reflection of atom beams by laser radiation.

Letokhov made decisive contributions to the development of methods of selective laser chemistry, including isotope-selective multiphoton dissociation of molecules by IR laser radiation and vibrationally mediated photochemistry. He developed several effective schemes of laser isotope separation and the first commercial plant for laser isotope separation was created in 1998. Letokhov and his coworkers performed groundbreaking experiments in laser mass spectroscopy of organic molecules, and they also developed methods of picosecond and femtosecond nonlinear laser spectroscopy for the investigation and control of ultrafast processes in condensed media. In recent years Letokhov was engaged in research on laser effects in stellar atmospheres, which he predicted at the beginning of his career.

The scientific results obtained by Letokhov and his coworkers were widely recognized. For his efforts, he was awarded the 1978 Lenin Prize, the 1998

Quantum Electronics Prize of the European Physical Society, the 2001 Rozhdestvensky Prize of the Russian Academy of Sciences, and the 2002 State Prize of the Russian Federation.

For many years Letokhov was involved in the publishing of international scientific journals. Among the publications he edited were *Laser Science and Technology* and the *Journal of Nonlinear Optics*. He also served on the editorial boards of the *Journal of Experimental and Theoretical Physics*, *Chemical Physics Letters*, *Applied Physics B*, and others. He was an author on more than 850 research papers, including 15 monographs.

Letokhov was a self-made man who, beginning in his school years, persistently used every possibility to broaden his educational and cultural knowledge. Although he was devoted to science and gave it considerable time, he also was deeply interested in literature, music, art, and history. He was an exceptionally interesting conversationalist and a man of great erudition.

He is sadly missed by his many colleagues and friends.

22nd International Conference on Atomic Physics IOP Publishing
Journal of Physics: Conference Series **264** (2011) 012001 doi:10.1088/1742-6596/264/1/012001

From optical pumping to quantum gases

Claude Cohen-Tannoudji

Collège de France and Laboratoire Kastler Brossel
Ecole Normale Supérieure, Paris, France

A full control of internal and external degrees of freedom of atoms and of their interactions can now be achieved in atomic physics experiments [1]. These achievements have been made possible by the development of optical pumping methods first proposed sixty years ago, by the availability of laser sources first realized fifty years ago, and by the invention of new methods, like laser cooling and trapping, evaporative cooling and Feshbach resonances. We will review in this paper a few breakthroughs in the recent evolution of atomic physics, showing the continuity between researches done at different times and emphasizing the new fruitful dialogue which is being established between atomic physics and other disciplines like theoretical physics, condensed matter physics and few body physics. The possibility to achieve ultracold gaseous samples of strongly interacting atoms and to control all experimental parameters allows one to explore new physical situations, to realize simple models of more complex quantum systems found in other fields of physics and to get a better understanding of their behaviour.

References

[1] The content of the paper is treated in more detail in *Advances in Atomic Physics. An Overview*, by Claude Cohen-Tannoudji and David Guéry-Odelin (World Scientific, Singapore), to be published in 2011. The pdf file of the talk can be found on the conference website: http:/www.swin.edu.au/icap2010

Journal of Physics: Conference Series **264** (2011) 012002 doi:10.1088/1742-6596/264/1/012002

Synthetic electric and magnetic fields for ultracold neutral atoms

W. D. Phillips[1], Y.-J. Lin[1], R. L. Compton[1], K. Jiménez-Garcia[1,2], A. R. Perry[1], J. V. Porto[1] and I. B. Spielman[1]

[1]Joint Quantum Institute, National Institute of Standards and Technology, and University of Maryland, Gaithersburg, Maryland, 20899-8424, USA

[2]Departamento de Física, Centro de Investigación y Estudios Avanzados del Instituto Politécnico Nacional, México D.F., 07360, México

Email: **ian.spielman@nist.gov**

Abstract. We use Raman coupling between magnetic sublevels of a ^{87}Rb Bose-Einstein condensate (BEC) to create an effective vector gauge field for the neutral atoms. They behave as if they were charged particles in a magnetic vector potential. With appropriate spatial gradients and time derivatives of this vector potential we produce synthetic magnetic and electric fields for the atoms. This paper is a brief summary of the presentation made by the first author at ICAP 2010 in Cairns, Australia. It is not intended as a complete report on the research described, nor does it attempt to be a fully referenced documentation of that research and the relevant literature. We do, however, provide references to the published work where a more complete description and more complete references to related literature may be found.

1. Introduction

Ultracold neutral atoms and quantum degenerate neutral atomic gases have proved to be useful in simulating quantum behaviour of a number of models, Hamiltonians, and physical systems. One reason for this is the ability to control the Hamiltonian for the atoms in ways that are difficult for other systems; another is the ability to measure certain quantities, like momentum distributions and correlations thereof, which are often not easily measured in other systems. Bose condensation, Fermi degeneracy, Cooper pairing, BCS/BEC (Bardeen-Cooper-Schrieffer/Bose-Einstein Condensation) crossover, and the behaviour of quantum particles in periodic potentials are among the areas where cold atoms exhibit behaviour that is hard to observe in analogous condensed matter systems. One of the difficulties facing the simulation of interesting Hamiltonians with neutral atoms is in simulating the response of charged particles to external electromagnetic fields, particularly to magnetic fields. One approach has been to use rotation, where the Coriolis force mimics the Lorentz force. Because of some difficulties with this approach, such as the introduction of a centrifugal force, the challenges associated with adding lattice potentials, as well as various technical imperfections, other techniques have been proposed involving the simulation of a magnetic vector potential.

2. Effective vector gauge potential for neutral atoms

The idea behind such an approach is that the Hamiltonian for a charged particle in a magnetic vector potential can be written as $H = (\mathbf{p} - q\mathbf{A})^2/2m$, where $\mathbf{p}$ is the particle's canonical momentum (the quantity canonically conjugate to position), m is its mass, q is its charge and $\mathbf{A}$ is the vector potential.

Evidently, the introduction of a vector potential is equivalent to displacing the origin of the particle's energy-momentum dispersion curve from having its minimum at $\mathbf{p} = 0$. A number of theoretical proposals [1-5] have suggested producing such a displacement via 2-photon Raman coupling between different spin states (internal states) and momentum states (external states) within the same electronic ground state manifold. Spatial variation of this Raman coupling can produce a curl of the effective vector potential. Here we outline a specific method [6] for producing an effective vector potential, its experimental realization, as well as the experimental demonstration of synthetic magnetic and electric fields arising from that vector potential. The method is related to an earlier idea [7] involving Raman coupling of spin and momentum states for a different purpose.

In the experimental realization [8] of this method we create an optically trapped Bose-Einstein condensate (BEC) of ^{87}Rb atoms in the $F = 1$ hyperfine level of the electronic ground state. The degeneracy of the magnetic levels is lifted by a magnetic field along the y-direction. We couple the sublevels using a pair of laser beams, counterpropagating along the x-direction, far-detuned from single-photon resonance with the electronically excited state, but with a frequency difference nearly resonant with the energy difference between the magnetic sublevels. The stimulated Raman transition from one m-state to another also transfers two units of photon momentum ηk. In a dressed atom picture where the atom's internal (spin) energy, centre-of-mass kinetic energy, and the photon energy are all considered, the Raman coupling produces dressed energy-eigenstates that are superpositions of different m-states and different centre-of-mass momentum states. For strong enough Raman coupling the lowest of these dressed states has a single minimum as a function of dressed-atomic (canonical) momentum, and with a curvature in $E(p)$ that is smaller than that of the free-space dispersion curve, implying a larger effective mass than the bare atomic mass. When the Raman detuning for the coupling between m-states is zero (Raman resonance), the minimum of $E(p)$ is at $p = 0$. For non-zero Raman detuning that minimum is shifted away from $p = 0$, resulting in an effective Hamiltonian equivalent to that of a charged particle in a magnetic vector potential. When the magnetic field applied along y is spatially uniform, the Raman detuning and hence the shift of the minimum of $E(p)$ are spatially uniform, so the effective vector potential, along x, is uniform, has no curl, and there is no effective magnetic field.

Reference [8] describes how we load a Bose-Einstein condensate of ^{87}Rb into the dressed state at the minimum of $E(p)$ and, using time-of-flight, measure the shift from zero of the minimum of $E(p)$ and hence measure the vector potential. We emphasize that while the atoms at the minimum have non-zero (canonical) momentum, their mechanical momentum is zero, as is their group velocity, given by $\mathrm{d}E(p)/\mathrm{d}p$.

3. Synthetic Magnetic and Electric Fields

To produce synthetic magnetic or electric fields from the effective vector potential arising from the Raman coupling, we must create either a curl of the vector potential (a spatial variation) or a time dependence.

In Ref. [9] we used a spatial dependence of the (real) magnetic field along y to create a y-derivative of the Raman detuning and hence of the effective vector potential. A linear gradient along y of the x-directed vector potential creates a curl equivalent to a synthetic magnetic field along z. When our BEC is subjected to a large enough synthetic magnetic field, we find that upon release (trap and Raman coupling turned off), the BEC is seen to contain many vortices, with the number of vortices increasing with the strength of the effective synthetic magnetic field. We also observe that there is a fairly large threshold value of the synthetic field, below which no vortices are seen in the BEC. This threshold behaviour is consistent with the energetic arguments detailed in [10], and in agreement with our own calculations. The vortices do not appear immediately upon the application of an appropriately large synthetic magnetic field, but over a time of some hundreds of milliseconds nucleate at the outside of the atom cloud and migrate to the interior.

Journal of Physics: Conference Series **264** (2011) 012002 doi:10.1088/1742-6596/264/1/012002

To create a uniform synthetic electric field, we need a time derivative of the uniform effective vector potential. The realization of this synthetic electric field is described in [11]. We produce the time derivative by changing the (real) magnetic field from one chosen value to another within a fraction of a millisecond. The resulting change in Raman detuning changes the vector potential, producing a pulse of synthetic electric field that impulsively accelerates the atoms. We turn off the Raman coupling and the trap and observe the velocity of the atoms after time-of-flight finding good agreement with the expected velocity change. We can also deliver a momentum impulse to the atoms, then allow the trapped, Raman-dressed atoms to oscillate rather than releasing them immediately to time-of-flight. We see that the atoms oscillate in the trap with a frequency determined by the effective mass appropriate to their dressed state.

4. Conclusions

We have demonstrated the creation of an effective vector gauge potential, created by Raman coupling between magnetic sublevels of a ^{87}Rb BEC. An appropriate spatial gradient of this vector potential produces a synthetic magnetic field, leading to a Lorentz-like force on the neutral atoms, as if they were charged particles in a magnetic field. A time variation of the vector potential produces a synthetic electric field that acts on the neutral particles as does a real electric field on charged particles. We hope to use the synthetic magnetic field to study quantum Hall and fractional quantum Hall effects in neutral atom systems.

5. Acknowledgements

This work was supported in part by the National Science Foundation through the Physics Frontier Center at the Joint Quantum Institute, the Office of Naval Research, and the Army Research Office with funds from the DARPA OLE program. R. L. C. acknowledges the support of the NIST/NRC postdoctoral fellowship program, and K.J.-G. thanks CONACYT.

References

[1] Juzeliūnas G, Ruseckas J, Öhberg P and Fleischhauer M 2006 *Phys. Rev. A* **73** 025602
[2] Zhu S-L, Fu H, Wu C-J, Zhang S-C and Duan L-M 2006 *Phys. Rev. Lett.* **97** 240401
[3] Liu X-J, Liu X, Kwek L C and Oh C H 2007 *Phys. Rev. Lett.* **98** 026602
[4] Cheneau M, Rath S P, Yefsah T, Gunter K J, Juzeliunas G and Dalibard J 2008 *Eur. Phys. J.* **83** 60001
[5] Günter K J, Cheneau M, Yefsah T, Rath S P and Dalibard J, *Phys. Rev. A* 2009 **79** 011604
[6] Spielman I B 2009 *Phys. Rev. A* **79** 063613
[7] Higbie J and Stamper-Kurn D M 2002 *Phys. Rev. Lett.* **88** 090401
[8] Lin Y-J, Compton R L, Perry A R, Phillips W D, Porto J V and Spielman I B 2009 *Phys. Rev. Lett.* **102** 130401
[9] Lin Y-J, Compton R L, Jiménez-García K, Porto J V and Spielman I B 2009 *Nature* **462** 628
[10] Lundh E, Pethick C J and Smith H 1997 *Phys. Rev. A* **55**, 2126
[11] Lin Y-J, Compton R L, Jimenez-Garcia K, Phillips W D, Porto J V and Spielman I B 2010 arXiv:1008.4864v1 [cond-mat.quant-gas]

Journal of Physics: Conference Series **264** (2011) 012003 doi:10.1088/1742-6596/264/1/012003

Non-Abelian Josephson effect in spinor Bose-Einstein condensates

Ran Qi, Xiao-Lu Yu and W. M. Liu

Beijing National Laboratory for Condensed Matter Physics, Institute of Physics, Chinese Academy of Sciences, Beijing 100080, China

E-mail: `wmliu@iphy.ac.cn`

Abstract. We investigate the non-Abelian Josephson effect in F=2 spinor Bose-Einstein condensates with double optical traps. We propose a real physical system which incorporates the non-Abelian Josephson effect and has very different density and spin tunneling characteristics compared with the Abelian case. We calculate the frequencies of the pseudo Goldstone modes in different phases between two traps, which are the crucial feature of the non-Abelian Josephson effect. We also give an experimental protocol to observe this novel effect in future experiments.

1. Introduction

The Josephson effect is a quantum tunneling phenomenon which occurs when a pair of superconducting or superfluid systems are weakly linked by some kind of physical barrier. Beginning with Josephson's original paper in 1962 [1], the Josephson effect has become a paradigm of the phase coherence manifestation in a macroscopic quantum system. With the rapid experimental progress in cold atom physics, the Josephson junction has been realized for trapped Bose-Einstein condensates (BEC) of ^{87}Rb [2] and ^{23}Na [3]. However, most of the extensive studies about this effect have focussed on the Abelian case so far, in terms of a junction of two systems with spontaneously broken Abelian symmetry [4, 5]. There are also some kinds of Josephson-type effect without junctions, such as the spin mixing process of a spin-1 condensate in a single well discussed as in Ref. [6].

Recently, F. Esposito *et al.* formulated the Josephson effect in a field theoretic language and provided a straightforward generalization to the non-abelian case [7]. The non-Abelian Josephson effect emerges in a junction of two weakly coupled systems with spontaneously broken non-Abelian symmetries, which often involves multi-component order parameters. The non-Abelian nature of the symmetry will induce more than one kind of tunneling mode in the Josephson effect. These different tunneling modes can be characterized by the excitation of so-called pseudo Goldstone bosons which have small but finite masses [7]. The emergence of pseudo Goldstone bosons is a consequence of the explicit symmetry breaking term due to the coupling between the two condensates. Theoretically speaking, the non-Abelian Josephson effect is ubiquitous in nature, covering many topics from particle physics to condensed matter physics. For example, it may be realized in the superfluid ^{3}He Josephson weak link [8], high density phases of QCD [9] and the artificial non-Abelian gauge field induced by nonlinear optics [10]. In Ref [7], the authors gave some samples in a relativistic quantum field system which involves

$O(N)$ symmetry in an N-component scalar field. However as far as we know, there have been no specific experimental constructions until now.

In this review, we provide an experimental protocol to observe this novel effect in future experiments. To our knowledge, this effect has not been explicitly spelled out in any real physical system. In order to generalize to the non-Abelian junction in experiments, we should consider a system with a multi-component order parameter which has a non-Abelian symmetry in the ground state. In contrast to the situation in a magnetic trap, the spin of alkali atoms is essentially free in an optical trap [12–14]. This spinor nature properly provides the scenario of our non-Abelian symmetry construction. We now briefly introduce the system consisting of a spinor atomic BEC in a double-well optical trap. Although the dynamical tunneling properties of spin-1 and "pseudo spin-1/2" bosonic systems have been calculated [15–19], the essence of the non-Abelian effect has not been captured yet. In the present paper, we focus on the pseudo Goldstone modes due to explicit non-Abelian symmetry breaking, which is at the heart of the Josephson effect. For a concrete construction, we propose a spin-2 BEC in double optical traps. A spin-2 system has possible advantages, compared with a spin-1 system, in the sense that the symmetry properties are much richer to explore the non-Abelian effect. We should note that in a spin-1 system some elements of the symmetry group are hidden in the ground state. For example, in the polar state of a spin-1 system, the symmetry group of the ground state is $U(1) \times S(2)$ which is a subgroup of the symmetry group of the Hamiltonian $U(1) \times SO(3)$ [14]. The reason this happens is that in a low spin system, some rotation in the symmetry group of the Hamiltonian will leave the ground state unchanged and does not contribute to Goldstone modes. In contrast, in the antiferromagnetic and cyclic phase of a spin-2 condensate, the full $U(1) \times SO(3)$ symmetry is preserved in the ground state configuration, except for some specific values of the parameter [20].

2. Ground state structure of a spin-2 BEC

Let us consider a system of a homogeneous spin-2 Bose gas with an s-wave interaction. This system can be described by the following mean field free energy:

$$F(\psi) = \frac{1}{2}[c_0(\psi^\dagger\psi)^2 + c_1\langle \mathbf{f}\rangle^2 + \frac{c_2}{5}|\Theta|^2] - \mu\psi^\dagger\psi, \tag{1}$$

where $\psi = (\psi_2, \psi_1, \psi_0, \psi_{-1}, \psi_{-2})^T$ is the order parameter of the system, c_0, c_1 and c_2 are interaction strengths related to the scattering length in different spin channels, $\langle \mathbf{f}\rangle = \psi^\dagger \mathbf{f} \psi$ is the mean value of the spin operator and $\Theta = \sum_{a=-2}^{2}(-1)^a \psi_a \psi_{-a}$ represents a single pair of identical spin-2 particles. The ground state configuration can be determined by minimizing this free energy. There are several distinct phases in this system [21, 22]. We will discuss these phases under zero magnetic field and analyze the symmetry and low-lying excitation spectrum of each phase.

2.1. Ferromagnetic phases

When $c_1 < 0$ and $c_1 - c_2/20 < 0$, two kinds of ferromagnetic phases are energetically favoured. The corresponding ground state configurations are given by $\psi = \sqrt{n}e^{i\theta}(1, 0, 0, 0, 0)$ or $\psi = \sqrt{n}e^{i\theta}(0, 1, 0, 0, 0)$, where $n = \mu/(c_0 + 4c_1)$ is the particle density and θ is an arbitrary global phase. It is obvious that these ground states have a $U(1)$ symmetry which leads to only one massless Goldstone mode. Therefore, two uncoupled systems have a $U(1) \bigotimes U(1)$ symmetry. This symmetry will softly break into a $U(1)$ diagonal symmetry when a weak coupling is applied. This pattern of symmetry breaking corresponds to an Abelian Josephson effect. The low-lying excitation spectrum of this state has been derived as $\omega_k = \sqrt{\epsilon_k(\epsilon_k + 2g_4 n)}$ [21], where $\epsilon_k = k^2$ and $g_4 = c_0 + 4c_1$. We should note that this Goldstone mode will break into two modes when coupling is applied: one zero energy mode and one pseudo Goldstone mode. This pseudo

Goldstone mode has a finite but small gap and leads to a density mode fluctuation in d.c. Josephson current.

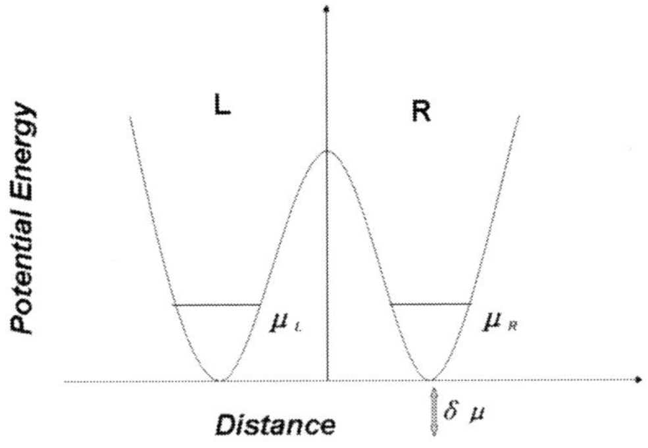

Figure 1. Experimental schematic of a spin-2 Bose gas trapped in a double well with chemical potentials μ_L of left and μ_R of right trap, which initially satisfy $\mu_L = \mu_R$. To drive the Josephson effect, we add a small distortion $\delta\mu$ to μ_R.

2.2. Antiferromagnetic phase

In the absence of a magnetic field, there is only one kind of antiferromagnetic phase when $c_2 < 0$ and $c_1 - c_2/20 > 0$ are satisfied. The corresponding ground state configuration is degenerate with respect to five continuous variables which lead to five massless Goldstone modes [22]. Four of them correspond to the $U(1) \times SO(3)$ symmetry of the free energy. The extra degeneracy besides the $U(1) \times SO(3)$ symmetry is correct only on the mean field level and will be removed when the quantum fluctuation is considered. In this paper, we will work at the mean field level and maintain all five Goldstone modes. The effect of quantum fluctuation will be included in future work.

2.3. Cyclic phase

When $c_1 > 0$ and $c_2 > 0$, the cyclic phase is energetically favoured. The ground state configuration is given by $\psi = \sqrt{n}(\frac{e^{i\theta_2}}{2}, 0, \frac{e^{i\theta_0}}{\sqrt{2}}, 0, \frac{e^{i\theta_{-2}}}{2})$ where $n = \mu/c_0$ is the particle density and the global phases $\theta_{\pm 2}$ and θ_0 satisfy $\theta_2 + \theta_{-2} - 2\theta_0 = \pi$ [22, 23]. Using the Schwinger boson representation [20], one can see that the ground state of the cyclic phase is mapped to a tetrahedron on the unit sphere. Therefore, the ground state of the cyclic phase has a full $U(1) \times SO(3)$ degeneracy and leads to four Goldstone modes. We will show that these modes also lead to the non-Abelian Josephson effect.

3. Non-Abelian Josephson effect and pseudo-Goldstone modes

We will analyze the Josephson effect of a spin-2 BEC system in a double-well optical trap, as shown in Fig. 1. We assume that the energy barrier between the two wells is strong enough so that the coupling between the Bose gas in each well is very weak and the overlap of the ground state wave functions in the left and right well (which we denote by $\varphi_L(\mathbf{x})$ and $\varphi_R(\mathbf{x})$) can be safely neglected. We will also use the single mode approximation which means we take the same mode function for all five spin components; this is a widely used approximation and is valid when the spin interaction is symmetric. Under these assumptions, the system can be described by the following potential

$$V_{couple} = F(\psi_L) + F(\psi_R) - J(\psi_L^\dagger \psi_R + \psi_R^\dagger \psi_L), \quad (2)$$

where ψ_L and ψ_R are the order parameters of the Bose system in the left and right wells, respectively, and J is the coupling parameter. It should be noted that we have taken the same chemical potential for the Bose gas in the right and left well, because we will only be interested in the d.c. Josephson effect which captures the essence of the non-Abelian symmetry breaking as simply as possible. Using the dynamic equations of the condensate $i\frac{d}{dt}\psi_{aL} = \frac{\delta V_{couple}}{\delta \psi^*_{aL}}$ and $i\frac{d}{dt}\psi_{aR} = \frac{\delta V_{couple}}{\delta \psi^*_{aR}}$, we can derive the equation of motion of the Josephson current in each phase and analyze the pseudo Goldstone modes. However, as we have mentioned above, the symmetry of the ground state in the ferromagnetic phase only leads to an Abelian Josephson effect which is not of interest in the present paper. Therefore, we will just analyze the antiferromagnetic phase and cyclic phase which are important realizations of the non-Abelian Josephson effect.

3.1. Antiferromagnetic phase

To obtain the pseudo Goldstone modes, we need to linearize the equation of motion around the ground state $\psi_L = \psi_R = \frac{\sqrt{n}}{\sqrt{2}}(1, 0, 0, 0, 1)$. By minimizing the potential V_{couple} under the above antiferromagnetic ground state, we get $n = (\mu + J)/(c_0 + c_2/5)$. With this ground state configuration, we can obtain the linearized equations of the fluctuations in this phase and analyze the excitation spectrum.

The $m = 0$ mode.

The equation of motion of the $m = 0$ mode is given as

$$i\frac{d}{dt}\phi_{0L} = (-\frac{c_2}{5}n + J)\phi_{0L} + \frac{c_2}{5}n\phi^*_{0L} - J\phi_{0R}, \tag{3}$$

and a similar equation for ϕ_{0R}. Since this mode is decoupled from the others, it corresponds to an Abelian Josephson current. By solving equation (3), we obtain the eigenenergies of this mode, one is zero corresponding to a massless Goldstone mode and the other is $\omega_0 = 2\sqrt{J(J + \frac{2|c_2|n}{5})}$ corresponding to a pseudo Goldstone mode.

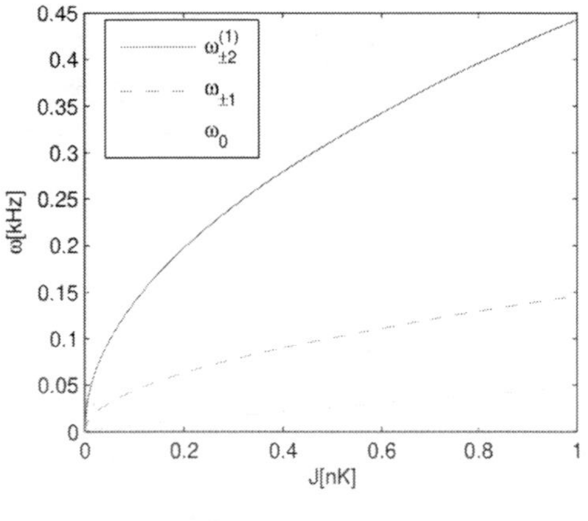

Figure 2. Frequencies of the pseudo Goldstone modes as a function of coupling parameter J for the case of the antiferromagnetic phase. All frequencies are proportional to $\sqrt{J}$ when J approaches zero and the dependence on J becomes linear when J is large compared to the interaction energy.

The $m = \pm 1$ coupled modes.

The fluctuaion of $m = \pm 1$ spin components are coupled in the following equations

$$\begin{aligned} i\frac{d}{dt}\phi_{1L} &= [n(c_1 - \frac{c_2}{5}) + J]\phi_{1L} + n(c_1 - \frac{c_2}{5})\phi^*_{-1L} - J\phi_{1R}, \\ i\frac{d}{dt}\phi_{-1L} &= [n(c_1 - \frac{c_2}{5}) + J]\phi_{-1L} + n(c_1 - \frac{c_2}{5})\phi^*_{1L} - J\phi_{-1R}, \end{aligned} \tag{4}$$

and a similar set of equations for $\phi_{\pm 1R}$. The solution involves two pseudo Goldstone modes with the same gap $\omega_{\pm 1} = 2\sqrt{n(c_1 - \frac{c_2}{5})J + J^2}$.

The $m = \pm 2$ coupled modes.

The equation of motion of $m = \pm 2$ spin components are coupled as follows

$$i\frac{d}{dt}\phi_{2L} = [\frac{n}{2}(c_0 + 4c_1) + J](\phi_{2L} + \phi_{2L}^*) + \frac{n}{2}(c_0 - 4c_1 + \frac{2}{5}c_2)(\phi_{-2L} + \phi_{-2L}^*) - J\phi_{2R}, \quad (5)$$

$$i\frac{d}{dt}\phi_{-2L} = [\frac{n}{2}(c_0 + 4c_1) + J](\phi_{-2L} + \phi_{-2L}^*) + \frac{n}{2}(c_0 - 4c_1 + \frac{2}{5}c_2)(\phi_{2L} + \phi_{2L}^*) - J\phi_{-2R}, \quad (6)$$

and a similar set of equations for $\phi_{\pm 2R}$. By solving the above equations, we obtain two pseudo Goldstone modes with energy gap $\omega_{\pm 2}^{(1)} = 2\sqrt{n(c_0 + \frac{c_2}{5})J + J^2}$ and $\omega_{\pm 2}^{(2)} = 2\sqrt{n(c_1 - \frac{c_2}{20})J + J^2}$. We can see that there are in total five pseudo Goldstone modes with four different gaps in this phase which is consistent with our analysis of the ground state degeneracy. Figure 2 shows the dependence of the above frequencies on coupling parameter J. Recently, a polar behavior has been observed in the F=2 ground state of the ^{87}Rb condensate [24]. We expect that the above modes of fluctuations can be observed in this system in future experiments. In the case of the ^{87}Rb system, the values of the interaction strengths under typical experimental conditions are given as [24]: $c_1n : 0 - 10$ nK, $c_2n : 0 - 0.2$ nK and c_0n about 150 nK. According to the weak coupling limit, we assume that the coupling parameter J is much smaller than the interaction energy of the condensate and given as about 0.1 nK. Under these conditions, we can obtain the frequencies of the fluctuation related to the antiferromagnetic phase, which is of order 100 Hz. The measurement of fluctuations on this characteristic time scale (about 10 ms) is accessible in current experiments.

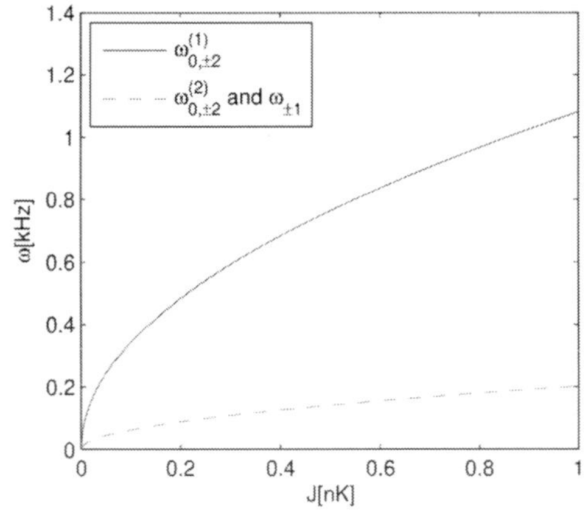

Figure 3. Frequencies of pseudo Goldstone modes as a function of coupling parameter J for the case of the cyclic phase. The dependence on J is similar to the case in the antiferromagnetic phase. However, there are two degenerate modes $\omega_{0,\pm 2}^{(2)}$ and $\omega_{\pm 1}$ in this case, which originate from the two degenerate Bogoliubov modes in the uncoupled system.

3.2. Cyclic phase

Following the same procedure as in the antiferromagnetic phase, the pseudo Goldstone modes for the cyclic phase can be obtained. We will not list the detailed equations of motion but just give the result in this section.

The $m = \pm 2, 0$ coupled modes.

Each Goldstone mode in the corresponding uncoupled system [21] breaks into one massless mode and one pseudo Goldstone mode. Since there are two Goldstone modes in the uncoupled system, they induce two pseudo Goldstone modes with energy $\omega_{0,\pm 2}^{(1)} = 2\sqrt{J^2 + c_0nJ}$ and $\omega_{0,\pm 2}^{(2)} = 2\sqrt{J^2 + 2c_1nJ}$.

The $m = \pm 1$ coupled modes.

In the uncoupled system, there are two massless Goldstone modes with the same energy [21]. By solving the equation of motions for this mode, it is found that each of them leads to a pseudo Goldstone mode with a gap $\omega_{\pm 1} = 2\sqrt{J^2 + 2c_1 nJ}$. There are four pseudo Goldstone modes in the cyclic phase which is consistent with the previous analysis on the symmetry of this phase. These kinds of fluctuations in cyclic phase are expected to be realized in a condensate of ^{85}Rb atoms [22]. Based on current estimates of scattering lengths, the value of interaction strengths under typical experimental condition are given as: $c_1 n : 0 - 20$ nK, $c_2 n : 0 - 0.6$ nK and $c_0 n$ about 600 nK. Under these conditions, we can also estimate the d.c. Josephson frequencies in the cyclic phase which are about 100-300 Hz. The dependence of the above frequencies on the coupling parameter J is shown in Fig. 3.

4. Experimental signatures of Non-Abelian Josephson effect

The experimental set up of a spin-2 Bose gas trapped in a double well is illustrated in Fig. 1. The d.c. non-Abelian Josephson current can be detected with the following steps. The first step is to initiate a density oscillation in the system. This can be realized by slightly changing the depth of one well, which will cause a small imbalance in chemical potential ($\mu_R \to \mu_R + \delta\mu$) between the two wells, and then tuning it back. The next step is to detect the time dependence of the particle numbers in the different spin components. Such kind of detection can be realized by first spatially separating different spin component with a Stern-Gerlach method during time of flight after switching off the trapping potential. Then, the number of atoms in each spin component will be related to the respective spatial density distributions which can be evaluated by the absorption imaging method. Following the above steps, one can measure the density oscillations in each spin component which are coupled together due to the non-Abelian symmetry of the system. A measurement of the dependence of the oscillation frequencies on J can be realized by varying the barrier between the two wells and repeating the above measurement.

In mean field theory, the condensates of ^{87}Rb atom in F=2 state are predicted to be polar ($c_1 - c_2/20 > 0$ and $c_2 < 0$), but close to the boundary of the cyclic phase ($c_1 > 0$ and $c_2 > 0$) [22]. Furthermore, polar behaviour in the F=2 ground state of the ^{87}Rb condensate has been observed in recent experiment [24]. As a result, we expect that the pseudo Goldstone modes of the antiferromagnetic phase could be observed in experiments. As we have mentioned, the values of interaction strengths under typical experimental conditions are given as [24]: $c_1 n : 0 - 10$ nK, $c_2 n : 0 - 0.2$ nK and $c_0 n$ about 150 nK, which leads to the fluctuation time scale of about 10 ms in this system. On this time scale, the measurement we proposed above is completely accessible within recent experimental techniques in F=2 spinor Bose-Einstein condensates of the ^{87}Rb system [24, 27]. In order to observe this dynamical oscillation clearly in an experiment, the temperature should be lower than the gap of the pseudo Goldstone modes, which is about 1-10 nk. Although there is still no such kind of measurement performed in a system with cyclic phase, we expect that it will be realized in a condensate of ^{85}Rb atoms in the near future [22].

In summary, we reveal a novel Josephson effect in spin-2 Bose system which involves non-Abelian symmetry and propose an experimental protocol to realize the so called non-Abelian Josephson effect in this system. It is found that the frequencies of the pseudo Goldstone modes are not only related to the coupling parameter but also to the interaction strength, which is a nonlinear effect due to the spin dependent interaction. These results are of particular significance for exploring the new features of the non-Abelian Josephson effect which are very distinct from the Abelian case.

Acknowledgments

This work was supported by the NSFC under Grant Nos. 10874235, 10934010 and 60978019 and the NKBRSF of China under Grant 2009CB930701, 2010CB922904 and 2011CB921500.

Journal of Physics: Conference Series **264** (2011) 012003 doi:10.1088/1742-6596/264/1/012003

References

[1] Josephson B D 1962 Phys. Lett. **1** 251
[2] Albiez M *et al.* 2005 Phys. Rev. Lett. **95** 010402
[3] Shin Y *et al.* 2005 Phys. Rev. Lett. **95** 170402
[4] Smerzi A, Fantoni S, Giovanazzi S and Shenoy S R 1997 Phys. Rev. Lett. **79** 4950
[5] Barone A and Paterno G 1982 *Physics and Applications of the Josephson Effect* (Wiley, New York)
[6] Chang M S, Qin Q, Zhang W, You L and Chapman M S 2005 Nature. Phys. **1** 111
[7] Esposito F P, Guay L P, MacKenzie R B, Paranjape M B and Wijewardhana L C R 2007 Phys. Rev. Lett. **98** 241602
[8] Davis J and Packard R 2002 Rev. Mod. Phys. **74** 741
[9] Rajagopal K and Wilczek F 2001 *At the Frontier of Particle Physics: Handbook of QCD: Boris Ioffe Festschrift*, edited by M. Shifman (World Scientific, Singapore), Ch. 35
[10] Ruseckas J, Juzeliunas G, Ohberg P and Fleischhauer M 2005 Phys. Rev. Lett. **95** 010404
[11] Jack M W and Yamashita M 2005 Phys. Rev. A **71** 033619
[12] Stamper-Kurn D M *et al.* 1998 Phys. Rev. Lett. **80** 2027
[13] Ohmi T and Machida K 1998 J. Phys. Soc. Jpn. **67** 1822
[14] Ho T L 1998 Phys. Rev. Lett. **81** 742
[15] Ashhab S and Lobo C 2002 Phys. Rev. A **66** 013609
[16] Ng H T, Law C K and Leung P T 2003 Phys. Rev. A **68** 013604
[17] Leggett A J 2001 Rev. Mod. Phys. **73** 307 (2001)
[18] Mustecaplioglu O E, Zhang M and You L 2005 Phys. Rev. A **71** 053616
[19] Mustecaplioglu O E, Zhang W and You L 2007 Phys. Rev. A **75** 023605
[20] Barnett R, Turner A and Demler E 2006 Phys. Rev. Lett. **97** 180412
[21] Ueda M and Koashi M 2002 Phys. Rev. A **65** 063602
[22] Ciobanu C V, Yip S K and Ho T L 2000 Phys. Rev. A **61** 033607
[23] Barnett R, Mukerjee S and Moore J E 2008 Phys. Rev. Lett. **100** 240405
[24] Schmaljohann H *et al.* 2004 Phys. Rev. Lett. **92** 040402
[25] Wu B and Niu Q 2001 Phys. Rev. A **64** 061603(R)
[26] Jin G R, Kim C K and Nahm K 2005 Phys. Rev. A **72** 045602
[27] Chang M S *et al.* 2004 Phys. Rev. Lett. **92** 140403

Journal of Physics: Conference Series **264** (2011) 012004 doi:10.1088/1742-6596/264/1/012004

Turbulence in a trapped Bose-Einstein condensate

J. A. Seman, R. F. Shiozaki, F. J. Poveda-Cuevas, E. A. L. Henn, K. M. F. Magalhães, G. Roati, G. D. Telles*, and V. S. Bagnato

Instituto de Física de São Carlos, Universidade de São Paulo, C.P. 369, 13560-970 São Carlos, SP, Brazil

E-mail: *gugs@ursa.ifsc.usp.br

Abstract. We have used an atomic ^{87}Rb BEC to study the emergence of quantum turbulence. The application of an external oscillating magnetic field gradient is used to nucleate vortices and anti-vortices spread over the cloud, in many directions, setting up the conditions for the turbulent regime to arise. Once the turbulence is established one may study it in a variety of its different aspects and we will present and discuss a few of these possibilities.

1. Introduction

Bose-Einstein condensates provide ideal experimental conditions for studying superfluidity [1, 2]. In such a coherent particle field, it follows that the superfluid component is irrotational. However, Nature finds its way to create rotation in these systems to form vortex filaments, which are lines along the condensate where the macroscopic wave function amplitude $\Psi(\mathbf{r})$ vanishes and has a curl-free circulation around it. For a free vortex filament, a stable condition implies that the quantized circulation must have a single quantum value [3]. A collection of vortex filaments in the superfluid may generate interesting spatial distributions, line vortices tangles, producing a turbulent state, named *quantum turbulence* (QT) [4]. The turbulent flow is necessarily rotational indicating that, inside the quantum fluid, there is only a single possible arrangement of vortex lines. This condition creates special features associated with QT, which makes this phenomenon somewhat different from its classical counterpart, and very interesting to be further investigated.

In the beginning, QT was only investigated in superfluid ^{4}He [5], evolving later to ^{3}He [6, 7], and, finally, observed in alkali BEC systems [8]. Because of the finite size and flexibility to control interactions in superfluid atomic samples, BECs represent a whole new window to explore the QT phenomenon.

We believe that better understanding of the features associated with QT phenomena may be an alternative path to clarify the overall understanding of its classical counterpart. Perhaps, it may be worthwhile to investigate the details of QT using condensates to help elucidate facts still unclear in dense systems like ^{3}He and ^{4}He. As stated by Davidson [9] in his recent book: "Despite the fact that the governing equations of turbulence have been known since 1845, there is still surprisingly little we can predict with relative certainty. The situation is reminiscent of the state of electromagnetism before it was transformed by Faraday and Maxwell. A myriad of tentative theories have been assembled, often centred around particular experiments, but there is not much in the way of a coherent theoretical framework".

In this paper, we review our recent work presenting the emergence of QT in a sample of ^{87}Rb BEC and some other interesting associated properties. We begin with a brief description of the experimental system used in our lab, and the procedure followed to produce the vortices and achieve the turbulent regime. A few characteristics related to the overall macroscopic behavior of the turbulent cloud are described. Some perspectives are also presented.

2. Experimental system

The experimental sequence to produce the BEC to generate the vortices in the condensate can be summarized as follows. First, we produce a ^{87}Rb BEC containing about 1×10^5 atoms with a small thermal fraction in a cigar-shaped magnetic trap with frequencies given by $\nu_r = 210$ Hz and $\nu_z = 23$ Hz. Once the condensate is obtained and still held in the trap, an oscillating external magnetic field gradient is applied. This field is produced by a pair of anti-Helmholtz coils placed with their axis closely, though not exactly, parallel to the weak trap axis. Since the coils are not perfectly aligned to the condensate axis, the external field gradient around the trapping region has components parallel to each trap eigen axis. This oscillatory scheme was developed to investigate coherent mode excitations in our Bose-condensates. Such excitations are also believed to produce vortices. A combination of shape modification, trap minimum displacement and rotation is coupled to the atoms. The quadrupolar field is applied for periods ranging from 20 to 60 ms, oscillating sinusoidally in time with a frequency of 200 Hz. This external magnetic field gradient has its offset chosen so that during a period of oscillation it goes from zero to a maximum value and then back to zero, never inverting its direction. The maximum value reached by the field gradient characterizes the strength of the excitation and has been varied from zero to 190 mG/cm along the vertical axis. The atoms are left inside the trap for an extra 20 ms wait time, as soon as the external perturbation stops. The measurements are performed via time-of-flight absorption images, with the atomic cloud in free fall expansion. For small amplitudes of the oscillating field as well as shorter excitation periods, we observe dipolar modes, quadrupolar modes, and scissor modes in our BEC but no appearance of vortices. Increasing the values of both parameters produced vortices growing in number with the amplitude and/or the excitation elapsed time.

3. Observation of vortex distribution and turbulence

During the experiments, it was observed that by increasing either the amplitude or perturbation time period we were able to nucleate many vortices [10]. Small amplitudes seem to be only able to make the cloud tilt as indicated in Figure 1. The cloud's axial motion is a clear indication that the oscillatory excitation is able to transfer angular momentum to atoms. The periodic axis tilting observed in the cloud is associated with the scissor modes. Originally, such a mode was investigated theoretically by Guéry-Odelin [11] and is generated by a sudden rotation of the confining trap. As already mentioned, the external coils produce a combination of rotation and translation of the trapping potential, inducing the modes. The mode frequencies are, naturally, closely related to the trap frequencies. In the case of a classical gas, the tilting oscillation does not present itself as a simple harmonic oscillation. The Thomas-Fermi cloud motion strongly indicates that angular momentum is being coupled to the atoms.

As the oscillation amplitude increases, the vortices start to nucleate. We have observed that, during the excitation, a combination of different collective modes is observed, as shown in Figure 2. Following the studies of Bradley and coworkers [12] the high excitation of collective modes promotes the formation of a thermal cloud which causes surface oscillations to become unstable. Long lived vortices are nucleated in the region between the condensed and the thermal cloud. Initially, they penetrate a small inward radial distance into the superfluid (on the order of the vortex size). As the cloud accumulates more (rotational) energy the vortices are able to fully enter into the condensed fraction. In Figure 3, an absorption image shows the process starting with the first inside vortex. We note that the surrounding peripheral vortices with little penetration into the condensed cloud are clearly visible [10].

The nucleation of vortices in the region interfacing the condensed and the thermal clouds may be equivalent to the Kelvin-Helmholtz instabilities phenomenon, well known in classical fluids. The particular point here is the instability generated by the excitation of the collective modes. Once the vortices are formed, they may be able to penetrate into the condensed fraction, when the conditions are energetically favorable. This has to take place while energy is still being pumped into the system

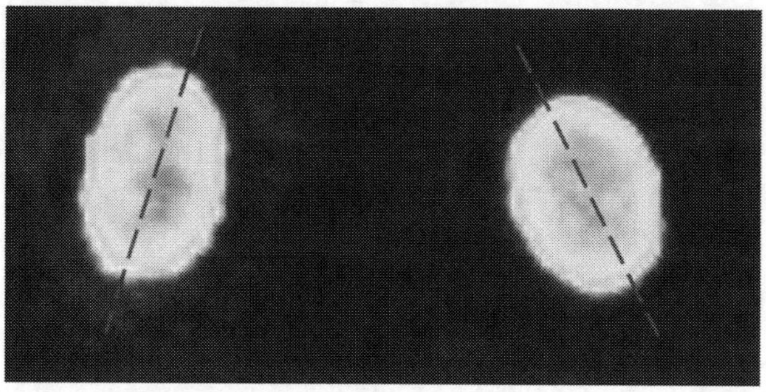

Figure 1. Axis tilt oscillations observed for 40 mG/cm of the external field gradient (low amplitude perturbation) after 15 ms of free expansion. The images were acquired from different experimental runs under identical conditions and are taken 0.5 ms apart.

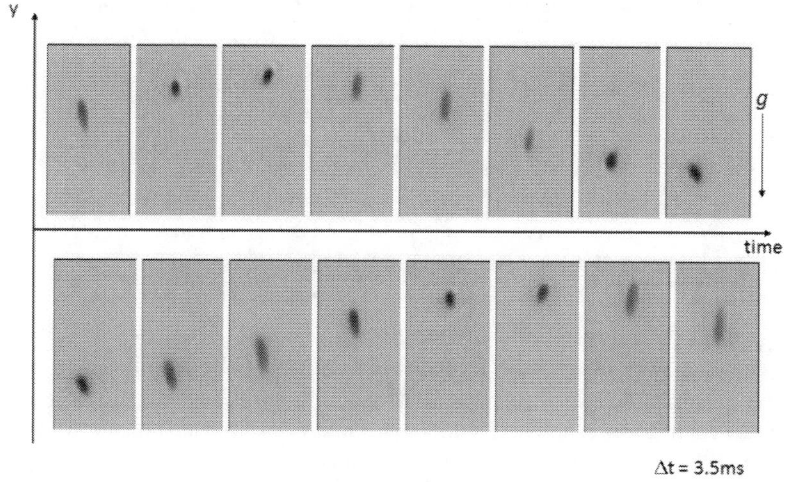

Figure 2. Combination of collective modes for different samples due to excitation coils and time of flight with a time interval of 3.5 ms.

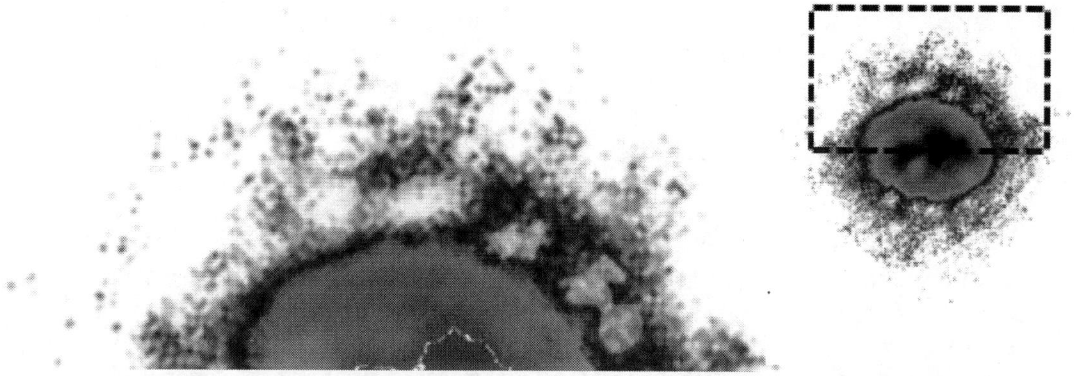

Figure 3. Zoomed boxed area of a typical absorption image (see right upper corner inset) containing vortices nucleated in the interface region between the condensed and the thermal fractions of the cloud.

by the external oscillatory field. We have made many observations in which the perturbed cloud had three singularities in the atomic density. We found out that when three vortices were being held together within the atomic cloud, they are most likely arranged as a triangle or a line. Theoretically, it is possible to justify that in the case of a spatial distribution close to a triangle the vorticity of each element presents the same signal while in the case of linear alignment they constitute a vortex tripole [13]. In the latter case, we have the two outer vortices with a given vorticity sign and the central one with the opposite sign. For the interaction regimes found in our experiment this configuration shall be considered metastable. The tripoles are expected to survive for about 20 ms in our setup [14]. The fast decay associated with the tripolar distributions makes their observation less frequent than the triangular ones for the case of equivalent vorticity for all three vortices. A detailed explanation of the observations of vortex anti-vortex clustering can be found in a recent publication [15].

An interesting fact is that the vortex formation in our experimental setup is not restricted to any preferential direction. In fact, the oscillatory generation of vortices corresponds to a combination of rotation, translation and deformation of the trapping potential in different planes [2]. The key idea is analogous to combining rotations in orthogonal directions. After nucleating orthogonal vortex filaments the distribution just established will evolve into turbulence. The evolution from a regular (superfluid)

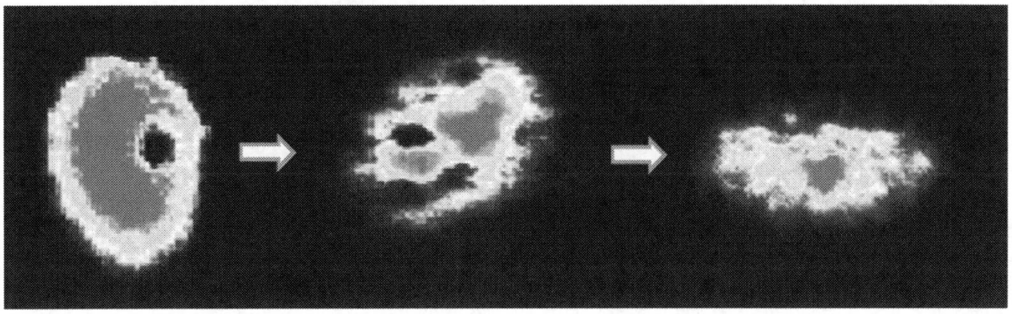

Figure 4. Two-dimensional atomic profiles showing the evolution from a regular condensate (with a single vortex) to the turbulent fluid (tangle cloud).

condensed cloud to the turbulent cloud is shown in Figure 4. The turbulent cloud is characterized by vortex distributions in different directions. Once the turbulent regime is reached there are many studies to be carried out. One of the studies corresponds to the investigation of the analogy of the turbulence onset in a trapped condensate to a flow of the superfluid through a geometry (trapping potential) whose dimension is on the same size as that of a typical BEC. In this case a quantum Reynolds number can be defined and evaluated.

4. Perspectives to turbulence in atomic superfluids

The possibility of studying turbulence in an alternative superfluid to liquid helium is very attractive and may still create new exciting possibilities. First of all the relatively low density of trapped atomic superfluids allows a clearer visualization of the vortex geometry as well as the direct observation of the distribution of vortices and other effects related to few vortices. One important class of these effects is the transverse excitations taking place in the vortices lines. Those oscillations, named Kelvin modes, have already been observed in trapped BECs [16] and are considered to be one of the proofs for the condition of seed sustained turbulence in a superfluid. Vortex reconnections [5], which can change the topology of the vortex lines, always play an important role in quantum turbulence and there is a lack of understanding about the microscopic details previously observed and studied in liquid helium.

Quantum turbulence is a strong non-equilibrium phenomenon representing an essentially non-uniform matter, which poses challenging problems for its theoretical description. But, as pointed out by Yukalov

[17], fully stationary turbulent states are likely to be obtained in trapped superfluids. This suggests that statistical methods may be used to model QT, creating new and exciting approaches to help understand this interesting phenomenon.

Finally, trapped atomic superfluids are systems of finite size and number. New effects due to the finite aspects of the superfluids will appear in these systems that might not be present in homogeneous systems like helium. These new effects might be remarkable concerning the evolution of the turbulent regime and its decay, generating thermal atoms, which can be easily detected and better studied in their dynamical aspects. The finite size effects may place limits on the maximum number of allowed vortices existing in the superfluid cloud before evolving to turbulence. This fact may have dramatic implications on the dynamical behavior of the turbulent states.

In conclusion, the study and understanding of turbulence in Bose-Einstein condensation of trapped neutral atoms is still at the very beginning, but has a great potential to become an important and exciting research field.

Acknowledgments
We wish to thank: A. Fetter, V. Yukalov, and M. Tsubota, for useful discussions; P. Castilho, C. Castelo Branco, P. E. S. Tavares, and S. R. Muniz for collaborating; and also the Brazilian funding agencies CAPES, CNPq, FAPESP.

References

[1] Maragò O M, Hopkins S A, Arlt J, Hodby E, Hechenblaikner G and Foot C J 2000 *Phys. Rev. Lett.* **84** 2056
[2] Kobayashi M and Tsubota M 2007 *Phys. Rev. A* **76** 045603
[3] Tsubota M 2009 *J. Phys.: Condens. Matter* **21** 164207
[4] Feynman R P 1955 *Prog. Low Temp. Phys.* **1** 17
[5] Vinen W F and Niemela J J 2002 *J. Low Temp. Phys.* **128** 167
[6] Fisher S N, Hale A J, Guénault A M and Pickett G R 2001 *Phys. Rev. Lett.* **86** 244
[7] Barenghi C F and Samuels D C 2002 *Phys. Rev. Lett.* **89** 155302
[8] Henn E A L, Seman J A, Roati G, Magalhães K M F and Bagnato V S 2009 *Phys. Rev. Lett.* **103** 045301
[9] Davidson P A 2004 *Turbulence: An Introduction for Scientists and Engineers* (Oxford University Press, USA) p vii
[10] Henn E A L, *et al* 2009 *Phys. Rev. A* **79** 043618
[11] Guéry-Odelin D, and Stringari S 1999 *Phys. Rev. Lett.* **83** 4452
[12] Bradley A S, Gardiner C W and Davis M J 2008 *Phys. Rev. A* **77** 033616
[13] Möttönen M, Virtanen S M M, Isoshima T and Salomaa M M 2005 *Phys. Rev. A* **71** 033626
[14] Pietilä V, Möttönen M, Isoshima T, Huhtamki J A M and Virtanen S M M 2006 *Phys. Rev. A* **74** 023603
[15] Seman J A, *et al* 2010 *Phys. Rev. A* **82** 033616
[16] Bretin V, Rosenbusch P, Chevy F, Shlyapnikov G V and Dalibard J 2003 *Phys. Rev. Lett.* **90** 100403
[17] Yukalov V I 2010 *Laser Phys. Lett.* **7** (6) 467

Journal of Physics: Conference Series **264** (2011) 012005 doi:10.1088/1742-6596/264/1/012005

Bose-Einstein condensation of photons in a 'white-wall' photon box

Jan Klärs, Julian Schmitt, Frank Vewinger and Martin Weitz
Institut für Angewandte Physik, Universität Bonn, Wegelerstr. 8, 53115 Bonn, Germany

Abstract. Bose-Einstein condensation, the macroscopic ground state occupation of a system of bosonic particles below a critical temperature, has been observed in cold atomic gases and solid-state physics quasiparticles. In contrast, photons do not show this phase transition usually, because in Planck's blackbody radiation the particle number is not conserved and at low temperature the photons disappear in the walls of the system. Here we report on the realization of a photon Bose-Einstein condensate in a dye-filled optical microcavity, which acts as a "white-wall" photon box. The cavity mirrors provide a trapping potential and a non-vanishing effective photon mass, making the system formally equivalent to a two-dimensional gas of trapped massive bosons. Thermalization of the photon gas is reached in a number conserving way by multiple scattering off the dye molecules. Signatures for a BEC upon increased photon density are: a spectral distribution that shows Bose-Einstein distributed photon energies with a macroscopically populated peak on top of a broad thermal wing, the observed threshold of the phase transition showing the predicted absolute value and scaling with resonator geometry, and condensation appearing at the trap centre even for a spatially displaced pump spot.

1. Introduction

When a gas of particles is either cooled or its density is increased up to the point where the de Broglie wavepackets overlap, quantum statistical effects come into play [1]. Specifically, for material particles with integer spin (bosons), the phenomenon of Bose-Einstein condensation is expected. Experimentally, this situation was achieved in the dilute gas limit in 1995, using laser and subsequent evaporative cooling of rubidium and sodium atoms, respectively [2, 3]. The signature of Bose-Einstein condensation was also observed with solid-state physics quasiparticles [4–7]. Other than material particles, photons in black-body radiation do not show a phase transition to a Bose-Einstein condensate [8]. Thermal photons usually have no chemical potential, corresponding to a non-conserved particle number, and at low temperature the photons disappear in the cavity walls instead of macroscopically populating the ground state mode. Early theoretical work has proposed to reach a Bose-Einstein condensate of photons using Compton scattering off a thermal electron gas to allow for a particle number conserving thermalization [9, 10]. After realization of the first atomic Bose-Einstein condensates, we have witnessed an increased interest in light sources, where a macroscopically populated photon mode is not the result of laser-like amplification, but is caused by an equilibrium phase transition. In a laser, both the state of the light field and that of the amplifying, usually inverted, medium is far removed from thermal equilibrium [11]. In an interesting work, Chiao et al. proposed to achieve a photon fluid in a two-dimensional nonlinear resonator configuration [12]. Thermal equilibrium of the photon gas was sought from photon-photon scattering, similar to atom-atom scattering

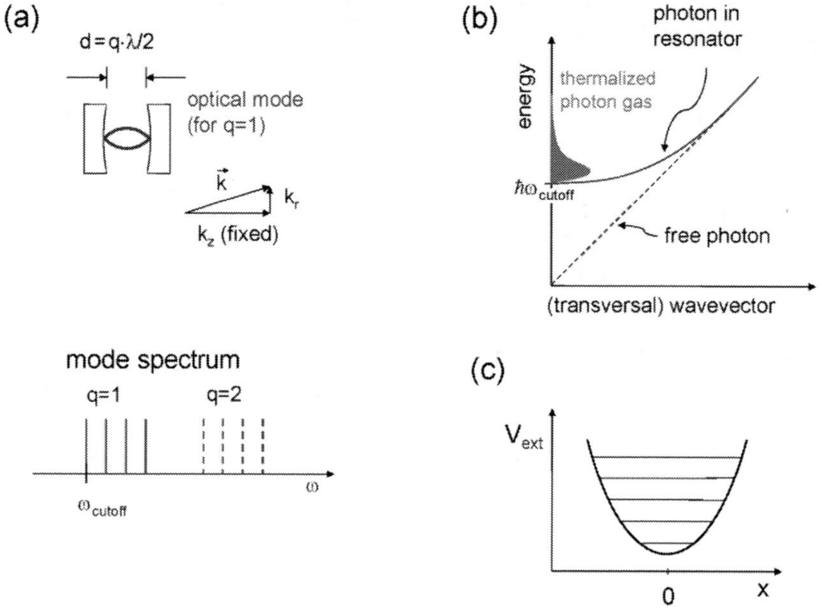

Figure 1. (a) Scheme of optical microresonator (top) and mode spectrum (bottom) for the case of a spacing between resonator mirrors of half an optical wavelength of the lowest optical mode ($q = 1$). The resonator is filled with dye solution and the photon gas thermalizes to the rovibrational temperature of the dye molecules. (b) Dispersion of a photon in the resonator (solid line) and the dispersion of a free photon (dashed line). (c) Trapping potential for the photon gas imposed by the curved mirrors.

in atomic physics BEC experiments. However the weak photon-photon interaction in available nonlinear materials has so far prevented thermalization with this method [13]. Experimentally, in a series of experiments exciton-polaritons, strongly coupled mixed states of matter and light, have been reported to allow a (quasi-) equilibrium Bose-Einstein condensation of polaritons [4–6]. The thermalization process used is based on interparticle interactions between polaritons, and in more recent experiments, superfluidity was observed [14, 15]. Our experiment described here is based on a dye molecular solution placed within an optical resonator. Thermalization of the photon gas with the dye medium is achieved by absorption and reemission of photons in the dye molecules. Rapid decoherence from frequent collisions of dye molecules with the solvent here prevents a coupling of the phases of dipole and photon [16, 17], so that we can assume the relevant particles to be well described by photons instead of polaritons. In initial experiments, a thermalized two-dimensional photon gas with freely adjustable chemical potential was realized [18]. More recently, we have observed a Bose-Einstein condensate of photons in the dye-filled optical microcavity setup [19].

2. Experimental Scheme and Setup

Our experiment traps photons in a dye solution filled optical microresonator, see Figure 1a for a schematic picture of the experiment. The distance between the two high reflecting curved resonantor mirrors is in the wavelength regime which leads to a large frequency spacing between longitudinal resonator modes - comparable with the emission width of the dye molecules - so that only photons of a fixed longitudinal mode number q can be found in the resonator. Then the longitudinal mode number is frozen out and the two remaining transverse modal quantum

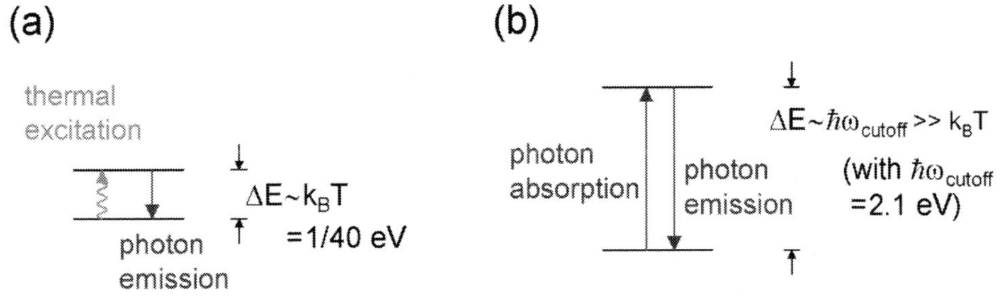

Figure 2. (a) Thermal emission in a blackbody radiator, for which the transition energy of the atoms in the cavity walls is of order of the thermal energy k_BT. Processes such as a thermal excitation of a wall atom followed by photon emission occur. The effect of this combined process is to create one photon. Clearly, the photon number here is not conserved. (b) In our experiment the photon energy is separated from the thermal energy. The used molecules have a quantized electronic structure with transition energy near or above the cavity cutoff. Thermal excitation of an electronically excited molecular state is suppressed in the limit of $\hbar\omega_{\mathrm{cutoff}} \gg k_BT$, and the (average) photon number is conserved.

numbers make the photon gas effectively two-dimensional. Moreover, the dispersion relation becomes modified with respect to free space, and acquires particle-like, i.e., quadratic, character. This is indicated in Figure 1b. The frequency of the transverse TEM_{00}-mode here becomes the lowest populated eigenfrequency, which acts as a cutoff frequency. Further, the mirror curvature leads to a harmonic confinement of the photon gas (Figure 1c). Thermal equilibrium of the photon gas is reached by repeated absorption and emission processes in the dye molecules. The photon gas thermalizes to the rovibrational temperatures of the molecules, which are at room temperature. We expect that the frequency distribution of the photon gas extends over a width of $k_BT/\hbar$ above the cutoff frequency. Essential to the experiment is that the thermalization process conserves the particle number, because thermal emission of photons is negligible in the limit of a cutoff frequency ($\hbar\omega_{\mathrm{cutoff}} \cong 2.1$ eV for light in the green spectral region) being much above the thermal energy (1/40 eV at room temperature). The dye molecules have quantized electronic excitation levels, with transition energy near or above the cutoff, whose thermal excitation is suppressed by a factor of order $\exp(-\hbar\omega_{\mathrm{cutoff}}/k_BT) \approx 10^{-36}$; see also Figure 2.

One can show that the photon gas confined in the resonator is formally equivalent to a harmonically confined two-dimensional gas of massive photons with effective mass $m_{\mathrm{eff}} = \hbar\omega_{\mathrm{cutoff}}/c^2 = \hbar k_z/c$, where ω_{cutoff} denotes the cutoff frequency [18]. The photon energy in the paraxial limit and in the limit of negligible interaction is

$$E \cong m_{\mathrm{eff}}c^2 + \frac{\hbar^2 k_r^2}{2m_{\mathrm{eff}}} + \frac{1}{2}m_{\mathrm{eff}}\Omega^2 r^2, \tag{1}$$

where k_r and k_z denote the transversal and longitudinal wavenumbers of the photon, c the speed of light in the medium and $\Omega = c/\sqrt{D_0 R/2}$ is the trap vibrational frequency, with R the mirror radius and D_0 the mirror spacing. For such a two-dimensional, harmonically confined system it is known that a Bose-Einstein condensate exists [20, 21]. This is expected if the particle number N exceeds a critical particle number

$$N_c = \frac{\pi^2}{3}\left(\frac{k_BT}{\hbar\Omega}\right)^2. \tag{2}$$

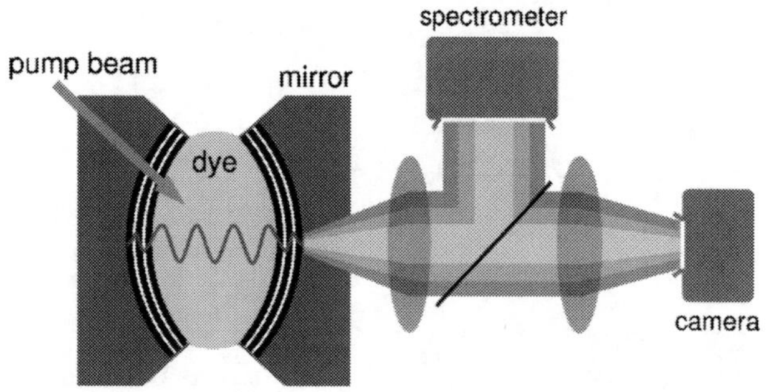

Figure 3. Experimental setup for observation of Bose-Einstein condensation of photons in a dye-filled optical microcavity. The typically used spacing between the mirrors is $D_0 \cong 1.46$ μm, corresponding to the $q = 7$ longitudinal mode, and the TEM_{00} transverse mode of this manifold acts as a low-frequency cutoff for the two-dimensional photon gas. The dye-filled resonator is pumped with radiation near 532 nm wavelength.

The typical trap vibrational frequency is $\Omega \cong 2\pi \cdot 4 \times 10^{10}$ Hz, and for $T = 300$ K we arrive at an expected critical particle number $N_c \cong 77,000$. The possibility to observe a BEC at room temperature can be understood from the effective mass $m_{\text{eff}} = \hbar\omega_{\text{cutoff}}/c^2 \cong 4 \times 10^{-36}$ kg being 10 orders of magnitude smaller than the mass of the rubidium atom, which significantly increases the phase transition temperature with respect to the atomic gas case.

Our experimental setup is shown in Figure 3. The optical resonator consists of two high reflecting dielectric mirrors with spherical curvature ($R = 1$ m). One of the mirrors is cut to 1 mm $\times$ 1 mm surface size to allow for a distance between mirrors in the micrometer regime ($D_0 \cong 1.46$ μm), as measured from the cavity free spectral range, despite the mirrors' curvature. The resonator is filled with a drop of liquid dye (rhodamine 6G or PDI) dissolved in an organic solvent. The resonator setup is pumped with a laser beam near 532 nm wavelength inclined at 45° to the cavity axis. The pumping is performed with an external laser beam because, other than in a perfect "photon box" [22], losses from coupling to optical modes not confined in the cavity, non-radiative decay and finite cavity finesse occur, which must be compensated for. We model the pumping as filling a reservoir of electronic excitations in the dye that can exchange particles with the photon gas. Thus the photon gas is understood as an open system, in the sense of a grand-canonical ensemble, with the thermalization originating from particle exchange with a reservoir of dye excitations being in equilibrium. The latter is characterized by rovibrational molecular states that are highly equilibrated both in the lower and in the upper electronic level due to subpicosecond relaxation from frequent collisions with solvent molecules [23]. This process efficiently decorrelates the states of absorbed and emitted photons within the nanosecond electronic lifetime of the dye, and leads to a temperature dependent absorption and emission spectral profile that is responsible for the thermalization. To relax both spatially and spectrally to an equilibrium distribution, a photon has to scatter several times off molecules before being lost.

3. Experimental results

In initial experiments with the dye-filled optical microresonator, we have tested for thermalization of the two-dimensional photon gas [18]. The observed photon spectra could be well described by Bose-Einstein distributed photon energies, showing that photon loss is sufficiently slow and equilibrium is reached. The chemical potential in this low density regime (well below the threshold to a BEC) is strongly negative. Moreover, a spatial concentration of light into the centre was observed, which is a consequence of the thermalization within the effective photon trapping potential, formed by curved mirrors. This behaviour is evident from the analogy with a trapped gas of material particles of finite temperature, and we expect that the effect can have prospects in the concentration of diffuse solar light [24].

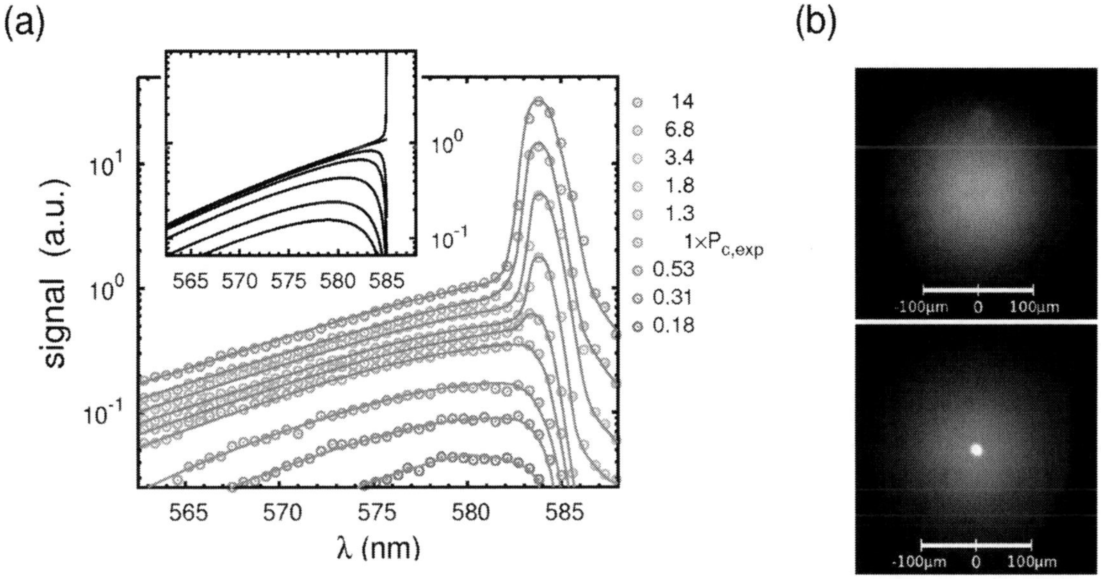

Figure 4. (a) The connected circles show measured spectral intensity distributions for different pump powers. The legend gives the optical intracavity power, determining the photon number. On top of a broad thermal wing, a spectrally sharp condensate peak at the position of the cavity cutoff is visible above a critical power $P_{c,\,\mathrm{exp}}$. The observed peak width is limited by the spectrometer resolution. The inset gives theoretical spectra based on Bose-Einstein distributed transverse excitations. (b) Images of the radiation emitted along the cavity axis, below (top) and above (bottom) the critical power. In the latter case, a condensate peak is visible in the centre.

More recently, we have operated the dye-filled microcavity in the high photon density regime, for which the optical pump beam was acoustooptically chopped to 0.5 μs long pulses (at least two orders of magnitude longer than the thermalization time) to avoid excessive population in the dye molecule triplet states. The repetition rate was 8 ms. Figure 4a shows typical spectra of the photon gas for different pumping rates [19]. While at small pumping rates the observed spectrum resembles a Boltzmann distribution above the cutoff frequency, near the expected phase transition a small shift of the spectral maximum towards the cutoff frequency is observed, and the spectrum resembles more that of a Bose-Einstein distribution. At optical intracavity powers above the critical value, in addition a spectrally sharp peak at the position of the cutoff frequency is visible, the Bose-Einstein condensate. The observed frequency width of the condensate peak is limited by the spectral resolution of our spectrometer. The experimental results are in good agreement with theoretical spectra based on a Bose-Einstein distribution of photon energies, see the inset of the figure. The observed critical photon number is $(6.3 \pm 2.4) \times 10^4$. Figure 4b shows spatial images of the photon gas below (top) and above (bottom) the critical power. Both images show a shift from the yellow spectral regime for light near the axis towards the green (with higher photon energy) for the radiation emitted off-axis. This is due to the higher energy of cavity resonances with large transverse momentum and correspondingly large mode diameter. Above the critical power, a bright spot is visible in the centre, indicating a macroscopically populated ground state (TEM_{00}) mode. The diameter of this condensate peak is (14 ± 3) μm, which is to be compared to an expected width of 12.2 μm for a noninteracting gas. Furthermore, we observe that the width of the condensate peak increases for larger photon numbers [19], which is attributed to a weak repulsive interaction

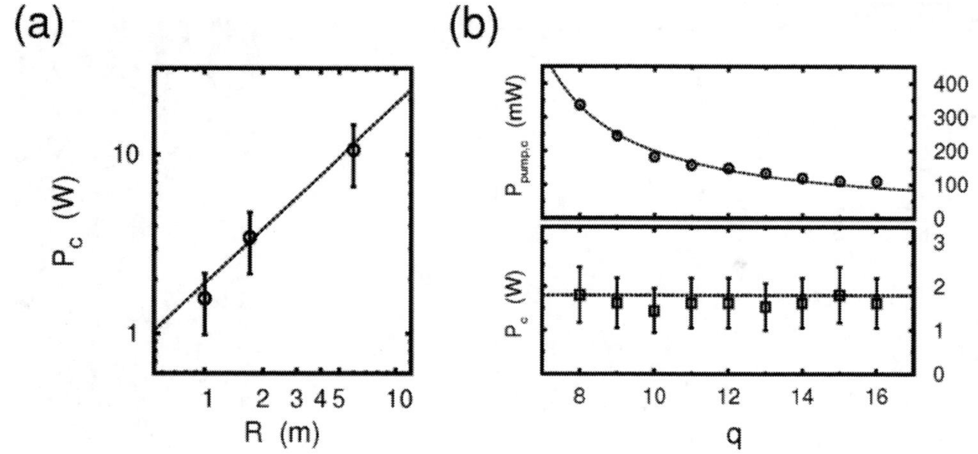

Figure 5. The data points give the optical intracavity power at criticality for three different curvatures of the cavity mirrors, and the line is the theoretical prediction based on Eq. 2. (b) Intracavity power at criticality (lower panel) versus the mirror separation, measured in units of the longitudinal mode number q. The data points in the upper panel give the required optical pump power (circles) along with a fit assuming an inverse proportionality to the absorption length in the dye.

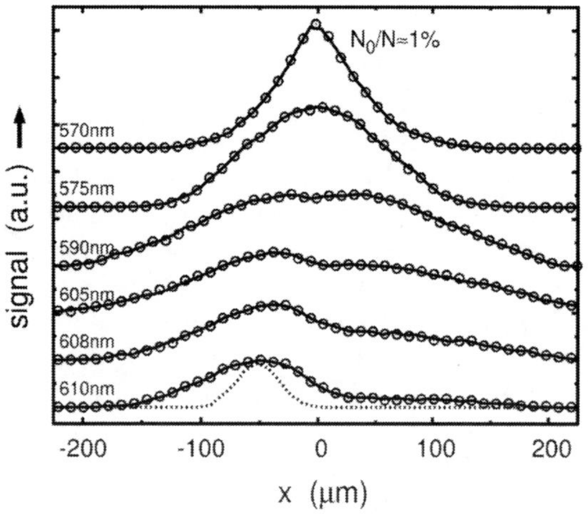

Figure 6. Spatial intensity profiles recorded with a pump beam spot spatially displaced by 50 μm from the trap centre, for different cutoff wavelengths. For data recorded with a cutoff wavelength of 610 nm (bottom profile), where reabsorption is weak, the emitted radiation is centred near the pump spot. The pump beam profile is shown by the dashed line. When tuning the cutoff to shorter wavelength values, where the reabsorption is increased, light is redistributed towards the trap centre. For data recorded with 570 nm cutoff wavelength (top profile), a sharpened peak appears, attributed to a partly condensed state with a BEC fraction of 1%.

of the photons. The origin of this is most likely thermal lensing in the dye, but in principle also a Kerr-nonlinearity would cause the same effect. Both effects can be modelled by a mean-field interaction, and from the observed increase of mode diameter, we can estimate a dimensionless interaction parameter of $\tilde{g} \approx (7 \pm 3) \times 10^{-4}$. This is much below the values $\tilde{g} \approx 10^{-2}...10^{-3}$ reported for two-dimensional atomic physics quantum gas experiments [25, 26] and also below the value $\tilde{g} \approx 10^{-3}$ given in Ref. [27], above which Kosterlitz-Thouless physics can be expected to be relevant in the harmonically trapped case. Experimentally, when directing the observed peak through a Michelson interferometer, we have not seen signatures for the phase blurring that was reported in two-dimensional atomic physics experiments [25]. This supports the conclusion that indeed a BEC is observed in the present experiment.

To test for the photon density dependence of the phase transition threshold, we have varied the resonator geometry. From Eq. 2 we arrive at an expected critical optical power $P_c = N_c \hbar\omega^2/2\pi q = (\pi^2/12)(k_B T)^2(\omega/\hbar c)R$, which grows linearly with the mirror radius R and is independent of the number of longitudinal modes q in the resonator. Figure 5a and the lower panel of Figure 5b show corresponding measurements of the critical power, with results in good agreement with the expected scaling. The upper panel of Figure 5b gives the required optical pump power to achieve the phase transition versus the number of longitudinal modes, which decreases for larger mirror spacings, as is expected from the stronger absorption of the pumping beam power in a dye film of larger thickness. The shown data is again in excellent agreement with a model assuming that the phase transition occurs at a constant value of the two-dimensional photon density, as anticipated for a photon BEC. This is in strong contrast to results from 'thresholdless' dye-based optical microlasers, for which an *increase* of the threshold with mirror separation was reported [28, 29]. For a macroscopic laser in the limit of low mirror losses, a fixed value of the pump intensity is required to reach inversion, and a threshold independent of the mirror separation is expected.

In further measurements, we have investigated the phase transition for a spatially mismatched pumping geometry to study the efficiency of the thermalization, which is expected to drive the photon gas into the trap centre. This effect is not known in laser models, but has been seen in the context of polariton condensation [6]. For our measurement, the pump beam was displaced 50 μm from the trap centre. Figure 6 shows a series of spatial intensity profiles recorded for a pumping power slightly above threshold, for different values of the cavity cutoff wavelength $\lambda_{\rm cutoff} = 2\pi c/\omega_{\rm cutoff}$, which tunes the efficiency of thermalization. The lower graph gives results recorded with 610 nm cutoff wavelength, for which the maximum of the fluorescence is centred at the position of the pump spot (shown by a dashed line). The weak reabsorption in this wavelength range prevents efficient photon thermalization into the trap centre. When the cutoff is moved to shorter wavelengths, the profiles become more symmetric around the trap centre, as expected from the increased reabsorption in this wavelength range. For a cavity cutoff near 570 nm, the photon gas appears fully thermalized and we observe a narrow peak at the position of the TEM_{00} mode on top of the thermal wing, which is attributed to a BEC with a 1% condensate fraction. The measurements indicate that, due to the photon thermalization process, Bose-Einstein condensation can be achieved even when the pumping intensity at the position of the ground state mode is essentially zero.

4. Conclusions

In summary, evidence for a Bose-Einstein condensation of photons was obtained from (i) Bose-Einstein distributed photon energies with a massively populated ground state mode on top of a broad thermal wing, (ii) the phase transition occurring both at the expected photon density and with the predicted cavity geometry dependence, and (iii) the ground state mode emerging even for a spatially displaced pump spot. It is instructive to discuss the relation of a photon BEC to optical microlasers, which also use high finesse cavities to capture the emission of excited state atoms and molecules in a small volume [28, 29]. The low lasing thresholds and the even possible inversion-less oscillation of microlasers, however, result from the coupling of spontaneous photons into a single cavity mode - which is not the case in the work reported here. This is evident from the observed highly multimodal emission below the phase transition. The main difference between a laser and Bose-Einstein condensation remains that the latter device in contrast to the former operates in thermal equilibrium. Note that the usual condition for the emergence of a coherent peak in a laser, of the gain coefficient exceeding the loss, is here replaced by Eq. (2), which essentially is equivalent to the condition $n\lambda_{\rm th}^2 \approx 1$, i.e. that the de Broglie wavepackets of the photon gas spatially overlap. Here n denotes the two-dimensional photon density, and $\lambda_{\rm th}$ is the de Broglie wavelength of the photon associated with the thermal motion in the resonator plane.

An intriguing possible consequence of the particle exchange between photons and the reservoir of excited state dye molecules is, in the limit of suitable relaxation times and small photon-photon interactions, a predicted realisation of a grand-canonical ensemble, which can give rise to unusually large number fluctuations in the condensed phase [30]. We expect that the concept of coherent light generation through condensation holds promise for new light sources, including coherent sources in wavelength regions where suitable laser media are rare or unavailable, as in the EUV regime [31]. A possible suitable medium for photon thermalization in this spectral region could be high pressure noble buffer gas samples, media that our group has recently started to investigate in the context of novel redistribution laser cooling techniques [32].

We acknowledge financial support from the Deutsche Forschungsgemeinschaft within the focused research unit FOR557.

References

[1] See, e.g., Bongs K and Sengstock K 2004 *Rep. Prog. Phys.* **67** 907
[2] Anderson M H et al. 1995 *Science* **269** 198
[3] Davis K B et al. 1995 *Phys. Rev. Lett.* **75** 3969
[4] Deng H, Weihs G, Santori C, Bloch J and Yamamoto Y 2002 *Science* **298** 199
[5] Kasprzak J et al. 2006 *Nature* **443** 409
[6] Balili R, Hartwell V, Snoke D, Pfeiffer L and West K 2007 *Science* **316** 1007
[7] Demokritov S O et al. 2006 *Nature* **443** 430
[8] See, e.g., Huang K 1987 *Statistical Mechanics*, 2nd edn, (New York: Wiley) pp. 293-294
[9] Zel'dovich Y B and Sunyaev R A 1972 *Sov. Phys. JETP* **81** 153
[10] Zel'dovich Y B and Levich E V 1969 *Sov. Phys. JETP* **28** 1287
[11] See, e.g., Siegman A E 1986 *Lasers* (Oxford: University Press)
[12] Chiao R Y 2000 *Opt. Commun.* **179** 157
[13] Mitchell M W, Hancox C I and Chiao R Y 2000 *Phys. Rev. A* **62** 043819
[14] Amo A et al. 2009 *Nature Phys.* **5** 805
[15] Sanvitto D et al. 2010 *Nature Phys.* **6** 527
[16] De Angelis E, De Martini F and Mataloni P 2000 *J. Opt. B* **2** 149
[17] Yokoyama H and Brorson S D 1989 *J. Appl. Phys.* **66** 4801
[18] Klaers J, Vewinger F and Weitz M 2010 *Nature Phys.* **6** 512
[19] Klaers J, Schmitt J, Vewinger F and Weitz M, 2010 *Nature* **468** 545
[20] Bagnato V and Kleppner D 1991 *Phys. Rev. A* **44** 7439
[21] Mullin W J 1997 *J. Low Temp. Phys.* **106** 615
[22] Bohr N 1949 in *Albert Einstein: Philosopher - Scientist*, P. A. Schilpp (ed.) (The Library of Living Philosophers. Vol. VII. Open Court. La Salle)
[23] See, e.g., Lakowicz J R 1999 *Principles of Fluorescence Spectroscopy*, 2nd edn, (Kluwer Academic/Plenum Publishers) p. 6
[24] Van Sark W G et al. 2008 *Optics Express* **16** 21773
[25] Hadzibabic Z et al. 2006 *Nature* **441** 1118
[26] Cladé P et al. 2009 *Phys. Rev. Lett.* **102** 170401
[27] Hadzibabic Z and Dalibard J 2009 *Preprint* arXiv:0912.1490
[28] De Martini F and Jacobovitz G R 1988 *Phys. Rev. Lett.* **60** 1711
[29] Yokoyama H et al. 1992 *Quant. Electron.* **24** S245
[30] Kocharovsky V V et al. 2006 *Adv. Atom. Mol. Opt. Phy.* **53** 291
[31] Jonkers J 2006 *Plasma Sources Sci. Technol.* **15** S8
[32] Vogl U and Weitz M 2009 *Nature* **461** 70

Journal of Physics: Conference Series **264** (2011) 012006 doi:10.1088/1742-6596/264/1/012006

Precision experiments with antihydrogen: an outlook

Michael Doser
CERN EP, 1211 Geneva 23, Switzerland
E-mail: michael.doser@cern.ch

Abstract. After a first generation of experiments has demonstrated the feasibility of forming - in a controlled manner - low-energy antihydrogen atoms via several different techniques, a second generation of experiments is now attempting to trap sufficiently cold atoms, or to form an atomic beam of antihydrogen atoms.

The goal of these experiments is to carry out comparative precision spectroscopy between hydrogen and antihydrogen, in view of testing the CPT theorem, either through 1S-2S spectroscopy or via a measurement of the hyperfine splitting of the ground state of antihydrogen. A related class of experiments combines techniques from these experiments with recent developments in the formation of positronium to test the gravitational interaction between matter and antimatter.

A significant number of challenges and limitations will still need to be overcome before precision measurements with antihydrogen become feasible, with the next significant milestones being either trapping of antihydrogen or the formation of a beam of antihydrogen.

1. Introduction

Precision measurements of antimatter hold promise of highly sensitive tests of CPT symmetry, as well as of a novel test of the weak equivalence principle with a new class of probe. The CPT theorem lies at the heart of much of 20th century physics, and rests on solid postulates. However, a number of extensions to the standard model have been proposed in recent years [1] that allow for the possibility of CPT violation. Specifically, a general framework has been developed [2] that parametrizes deviations from the standard model Lagrangian through CPT and Lorenz-violating terms.

Also the gravitational interaction between matter and antimatter is an open question. While numerous tests of the weak equivalence principle have been carried out, constraints derived from these tests [3] are often assumption dependent [4], and can be avoided through specifically constructed models, such as those incorporating gravivector and graviscalar components [5].

Antihydrogen is a highly sensitive test-bed for CPT, if the same spectroscopic precision as that obtained in the case of hydrogen [6, 7] can be achieved, while in the case of the gravitational interaction between matter and antimatter, it is one of the very few antimatter systems in which measurements can be envisaged (proposals have also been put forward to study two other neutral systems, positronium (Ps) [8] and most recently, muonium [9]) since their sensitivity to electric and magnetic fields excludes charged (anti)particles from such tests.

Should any asymmetry in the properties of matter and antimatter become apparent, then these tests may in turn lead to a better understanding of the origin of the baryon asymmetry of the Universe.

2. Antihydrogen formation experiments

Among a number of techniques to form antihydrogen [10], two have been focused on, experimentally implemented and studied. Both methods require prior trapping and cooling of antiprotons and positrons in Penning traps [11]. The first method [12] relies on nested Penning traps to allow antiprotons and positrons to be brought into proximity and be allowed to interact. The resulting formation [13] of antihydrogen proceeds via the three-body process
$e^+ + e^+ + \bar{p} \rightarrow \bar{H} + e^+$
and leads to weakly bound states susceptible to be re-ionized both collisionally and through field-ionization. These effects significantly modify the expected dependency of the formation rates on plasma density and temperature [14], and represent a transport mechanism for antiprotons from the core of the positron plasma (assuming on-axis injection of the antiprotons into the positrons) to its surface, particularly in the case of high density positron plasmas. Such a transport mechanism is problematic for attempts to produce trappable antihydrogen, since not only does it reduce the formation rate through a rapid reduction of the available antiprotons, but in addition, the velocity of antihydrogen formed at the surface of the positron plasma will be determined by the **E x B** drift of the antiproton prior to formation. According to simulations [14], the resulting antiproton energies would be in excess of the depth of achievable magnetic multi-polar traps. The same simulations also underline the importance of controlling the density of the positron plasma in this formation scheme to tune the antihydrogen velocity distribution, and thus optimize their trappability.

A second production technique [15] relies on interacting antiprotons in a Penning trap with Ps, that is produced by interacting Rydberg Cs atoms with positrons in a second nearby Penning trap. The cross section for this double-charge exchange process is lower than the three-body process, but - assuming sufficiently low-energy antiprotons - should result in low energy, and thus potentially trappable, antihydrogen atoms, given the very low momentum transfer from the (low mass with respect to antiprotons) Ps to the formed antihydrogen atoms.

Experimental evidence for production of high-temperature (and thus untrappable) antihydrogen atoms via the three-body formation process comes from a transmission measurement of antihydrogen atoms through an oscillatory electric field, which is able to field ionize specific Rydberg antihydrogen atoms [16]. The resulting velocity distribution is in good agreement with simulations [17].

3. Attempts at trapping antihydrogen

The trapping of antihydrogen is rendered difficult by two constraints. On one hand, the currently employed formation mechanisms require trapping antihydrogen atoms in the same volume in which they are produced, and in which both positrons and antiprotons must be confined, with current designs superimposing magnetic multipole traps (for the confinement of the produced antihydrogen) and nested Penning traps (for the formation of antihydrogen) [18].

On the other hand, optically trapping antihydrogen immediately after production is rendered impossible by the relatively high temperature (at best O(10 K)) of the formed antihydrogen, the excited state in which the Antihydrogen is formed, and the non-pulsed production mechanism. Even if antihydrogen were formed in the ground state, direct trapping would still require a currently unrealistically high power cw laser to provide the required cooling into the trap.

The solution chosen by the ATRAP and ALPHA collaborations is to form antihydrogen at as low a temperature as possible within a magnetic multipole that, when combined with the surrounding solenoidal fields, produces a confining potential in 3D. The requirement of confining positrons and antiprotons in the same volume imposes a minimal solenoidal field of 1T, onto which the magnetic multipole is superimposed. The depth of the potential is then sufficient to confine antihydrogen atoms with a temperature of less than 0.5 K. The defining challenge of this approach is thus that of producing antihydrogen at $\sim$ K temperatures.

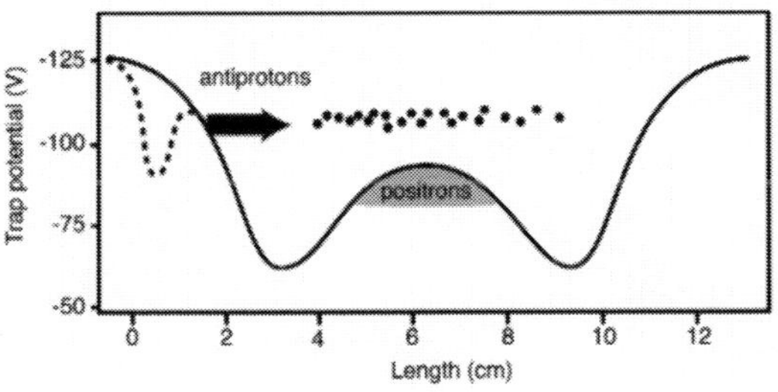

Figure 1. Nested trap for the formation of $\bar{H}$ atoms. Both positrons and antiprotons are held in Penning traps, but of opposite polarity. While the trap potentials indicated here correspond to those used in the first experiments to form antihydrogen by injecting antiprotons from the dotted well into the central well, current experiments employ greatly reduced potential values.

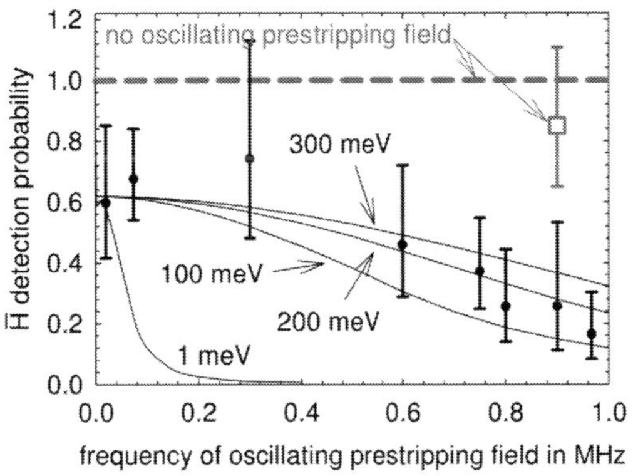

Figure 2. From [16]: the fraction of $\bar{H}$ atoms detected via field ionization in a detection well as a function of the frequency of the stripping field, and comparison to models (solid lines, detailed in [16]).

Although operating in a cryogenic environment, initial experiments that mixed antiprotons and positron plasmas resulted in high temperature (O(1000 K)) antihydrogen [16, 19]. A much deeper understanding of the underlying processes [20] has modified the approaches taken by ATRAP and ALPHA, who have emphasized both the initial temperature of the antiprotons and positrons, as well as the method employed to bring them into contact. ATRAP has reported [21] results on achieving ultra-cold positron and electron plasmas, with temperatures in the few K region. Recently, ALPHA has applied the technique of forced evaporative cooling to cold charged plasmas for the first time [22]. By slowly lowering the electrostatic potential well of a Penning trap containing antiprotons at a density of $\sim 10^6 cm^{-3}$, and allowing sufficient time for elastic collisions between antiprotons to scatter the highest-energy particles out of the well, a decrease of the temperature from 1000 K to antiproton temperatures as low as (9 $\pm$ 4) K has been demonstrated (Fig. 3). Evaporative cooling is by its nature a lossy procedure: from an initial population of $\sim$ 50,000 antiprotons, (6 $\pm$ 1) % survive to the lowest temperature, corresponding to a potential depth of (10 $\pm$ 4) mV.

4. Formation of an atomic beam of antihydrogen

A number of methods [23-27] have been proposed to form a beam of atomic antihydrogen, of which two are being experimentally evaluated. In a first scheme, antihydrogen atoms are formed in an environment that almost, but not completely, confines them, allowing individual atoms to escape along a well-defined path, while in a second scheme, Rydberg antihydrogen atoms are accelerated via the Stark effect.

The standard method of forming antihydrogen atoms utilizes Penning traps, upon which - if trapping is desired - magnetic bottles are superimposed. The ASACUSA collaboration has proposed two alternatives to this scheme, in which the charged components are trapped either through a Paul trap [28] or in a new type of trap, a cusp trap [24] (see Fig. 4). The latter traps positrons at the centre of a field consisting of a magnetic quadrupole and an electric octupole. The proponents expect that the total electric field of this octupole, together with the space charge of the trapped positrons themselves, will be strong enough to confine antiprotons inside the positron cloud, where the two particle species will combine to produce antihydrogen atoms

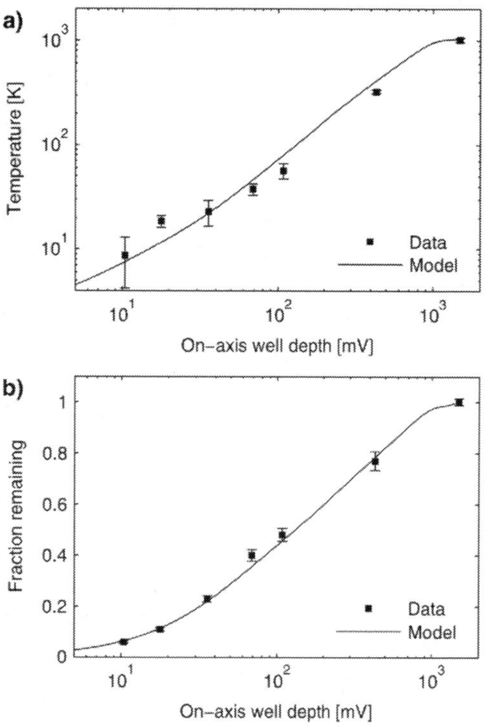

Figure 3. From [22]: Temperature of antiprotons as obtained through evaporative cooling by the ALPHA collaboration. The model calculation in [22] is shown as solid lines. (a) Temperature vs. the on-axis depth of the Penning trap holding the antiprotons. (b) Fraction of antiprotons remaining after evaporative cooling vs on-axis well depth.

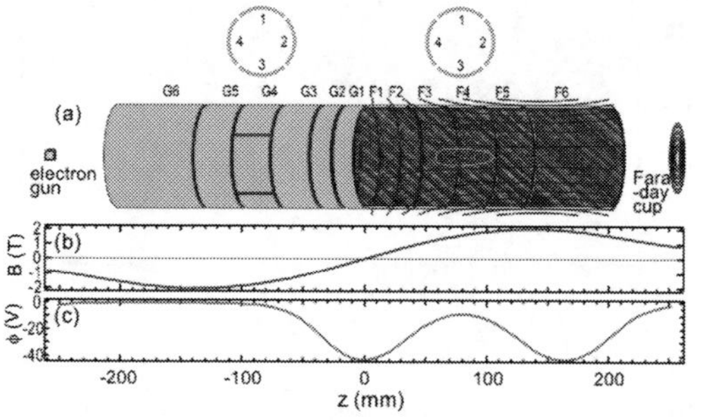

Figure 4. From [29]: a) Schematic view of the trap electrodes, field lines and confined plasma in a test setup of a cusp trap as proposed by the ASACUSA experiment. b) Magnetic field strength profile. c) Electrostatic potential profile on the rotational symmetry axis.

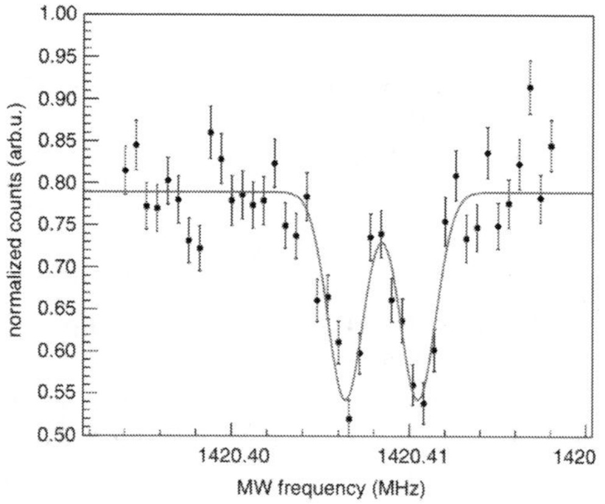

Figure 5. From [34] Simulated microwave resonance scan of the σ_1 hyperfine transition in antihydrogen for the proposed measurement by the ASACUSA experiment.

spontaneously via radiative recombination. Of particular relevance to the formation of a beam of antihydrogen atoms is the observation, in a numerical simulation, that a considerable fraction of antihydrogen atoms in low-field-seeking states formed at 5 K near the centre of the trap may be transported as a 99% polarised beam.

While the beam resulting from these methods would be continuous, it can be advantageous to know the exact moment of formation of antihydrogen atoms, both for spectroscopy with untrapped atoms, as well as to produce a pulsed beam. The formation method proposed by the AEGIS collaboration [23] relies on the interaction between antiprotons that are cooled to the lowest possible temperature by one of a number of techniques (evaporative cooling [22], electron cooling [30] or sympathetic cooling [31]) and positronium that has been laser-excited into a Rydberg state [32]. The resulting Rydberg antihydrogen atoms are then expected to have a

Maxwellian velocity distribution peaked at around 50 m/s (corresponding to 120 mK). Because these atoms are in a Rydberg state, they are amenable to being Stark-accelerated along the axis of the magnet confining the antiprotons in the Penning trap to several 100 m/s, analogously to Stark deceleration of hydrogen [33], to form a pulsed beam.

4.1. Hyperfine splitting

In the specific framework of the Standard Model Extension formulated by Kostelecký [2], the most sensitive test of CPT violation is not in the comparison of the 1s-2s transitions between hydrogen and antihydrogen, but rather in the comparison of the hyperfine splitting of the 1s state. In this model, the energy levels of the 1s (and 2s) hyperfine sublevels are potentially different for Hydrogen and Antihydrogen. The ASACUSA collaboration is preparing a measurement of these transitions using their proposed continuous antihydrogen beam [34] and a scheme based on a Stern-Gerlach-based apparatus. Both high-field and low-field seekers, depending on their magnetic moment, will be present in the beam; when traversing a sextupole magnet on their trajectory, the former will be defocussed, while the latter will be focused and can be directed towards a radiofrequency cavity, which is then again followed by a (sextupole) analyzing magnet. Spin-flips induced on resonance in the microwave cavity will cause the second magnet to act as a defocusing magnet, and thus reduce the flux of antihydrogen atoms arriving at a downstream detector. A simulation of the expected signal is shown in Fig. 5.

4.2. Measurement of the gravitational interaction between matter and antimatter

Any measurement of the gravitational fall of antihydrogen will be confronted with the thermal motion of the atoms at a temperature T, which must be compared to the scale of the gravitational potential energy over the distance of fall. The figure of merit is Mgh/k_BT. In the case of (anti)hydrogen, k_B/Mg is approximately 85 cm/mK. Attempting to measure the gravitational interaction of trapped antihydrogen thus requires cooling the atoms to $\sim$ 1 mK, within a $\sim$ 1m apparatus.

However, it is possible to circumvent the requirement to trap and cool antihydrogen if, instead, it is transformed into a sufficiently cold ($\sim$ 100 mK), pulsed horizontal beam. This is the approach proposed [23] by the AEGIS collaboration (Fig. 6). After having been produced and horizontally accelerated (by switching the electrodes of the Penning trap from the configuration used to confine the antiprotons into a Stark-accelerator configuration), the antihydrogen atoms freely fly through two gratings of a Moiré deflectometer before impinging on a position- and time-sensitive detector (Fig. 7). The gratings function similarly to an atom interferometer, albeit in the classical regime, and produce a periodic pattern of the antihydrogen impact points in the plane of the detector. The gravitational coupling will then be extracted from a measurement of the time-of-flight-dependent downward shift of the periodic pattern. This technique, which has been used to measure the gravitational fall of Ar atoms to 10^{-4} [35], will allow reaching a precision of $\sim 1\%$ with a few months of antiproton beam time, underlining the importance of the control of systematics over extended periods of time.

5. Outlook

Since the first formation of low temperature antihydrogen in 2002, steady progress has been made in understanding the formation mechanisms and in developing the techniques needed to produce antihydrogen that is sufficiently cold that it can be trapped, such that two experiments, ATRAP and ALPHA, are very close to achieving this goal. Spectroscopy of antihydrogen should follow closely on the heels of this expected break-through, and should rapidly lead to the first laser-cooling of trapped antihydrogen atoms, and subsequently to the first moderate precision comparisons between the spectra of hydrogen and antihydrogen, thus opening the way to a new class of highly-precise CPT tests.

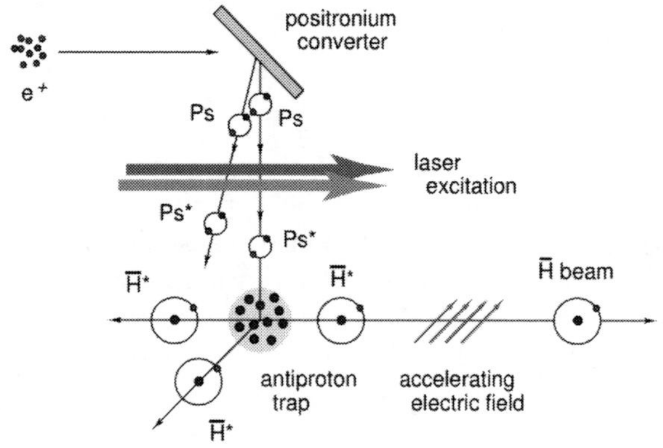

Figure 6. Proposed method for the production of a pulsed beam of cold antihydrogen atoms (AE$\bar{g}$IS experiment).

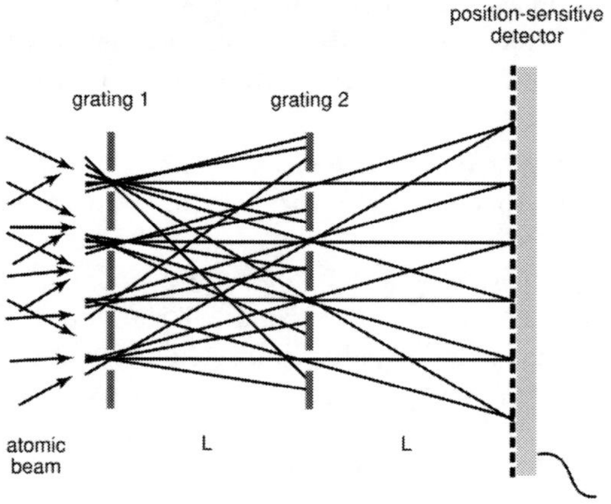

Figure 7. Schematic of the principle of the Moiré deflectometer (consisting of two gratings coupled to a position-sensitive detector) for the AE$\bar{g}$IS measurement of the gravitational coupling between matter and antimatter.

In parallel, new developments and concepts have entered the field, and have prepared the technical basis to attempt to form low intensity beams of cold antihydrogen atoms, opening the possibility of a highly sensitive measurement of the HFS of antihydrogen, as well as a first measurement of the gravitational interaction between matter and antimatter.

Finally, a number of recent initiatives on further paths to produce and study antihydrogen [36, 25] or the gravitational interaction of antimatter [37, 38] are a good indication of the increasing interest that this field at the intersection between high-energy physics and atomic physics is attracting.

(Note added in proof: As hoped for, the first breakthrough - that of trapping cold antihydrogen atoms - has just been achieved: the publication [39] by the ALPHA collaboration on trapping small numbers of antihydrogen atoms in their octupolar magnetic trap raises hopes that first spectroscopy of antihydrogen will soon be within reach.)

References

[1] Kostelecky A et al. 1997 *Phys. Rev. D* **55** 6760
[2] Blum R, Kostelecky A and Russel N 1999 *Phys. Rev. Lett.* **82** 2254
[3] Alves D, Jankowiak M and Saraswat P 2009 *Preprint* hep-ph/0907.4110v1
[4] Nieto M and Goldman T 1992 *Phys. Rep.* **205** 5 221-281
[5] Sherk J 1979 *Phys. Lett. B* **88** 265
[6] Hänsch T et al. 2000 *Phys. Rev. Lett.* **84** 54965499
[7] Ramsey N F, Nobel Lecture (1989); Ramsey N F 1995 *Physica Scripta* **T59** 323
[8] Oberthaler M 2002 *Nucl. Inst. Meth. B* **192** 129
[9] Kirch K *Preprint* hep-ex/0702143
[10] Holzscheiter M and Charlton M 1999 *Rep. Prog. Phys.* **62** 1-60
[11] Gabrielse G et al. 1986 *Phys. Rev. Lett.* **57** 2504
[12] Gabrielse G, Rolston S, Haarsma L and Kells W 1988 *Phys. Lett. A* **129** 38
[13] Amoretti M et al. (ATHENA collaboration) 2002 *Nature* **419** 202;
Gabrielse G et al. 2002 *Phys. Rev. Lett.* **89** 213401
[14] Jonsell J, van der Werf D P, Charlton M and Robicheaux F 2009 *J. Phys. B: At. Mol. Opt. Phys.* **42** 215002
[15] Storry C et al. (ATRAP) 2004 *Phys. Rev. Lett.* **93** 263401

Journal of Physics: Conference Series **264** (2011) 012006 doi:10.1088/1742-6596/264/1/012006

[16] Gabrielse G et al. 2003 *Phys. Rev. Lett.* **93** 073401
[17] Robicheaux F 2004 *Phys. Rev. A* **70** 022510
[18] Walraven J 1993 *Hyperfine Int.* **76** 203
[19] Madsen N et al. (ATHENA collaboration) 2004 *Phys. Rev. Lett.* **94** 033403
[20] Robicheaux F 2008 *J. Phys. B: At. Mol. Opt. Phys.* **41** 192001
[21] Gabrielse G et al. 2010 *SPSC report* *http://cdsweb.cern.ch/record/1234581/files/SPSC-SR-057.pdf?version=1*
[22] Andresen G B et al. 2010 *Phys. Rev. Lett.* **105** 013003
[23] Kellerbauer A et al. 2008 *Nucl. Instrum. Methods Phys. Res. B* **266** 351
[24] Mohri A and Yamazaki T 2003 *Europhys. Lett.* **63** 207
[25] Kaplan D 2009 *Hyperfine Int.* **194** 145; Cronin A D et al. *AGE Letter of Intent http://www.fnal.gov/directorate/program_planning/Mar2009PACPublic/AGELOIFeb2009.pdf*
[26] Ordonez C A 2009 *J. App. Phys.* **106** 024905
[27] Robicheaux F 2010 *J. Phys. B: At. Mol. Opt. Phys.* **43** 015202
[28] ASACUSA proposal addendum 2005 *CERN/SPSC 2005-002, SPS P-307 Add. 1*
[29] Saitoh H, Mohri A, Enomoto Y, Kanai Y and Yamazaki Y 2008 *Phys. Rev. A* **77** 051403
[30] Gabrielse G et al. 1989 *Phys. Rev. Lett.* **63** 1360
[31] Kellerbauer A and Walz J 2006 *New Journ. Physics* **8** 45
[32] Charlton M 1990 *Phys. Lett. A* **143** 143
[33] Vliegen E, Hogan S, Schmutz H and Merkt F 2007 *Phys. Rev. A* **76** 023405
[34] Juhász B and Widmann E 2009 *Hyperfine Interactions* **193** 305
[35] Oberthaler M, Bernet S, Rasel E M, Schmiedmayer J and Zeilinger A 1996 *Phys. Rev. A* **54** 3165
[36] Perez P and Rosowsky A 2005 *Nucl. Inst. Meth.* **545** 20
[37] Phillips T 1996 *Hyperfine Int.* **100** 163
[38] Walz J and Hänsch T 2004 *General Relativity and Gravitation* **36** 561
[39] Andresen G B et al. 2010 *Nature* **468** 355

22nd International Conference on Atomic Physics IOP Publishing
Journal of Physics: Conference Series **264** (2011) 012007 doi:10.1088/1742-6596/264/1/012007

Helium fine structure theory for determination of α

Krzysztof Pachucki

Institute of Theoretical Physics, University of Warsaw, Hoża 69, 00–681 Warsaw, Poland

E-mail: krp@fuw.edu.pl

Vladimir A. Yerokhin

Center for Advanced Studies, St. Petersburg State Polytechnical University, Polytekhnicheskaya 29, St. Petersburg 195251, Russia

E-mail: yerokhin@pcqnt1.phys.spbu.ru

Abstract. We present recent progress in the calculation of the helium fine-structure splitting of the 2^3P_J states, based on the quantum electrodynamic theory. Apart from the complete evaluation of $m\,\alpha^7$ and $m^2/M\,\alpha^6$ corrections, we have performed extensive tests by comparison with all experimental results for light helium-like ions and with the known large nuclear charge asymptotics of individual corrections. Our theoretical predictions are still limited by the unknown $m\,\alpha^8$ term, which is conservatively estimated to be 1.7 kHz. However, comparison with the latest experimental result for the $2^3P_0 - 2^3P_2$ transition [M. Smiciklas and T. Shiner, Phys. Rev. Lett. **105**, 123001 (2010)] suggests that the higher-order contribution is in fact much smaller than the theoretical estimate. This means that the spectroscopic determination of α can be significantly improved if another measurement of the $2^3P_0 - 2^3P_2$ transition in helium-like Li^+ or Be^{2+} ion is performed.

1. Introduction

The quantum electrodynamic (QED) theory of atomic energy levels has achieved a precision level that makes possible the determination of nuclear properties, like the charge radius, the magnetic dipole, or even the nuclear polarizability from measured atomic spectra. If the nuclear structure effects are negligible or can be eliminated, one may obtain fundamental constants from comparison of theoretical predictions with experimental results. The most important examples include the Rydberg constant determined from hydrogen spectroscopy, the electron mass derived from the bound-electron g factor in hydrogen-like ions, and α obtained from the helium fine structure. As first pointed out by Schwartz in 1964 [1], the splitting of the 2^3P_J levels in helium can be used for an accurate determination of the fine structure constant α. The attractive features of the fine structure in helium as compared to other atomic transitions are, first, the long lifetime of the metastable 2^3P_J levels (roughly two orders of magnitude longer than that of the $2p$ state in hydrogen) and, second, the relative simplicity of the theory. Schwartz's suggestion stimulated a sequence of calculations [2–5], which resulted in a theoretical description of the helium fine structure complete up to order $m\alpha^6$ (or α^4 Ry) and a value of α accurate to 0.9 ppm [6].

The present experimental precision for the fine-structure intervals in helium is sufficient for a determination of α with an accuracy of 14 ppb from Refs. [7, 8] and even 5 ppb from Ref. [9].

In order to match this level of accuracy in the theoretical description of the fine structure, the complete calculation of the next-order, $m\alpha^7$ contribution and an estimation of the higher-order effects is needed. Work towards this end started in the 1990s and extended over two decades [10–19]. In 2006 the first complete evaluation of the $m\alpha^7$ correction to the helium fine structure was reported by one of us (KP) [20]. However, the numerical results presented there were in disagreement with the experimental values by more than 10 standard deviations (σ).

In our recent investigations [21, 22], we recalculated, using formulae from Ref. [20], all effects up to order $m\alpha^7$ to the fine structure of helium and performed calculations for helium-like ions with nuclear charges Z up to 10. The calculations were extensively checked by studying the hydrogenic ($Z \to \infty$) limit of individual corrections and by comparing them with the results known from the hydrogen theory. We found several problems in previous numerical calculations and, in the meantime, the experimental value of the $2^3P_1 - 2^3P_2$ transition was changed by 3σ [8]. As a result, the present theoretical predictions are in agreement with the latest experimental data for the fine-structure intervals in helium, as well as with most of the experimental data available for light helium-like ions. Our calculation of the $m\alpha^7$ correction for the fine-structure splitting in light helium-like atoms was reported in Refs. [22, 23]. In this paper, we present a detailed description of all corrections to helium fine structure and a summary of the numerical results.

2. QED theory of the helium fine structure

According to the quantum electrodynamic theory (QED) the energy levels of an atomic system are a function of the fine structure constant α and the electron-nucleus mass ratio. We omit possible nuclear structure effects, as their contribution to the helium fine structure is negligible. The fine-structure splitting $E_{\rm fs}(\alpha)$ can be expanded in powers of α,

$$E_{\rm fs} = E_{\rm fs}^{(4)} + E_{\rm fs}^{(5)} + E_{\rm fs}^{(6)} + E_{\rm fs}^{(7)} + O(\alpha^8)\,. \tag{1}$$

The expansion terms $E_{\rm fs}^{(n)} \equiv m\,\alpha^n\,\mathcal{E}^{(n)}$ are of order $m\,\alpha^n$. They implicitly depend on the electron-nucleus mass ratio and may additionally involve powers of $\ln\alpha$. The advantage of this approach is that each of the expansion terms is expressed as the expectation value of some effective Hamiltonian, as presented in the following. For convenience, we first consider the infinite nuclear mass limit, and then account for the finite nuclear mass corrections separately.

The dominant contribution to the helium fine structure is induced by the spin-dependent part of the Breit-Pauli Hamiltonian, which is, for an infinitely heavy nucleus,

$$\begin{aligned}
H_{\rm fs} &= \frac{1}{4}\left(\frac{\vec\sigma_1\cdot\vec\sigma_2}{r^3} - 3\,\frac{\vec\sigma_1\cdot\vec r\,\vec\sigma_2\cdot\vec r}{r^5}\right)(1+a_e)^2 \\
&+ \frac{Z}{4}\left[\frac{1}{r_1^3}\,\vec r_1\times\vec p_1\cdot\vec\sigma_1 + \frac{1}{r_2^3}\,\vec r_2\times\vec p_2\cdot\vec\sigma_2\right](1+2a_e) \\
&+ \frac{1}{4\,r^3}\bigg\{\big[(1+2\,a_e)\,\vec\sigma_2 + 2\,(1+a_e)\,\vec\sigma_1\big]\cdot\vec r\times\vec p_2 \\
&\qquad -\big[(1+2\,a_e)\,\vec\sigma_1 + 2\,(1+a_e)\,\vec\sigma_2\big]\cdot\vec r\times\vec p_1\bigg\}\,,
\end{aligned} \tag{2}$$

where $\vec r = \vec r_1 - \vec r_2$. The above Hamiltonian includes the effect of the anomalous magnetic moment (amm) a_e, which is given by [24] (neglecting small vacuum-polarization corrections coming from particles heavier than an electron)

$$a_e = \frac{\alpha}{2\pi} - 0.328\,478\,966\left(\frac{\alpha}{\pi}\right)^2 + 1.181\,241\,457\left(\frac{\alpha}{\pi}\right)^3 - 1.914\,4(35)\left(\frac{\alpha}{\pi}\right)^4 + \ldots\,. \tag{3}$$

Journal of Physics: Conference Series **264** (2011) 012007 doi:10.1088/1742-6596/264/1/012007

Expanding the amm prefactors in Eq. (2), $H_{\rm fs}$ can be written as a sum of operators contributing to different orders in α

$$H_{\rm fs} = H_{\rm fs}^{(4)} + \alpha\, H_{\rm fs}^{(5)} + \alpha^2\, H_{\rm fs,amm}^{(6)} + \alpha^3\, H_{\rm fs,amm}^{(7)} + \ldots . \tag{4}$$

Here, $H_{\rm fs}^{(4)}$ and $H_{\rm fs}^{(5)}$ are the complete effective Hamiltonians to order $m\,\alpha^4$ and $m\,\alpha^5$, respectively, whereas $H_{\rm fs,amm}^{(6)}$ and $H_{\rm fs,amm}^{(7)}$ are the amm parts of the corresponding higher-order operators. The contributions to the fine structure are

$$\mathcal{E}^{(4)} = \langle H_{\rm fs}^{(4)}\rangle + O(m/M)\,, \tag{5}$$

$$\mathcal{E}^{(5)} = \langle H_{\rm fs}^{(5)}\rangle + O(m/M)\,, \tag{6}$$

where the expectation values are calculated with the corresponding eigenstate of the nonrelativistic Hamiltonian H_0

$$H_0 = \frac{p_1^2 + p_2^2}{2} - \frac{Z}{r_1} - \frac{Z}{r_1} + \frac{1}{r}\,. \tag{7}$$

The finite nuclear mass corrections up to order $m\alpha^5$ are conveniently divided into three parts, termed the mass scaling, the mass polarization, and the recoil operators. The effect of the mass scaling is accounted for by including the prefactor $(m_r/m)^3$ in the operator $H_{\rm fs}$, where m_r is the reduced mass for the electron-nucleus system. The effect of the mass polarization can be accounted for to all orders by evaluating expectation values of all operators on the eigenfunctions of the Schrödinger Hamiltonian with the mass-polarization operator $(m_r/M)\,\vec{p}_1\cdot\vec{p}_2$ included. The third effect is induced by the recoil addition to the Breit-Pauli Hamiltonian

$$H_{\rm fs,rec} = \frac{Z}{2}\frac{m}{M}\left[\frac{\vec{r}_1}{r_1^3}\times(\vec{p}_1+\vec{p}_2)\cdot\vec{\sigma}_1 + \frac{\vec{r}_2}{r_2^3}\times(\vec{p}_1+\vec{p}_2)\cdot\vec{\sigma}_2\right](1+a_e)\,. \tag{8}$$

3. The spin-dependent $m\,\alpha^6$ contribution

The $m\,\alpha^6$ contribution to the helium fine structure is a sum of the second-order perturbation corrections induced by the Breit-Pauli Hamiltonian and the expectation value of the effective fine-structure Hamiltonian to this order, $H_{\rm fs}^{(6)}$,

$$\mathcal{E}^{(6)} = \left\langle H_{\rm fs}^{(4)}\frac{1}{(E_0-H_0)'}H_{\rm fs}^{(4)}\right\rangle + 2\left\langle H_{\rm nfs}^{(4)}\frac{1}{(E_0-H_0)'}H_{\rm fs}^{(4)}\right\rangle + \left\langle H_{\rm fs}^{(6)} + H_{\rm fs,amm}^{(6)}\right\rangle\,. \tag{9}$$

Here, $1/(E_0-H_0)'$ is the reduced Green function and $H_{\rm nfs}^{(4)}$ is the spin-independent part of the Breit-Pauli Hamiltonian,

$$H_{\rm nfs}^{(4)} = -\frac{1}{8}\,(p_1^4+p_2^4) + \frac{Z\,\pi}{2}\,[\delta^3(r_1)+\delta^3(r_2)] - \frac{1}{2}\,p_1^i\left(\frac{\delta^{ij}}{r}+\frac{r^i\,r^j}{r^3}\right)p_2^j\,, \tag{10}$$

where we have omitted a term with $\delta^3(r)$ since it vanishes for the triplet states. $H_{\rm fs}^{(6)}$ consists of 15 operators first derived by Douglas and Kroll (DK) [2] in the framework of the Salpeter equation. These operators were later rederived using the much simpler effective field method in Ref. [15]. The result is

$$H_{\rm fs}^{(6)} = \sum_{i=1}^{15} B_i\,, \tag{11}$$

Table 1. Effective operators contributing to $H_{\rm fs}^{(6)}$ (left column) and H_H (right column)

Operator $\times m\,\alpha^6$	Operator $\times m\,\alpha^7/\pi$
$B_1 = -\frac{3Z}{8}\,p_1^2\,\frac{\vec{r}_1}{r_1^3}\times\vec{p}_1\cdot\vec{\sigma}_1$	$H_1 = -\frac{Z}{4}\,p_1^2\,\frac{\vec{r}_1}{r_1^3}\times\vec{p}_1\cdot\vec{\sigma}_1$
$B_2 = -Z\,\frac{\vec{r}_1}{r_1^3}\times\frac{\vec{r}}{r^3}\cdot\vec{\sigma}_1\,(\vec{r}\cdot\vec{p}_2)$	$H_2 = -\frac{3Z}{4}\,\frac{\vec{r}_1}{r_1^3}\times\frac{\vec{r}}{r^3}\cdot\vec{\sigma}_1\,(\vec{r}\cdot\vec{p}_2)$
$B_3 = \frac{Z}{2}\,\frac{\vec{r}}{r^3}\cdot\vec{\sigma}_1\,\frac{\vec{r}_1}{r_1^3}\cdot\vec{\sigma}_2$	$H_3 = \frac{3Z}{4}\,\frac{\vec{r}}{r^3}\cdot\vec{\sigma}_1\,\frac{\vec{r}_1}{r_1^3}\cdot\vec{\sigma}_2$
$B_4 = \frac{1}{2\,r^4}\,\vec{r}\times\vec{p}_2\cdot\vec{\sigma}_1$	$H_4 = \frac{1}{2\,r^4}\,\vec{r}\times\vec{p}_2\cdot\vec{\sigma}_1$
$B_5 = -\frac{1}{2\,r^6}\,\vec{r}\cdot\vec{\sigma}_1\,\vec{r}\cdot\vec{\sigma}_2$	$H_5 = -\frac{3}{4\,r^6}\,\vec{r}\cdot\vec{\sigma}_1\,\vec{r}\cdot\vec{\sigma}_2$
$B_6 = \frac{5}{8}\,p_1^2\,\frac{\vec{r}}{r^3}\times\vec{p}_1\cdot\vec{\sigma}_1$	$H_6 = \frac{1}{4}\,p_1^2\,\frac{\vec{r}}{r^3}\times\vec{p}_1\cdot\vec{\sigma}_1$
$B_7 = -\frac{3}{4}\,p_1^2\,\frac{\vec{r}}{r^3}\times\vec{p}_2\cdot\vec{\sigma}_1$	$H_7 = -\frac{1}{4}\,p_1^2\,\frac{\vec{r}}{r^3}\times\vec{p}_2\cdot\vec{\sigma}_1$
$B_8 = -\frac{i}{4}\,p_1^2\,\frac{1}{r}\,\vec{\sigma}_1\cdot(\vec{p}_1\times\vec{p}_2)$	$H_8 = -\frac{Z}{4\,r}\,\frac{\vec{r}_1}{r_1^3}\times\vec{p}_2\cdot\vec{\sigma}_1$
$B_9 = -\frac{3i}{4}\,p_1^2\,\frac{1}{r^3}\,\vec{r}\cdot\vec{p}_2\,\vec{r}\times\vec{p}_1\cdot\vec{\sigma}_1$	$H_9 = -\frac{i}{2}\,p_1^2\,\frac{1}{r^3}\,\vec{r}\cdot\vec{p}_2\,\vec{r}\times\vec{p}_1\cdot\vec{\sigma}_1$
$B_{10} = \frac{3i}{8\,r^5}\,\vec{r}\times(\vec{r}\cdot\vec{p}_2)\,\vec{p}_1\cdot\vec{\sigma}_1$	$H_{10} = \frac{3i}{4\,r^5}\,\vec{r}\times(\vec{r}\cdot\vec{p}_2)\,\vec{p}_1\cdot\vec{\sigma}_1$
$B_{11} = -\frac{3}{16\,r^5}\,\vec{r}\times(\vec{r}\times\vec{p}_1\cdot\vec{\sigma}_1)\,\vec{p}_2\cdot\vec{\sigma}_2$	$H_{11} = -\frac{3}{8\,r^5}\,\vec{r}\times(\vec{r}\times\vec{p}_1\cdot\vec{\sigma}_1)\,\vec{p}_2\cdot\vec{\sigma}_2$
$B_{12} = -\frac{1}{16\,r^3}\,\vec{p}_1\cdot\vec{\sigma}_2\,\vec{p}_2\cdot\vec{\sigma}_1$	$H_{12} = -\frac{1}{8\,r^3}\,\vec{p}_1\cdot\vec{\sigma}_2\,\vec{p}_2\cdot\vec{\sigma}_1$
$B_{13} = \frac{3}{2}\,p_1^2\,\frac{1}{r^5}\,\vec{r}\cdot\vec{\sigma}_1\,\vec{r}\cdot\vec{\sigma}_2$	$H_{13} = \frac{21}{16}\,p_1^2\,\frac{1}{r^5}\,\vec{r}\cdot\vec{\sigma}_1\,\vec{r}\cdot\vec{\sigma}_2$
$B_{14} = -\frac{i}{4}\,p_1^2\,\frac{\vec{r}}{r^3}\cdot\vec{\sigma}_1\,\vec{p}_1\cdot\vec{\sigma}_2$	$H_{14} = -\frac{3i}{8}\,p_1^2\,\frac{\vec{r}}{r^3}\cdot\vec{\sigma}_1\,\vec{p}_1\cdot\vec{\sigma}_2$
$B_{15} = \frac{i}{8}\,p_1^2\,\frac{\vec{r}}{r^3}\cdot\vec{\sigma}_1\,\vec{p}_2\cdot\vec{\sigma}_2$	$H_{15} = \frac{i}{8}\,p_1^2\,\frac{1}{r^3}\,\big(\vec{r}\cdot\vec{\sigma}_2\,\vec{p}_2\cdot\vec{\sigma}_1 + \vec{r}\cdot\vec{\sigma}_1\,\vec{p}_2\cdot\vec{\sigma}_2$
	$-\frac{3}{r^2}\,\vec{r}\cdot\vec{\sigma}_1\,\vec{r}\cdot\vec{\sigma}_2\,\vec{r}\cdot\vec{p}_2\big)$
	$H_{16} = -\frac{1}{4}\,\vec{p}_1\cdot\vec{\sigma}_1\,\vec{p}_1\times\frac{\vec{r}}{r^3}\cdot\vec{p}_2$
	$H_{17} = \frac{1}{8}\,\vec{p}_1\cdot\vec{\sigma}_1\,\big(-\vec{p}_1\cdot\vec{\sigma}_2\,\frac{1}{r^3} + 3\vec{p}_1\cdot\vec{r}\,\frac{\vec{r}}{r^5}\cdot\vec{\sigma}_2\big)$

where the B_i are given in Table 1.

The finite nuclear mass corrections to the $m\,\alpha^6$ contribution can be divided into the mass scaling, the mass polarization, and the operator parts. The mass scaling prefactor is $(m_r/M)^4$ for the B_2, B_3, B_4, and B_5, $(m_r/M)^5$ for the other B_i operators, $(m_r/M)^6$ for the second-order corrections involving the first term in Eq. (10), and $(m_r/M)^5$ for all other second-order corrections. The mass polarization effect is most easily accounted for by including the mass polarization operator in the zeroth-order Hamiltonian. The operator part comes from recoil corrections to $H_{\rm fs}^{(4)}$, $H_{\rm nfs}^{(4)}$, and $H_{\rm fs}^{(6)}$. The recoil part of $H_{\rm fs}^{(4)}$ is given by Eq. (8). The spin-independent recoil part of the Breit-Pauli Hamiltonian is

$$H_{\rm nfs,rec}^{(4)} = -\frac{Z}{2}\,\frac{m}{M}\sum_{a=1,2} p_a^i\left(\frac{\delta^{ij}}{r_a} + \frac{r_a^i r_a^j}{r_a^3}\right)(p_1^j + p_2^j)\,. \tag{12}$$

Recoil corrections to the DK operators were studied by Zhang [14] and by Pachucki and

Sapirstein [19]. The result is given by the effective Hamiltonian

$$\begin{aligned}
H^{(6)}_{\rm fs,rec} =& \frac{m}{M}\biggl[\frac{iZ}{4}p_1^2\frac{1}{r_1}\vec\sigma_1\cdot(\vec p_1\times\vec p_2)-\frac{iZ}{4}p_1^2\frac{\vec r_1}{r_1^3}(\vec\sigma_1\cdot\vec r_1\times\vec p_1)\cdot(\vec p_1+\vec p_2)\\
&-\frac{3Z}{4}p_1^2\vec\sigma_1\cdot\frac{\vec r_1}{r_1^3}\times(\vec p_1+\vec p_2)+Z\,\vec\sigma_1\cdot\frac{\vec r}{r_1\,r^3}\times(\vec p_1+\vec p_2)+Z\,\vec\sigma_1\cdot\frac{\vec r}{r^3}\times\frac{\vec r_1}{r_1^3}(\vec r_1\cdot(\vec p_1+\vec p_2))\\
&+Z^2\,\vec\sigma_1\cdot\frac{\vec r_1}{r_1^3}\times\frac{\vec r_2}{r_2^3}(\vec r_1\cdot\vec p_1)-\frac{Z^2}{2}\vec\sigma_1\cdot\frac{\vec r_1}{r_1^4}\times(\vec p_1+\vec p_2)-\frac{Z^2}{4}\vec\sigma_1\cdot\frac{\vec r_2}{r_2^3}\vec\sigma_2\cdot\frac{\vec r_1}{r_1^3}\biggr]. \qquad (13)
\end{aligned}$$

4. The spin-dependent $m\alpha^7$ correction

The $m\alpha^7$ correction to the helium fine structure can be conveniently separated into four parts

$$\mathcal{E}^{(7)}=\mathcal{E}^{(7)}_{\rm log}+\mathcal{E}^{(7)}_{\rm first}+\mathcal{E}^{(7)}_{\rm sec}+\mathcal{E}^{(7)}_{L}\,. \qquad (14)$$

The first term above combines all terms with $\ln Z$ and $\ln\alpha$ [11–13,15,20],

$$\begin{aligned}
\mathcal{E}^{(7)}_{\rm log} =& \ln[(Z\,\alpha)^{-2}]\biggl[\Bigl\langle\frac{2Z}{3}\,i\,\vec p_1\times\delta^3(r_1)\,\vec p_1\cdot\vec\sigma_1\Bigr\rangle-\Bigl\langle\frac{1}{4}(\vec\sigma_1\cdot\vec\nabla)\,(\vec\sigma_2\cdot\vec\nabla)\delta^3(r)\Bigr\rangle\\
&-\Bigl\langle\frac{3}{2}\,i\,\vec p_1\times\delta^3(r)\,\vec p_1\cdot\vec\sigma_1\Bigr\rangle+\frac{8Z}{3}\Bigl\langle H^{(4)}_{\rm fs}\frac{1}{(E_0-H_0)'}[\delta^3(r_1)+\delta^3(r_2)]\Bigr\rangle\biggr]\,. \qquad (15)
\end{aligned}$$

The second part of $\mathcal{E}^{(7)}$ is induced by effective Hamiltonians to order $m\alpha^7$. They were derived by one of us (K.P.) in Refs. [20, 21]. (The previous derivation of this correction by Zhang [11, 12] turned out to be not entirely consistent.) The result is

$$\mathcal{E}^{(7)}_{\rm first}=\Bigl\langle H_Q+H_H+H^{(7)}_{\rm fs,amm}\Bigr\rangle\,. \qquad (16)$$

The Hamiltonian H_Q is induced by the two-photon exchange between the electrons, the electron self-energy and the vacuum polarization. It is given by [20]

$$\begin{aligned}
H_Q =& Z\,\frac{91}{180}\,i\,\vec p_1\times\delta^3(r_1)\,\vec p_1\cdot\vec\sigma_1-\frac{1}{2}\,(\vec\sigma_1\cdot\vec\nabla)\,(\vec\sigma_2\cdot\vec\nabla)\,\delta^3(r)\left[\frac{83}{30}+\ln Z\right]\\
&+3\,i\,\vec p_1\times\delta^3(r)\,\vec p_1\cdot\vec\sigma_1\left[\frac{23}{10}-\ln Z\right]-\frac{15}{8\,\pi}\frac{1}{r^7}\,(\vec\sigma_1\cdot\vec r)\,(\vec\sigma_2\cdot\vec r)-\frac{3}{4\,\pi}\,i\,\vec p_1\times\frac{1}{r^3}\,\vec p_1\cdot\vec\sigma_1\,. \qquad (17)
\end{aligned}$$

Here, the terms with $\ln Z$ compensate the logarithmic dependence implicitly present in the expectation values of the singular operators $1/r^3$ and $1/r^5$, so that matrix elements of H_Q do not have any logarithms in their $1/Z$ expansion. The singular operators are defined through their integrals with the arbitrary smooth function f

$$\int d^3r\,\frac{1}{r^3}\,f(\vec r)\equiv\lim_{\epsilon\to0}\int d^3r\left[\frac{1}{r^3}\,\theta(r-\epsilon)+4\,\pi\,\delta^3(r)\,(\gamma+\ln\epsilon)\right]f(\vec r) \qquad (18)$$

and

$$\begin{aligned}
&\int d^3r\,\frac{1}{r^7}\left(r^i\,r^j-\frac{\delta^{ij}}{3}\,r^2\right)f(\vec r)\equiv \qquad (19)\\
&\lim_{\epsilon\to0}\int d^3r\left[\frac{1}{r^7}\left(r^i\,r^j-\frac{\delta^{ij}}{3}\,r^2\right)\theta(r-\epsilon)+\frac{4\,\pi}{15}\,\delta^3(r)\,(\gamma+\ln\epsilon)\left(\partial^i\,\partial^j-\frac{\delta^{ij}}{3}\,\partial^2\right)\right]f(\vec r)\,,
\end{aligned}$$

where γ is the Euler constant. The effective Hamiltonian H_H represents the anomalous magnetic moment (amm) correction to the Douglas-Kroll $m\alpha^6$ operators and is given by [20]

$$H_H = \sum_{i=1}^{17} H_i\,, \tag{20}$$

where the H_i are presented in Table 1. The last term of $\mathcal{E}^{(7)}_{\rm first}$ in Eq. (16), the Hamiltonian $H^{(7)}_{\rm fs,amm}$ is the $m\alpha^7$ amm correction to the Breit-Pauli Hamiltonian, see Eq. (2).

The third part of $\mathcal{E}^{(7)}$ is given by the second-order matrix elements of the form [20]

$$\mathcal{E}^{(7)}_{\rm sec} = 2\left\langle H^{(4)}_{\rm fs}\frac{1}{(E_0-H_0)'}H^{(5)}_{\rm nlog}\right\rangle + 2\left\langle\left[H^{(4)}_{\rm fs}+H^{(4)}_{\rm nfs}\right]\frac{1}{(E_0-H_0)'}H^{(5)}_{\rm fs}\right\rangle\,, \tag{21}$$

where $H^{(5)}_{\rm nlog}$ is the effective Hamiltonian responsible for the nonlogarithmic $m\alpha^5$ correction to the energy

$$H^{(5)}_{\rm nlog} = -\frac{7}{6\,\pi\,r^3} + \frac{38Z}{45}\left[\delta^3(r_1)+\delta^3(r_2)\right]\,. \tag{22}$$

$H^{(4)}_{\rm nfs}$ is the spin-independent part of the Breit-Pauli Hamiltonian given by Eq. (10), and $H^{(5)}_{\rm fs}$ is the $m\alpha^5$ amm correction to $H^{(4)}_{\rm fs}$, see Eq. (2).

The fourth part of $\mathcal{E}^{(7)}$ is the contribution induced by the emission and reabsorption of virtual photons of low energy. It is denoted as $\mathcal{E}^{(7)}_L$ and interpreted as the relativistic correction to the Bethe logarithm. The expression for $\mathcal{E}^{(7)}_L$ reads [16]

$$\begin{aligned}\mathcal{E}^{(7)}_L = &-\frac{2}{3\,\pi}\,\delta\left\langle(\vec{p}_1+\vec{p}_2)\cdot(H_0-E_0)\ln\left[\frac{2(H_0-E_0)}{Z^2}\right](\vec{p}_1+\vec{p}_2)\right\rangle\\ &+\frac{i\,Z^2}{3\,\pi}\left\langle\left(\frac{\vec{r}_1}{r_1^3}+\frac{\vec{r}_2}{r_2^3}\right)\times\frac{\vec{\sigma}_1+\vec{\sigma}_2}{2}\ln\left[\frac{2(H_0-E_0)}{Z^2}\right]\left(\frac{\vec{r}_1}{r_1^3}+\frac{\vec{r}_2}{r_2^3}\right)\right\rangle\,,\end{aligned} \tag{23}$$

where $\delta\,\langle\ldots\rangle$ denotes the first-order perturbation of the matrix element $\langle\ldots\rangle$ by $H^{(4)}_{\rm fs}$, implying perturbations of the reference-state wave function, the reference-state energy, and the electron Hamiltonian.

5. Results for helium fine-structure

Summary of the individual contributions to the fine-structure intervals of helium is given in Table 2. Numerical results are presented for the large ν_{01} and the small ν_{12} intervals, defined by

$$\nu_{01} = \left[E(2^3P_0)-E(2^3P_1)\right]/h \tag{24}$$

$$\nu_{12} = \left[E(2^3P_1)-E(2^3P_2)\right]/h\,. \tag{25}$$

We note that the style of breaking the total result into separate entries used in Table 2 differs from that used in the summary tables of the previous papers by Pachucki *et al.* [20, 21]. In particular, the lower-order terms listed in Table III of Ref. [20] and in Table II of Ref. [21] contained contributions of higher orders, whereas in the present work the entries in Table 2 contain only the contributions of the order specified.

A term-by-term comparison with the independent calculation by Drake [18] was performed in Ref. [23]. We observe good agreement between the two calculations for the lower-order terms,

Table 2. Summary of individual contributions to the fine-structure intervals in helium, in kHz. The parameters [25] are $\alpha^{-1} = 137.035\,999\,679(94)$, $cR_\infty = 3\,289\,841\,960\,361(22)$ kHz, and $m/M = 1.370\,933\,555\,70 \times 10^{-4}$. The label $(+m/M)$ indicates that the corresponding entry comprises both the non-recoil and recoil contributions of the specified order in α.

Term	ν_{01}	ν_{12}	ν_{02}
$m\alpha^4(+m/M)$	29 563 765.45	2 320 241.43	
$m\alpha^5(+m/M)$	54 704.04	−22 545.00	
$m\alpha^6$	−1 607.52(2)	−6 506.43	
$m\alpha^6 m/M$	−9.96	9.15	
$m\alpha^7 \log(Z\alpha)$	81.43	−5.87	
$m\alpha^7$, nlog	18.86	−14.38	
$m\alpha^8$	±1.7	±1.7	
Total theory	29 616 952.29 ± 1.7	2 291 178.91 ± 1.7	31 908 131.20 ± 1.7
Experiment	29 616 951.66(70)^a	2 291 177.53(35)^d	31 908 131.25(30)^f
	29 616 952.7(10)^b	2 291 175.59(51)^a	31 908 126.78(94)^a
	29 616 950.9(9)^c	2 291 175.9(10)^e	

^{a} Ref. [7], ^{b} Ref. [26], ^{c} Ref. [27], ^{d} Ref. [8], ^{e} Ref. [28], ^{f} Ref. [9].

namely, for the $m\alpha^4$, $m\alpha^5$, and $m\alpha^6$ corrections. However, for the recoil correction to order $m\alpha^6$, our results differ from those of Drake by about 0.5 kHz for both intervals. The reason for this disagreement seems to be different for the large and the small intervals. For the large interval, the deviation is due to the recoil operator part, whereas for the small interval, it is mainly due to the mass polarization part (see discussion in Ref. [21]).

Our present estimates of the uncalculated higher-order effects for helium are larger than those in the previous studies [17, 18]. The previous estimates were significantly less than 1 kHz. They were based on logarithmic contributions to order $m\alpha^8$ corresponding to the hydrogen fine structure. However, a larger contribution might originate from the nonlogarithmic relativistic corrections. So our present estimate is obtained by multiplying the $m\alpha^6$ contribution for the $\nu_{02} = \nu_{01} + \nu_{12}$ interval by the factor of $(Z\alpha)^2$, which yields a conservative estimate of ±1.7 kHz for all ν_{01}, ν_{12}, and ν_{02} intervals. All nuclear structure effects are completely negligible at the current precision level. The finite nuclear size correction is estimated to yield 18 Hz for ν_{01} and 6 Hz for ν_{12}.

Our result for the ν_{01} interval of helium agrees well with all recent experimental values [7, 26, 27]. For the ν_{12} interval, theoretical result is by about 2σ larger than the values obtained in Refs. [7, 28] but in agreement with the latest measurement by Hessels and coworkers [8]. Our theoretical prediction for the ν_{02} interval is in excellent agreement with the very recent measurement of Smiciklas and Shiner [9]. Comparison with this experimental result suggests that the higher-order contribution might in fact be much smaller than our conservative estimate. This means that, if an independent measurement on Li^+ or Be^{2+} confirms the smallness of the $m\alpha^8$ terms, the helium determination of α will be significantly improved. The measurement should be performed for the $2^3P_0 - 2^3P_2$ transition, since it is not affected by the singlet-triplet mixing effects, which strongly depend on Z.

In summary, the theory of the fine structure of helium and light helium-like ions is now complete up to orders $m\alpha^7$ and $\alpha^6 m^2/M$. The theoretical predictions agree with the latest experimental results for helium, as well as with most of the experimental data for light helium-like ions. A combination of the theoretical and experimental results [9] for the $2^3P_0 - 2^3P_2$ interval in helium yields an independent determination of the fine structure constant α

$$\alpha^{-1} = 137.035\,999\,55(64)(4)(368)\,, \tag{26}$$

where the first error is the experimental uncertainty, the second one is the numerical uncertainty, and the third comes from the estimate of the $m\,\alpha^8$ term (± 1.7 kHz). The result (26) is accurate to 27 ppb and in agreement with the recent value obtained from the electron g factor [29].

Acknowledgments
Support by NIST through Precision Measurement Grant PMG 60NANB7D6153 is gratefully acknowledged.

References

[1] Schwartz C 1964 Phys. Rev. **134** A1181
[2] Douglas M and Kroll N 1974 Ann. Phys. (NY) **82** 89
[3] Hambro L 1972 Phys. Rev. A **5** 2027
[4] Hambro L 1972 Phys. Rev. A **6** 865
[5] Hambro L 1973 Phys. Rev. A **7** 479
[6] Lewis M L and Serafino P H 1978 Phys. Rev. A **18** 867
[7] Zelevinsky T, Farkas D and Gabrielse G 2005 Phys. Rev. Lett. **95** 203001
[8] Borbely J S, George M C, Lombardi L D, Weel M, Fitzakerley D W and Hessels E A 2009 Phys. Rev. A **79** 0605030(R)
[9] Smiciklas M and Shiner D 2010 Phys. Rev. Lett. **105** 123001
[10] Yan Z -C and Drake G W F 1995 Phys. Rev. Lett. **74** 4791
[11] Zhang T 1996 Phys. Rev. A **54** 1252
[12] Zhang T 1996 Phys. Rev. A **53** 3896
[13] Zhang T, Yan Z -C and Drake G W F 1996 Phys. Rev. Lett. **77** 1715
[14] Zhang T 1997 Phys. Rev. A **56** 270
[15] Pachucki K 1999 J. Phys. B **32** 137
[16] Pachucki K and Sapirstein J 2000 J. Phys. B **33** 5297
[17] Pachucki K and Sapirstein J 2002 J. Phys. B **35** 1783
[18] Drake G W F 2002 Can. J. Phys. **80** 1195
[19] Pachucki K and Sapirstein J 2003 J. Phys. B **36** 803
[20] Pachucki K 2006 Phys. Rev. Lett. **97** 013002
[21] Pachucki K and Yerokhin V A 2009 Phys. Rev. A **79** 062516; *ibid.* 2009 **80** 019902(E); *ibid.* 2010 **81** 039903(E)
[22] Pachucki K and Yerokhin V A 2010 Phys. Rev. Lett. **104** 070403
[23] Pachucki K and Yerokhin V A 2010 Can. J. Phys. *in print*
[24] Aoyama T, Hayakawa M, Kinoshita T and Nio M 2007 Phys. Rev. Lett. **99** 110406
[25] Mohr P J, Taylor B N and Newell D B 2008 Rev. Mod. Phys. **80** 633
[26] Giusfredi G, Pastor P C, Natale P D, Mazzotti D, de Mauro C, Fallani L, Hagel G, Krachmalnicoff V and Inguscio M 2005 Can. J. Phys. **83** 301
[27] George M C, Lombardi L D and Hessels E A 2001 Phys. Rev. Lett. **87** 173002
[28] Castillega J, Livingston D, Sanders A and Shiner D 2000 Phys. Rev. Lett. **84** 4321
[29] Hanneke D, Fogwell S and Gabrielse G 2008 Phys. Rev. Lett. **100** 120801

The size of the proton and the deuteron

Randolf Pohl[1,13], Fernando D. Amaro[3], Aldo Antognini[1,12], François Biraben[2], João M. R. Cardoso[3], Daniel S. Covita[3,4], Andreas Dax[5], Satish Dhawan[5], Luis M. P. Fernandes[3], Adolf Giesen[6,10], Thomas Graf[6], Theodor W. Hänsch[1,13], Paul Indelicato[2], Lucile Julien[2], Cheng-Yang Kao[7], Paul Knowles[8], Eric-Olivier Le Bigot[2], Yi-Wei Liu[7], José A. M. Lopes[3], Livia Ludhova[8], Cristina M. B. Monteiro[3], Françoise Mulhauser[8], Tobias Nebel[1], François Nez[2], Paul Rabinowitz[9], Joaquim M. F. dos Santos[3], Lukas A. Schaller[8], Karsten Schuhmann[10], Catherine Schwob[2], David Taqqu[11], João F. C. A. Veloso[4] and Franz Kottmann[12]

[1] Max–Planck–Institut für Quantenoptik, Garching, Germany.
[2] Laboratoire Kastler Brossel, École Normale Supérieure, CNRS and Université P. et M. Curie, Paris, France.
[3] Universidade de Coimbra, Portugal.
[4] Universidade de Aveiro, Portugal.
[5] Physics Department, Yale University, USA.
[6] IFSW, Universität Stuttgart, Germany.
[7] National Tsing Hua University, Taiwan.
[8] Université de Fribourg, Switzerland.
[9] Department of Chemistry, Princeton University, USA.
[10] Dausinger & Giesen GmbH, Stuttgart, Germany.
[11] Paul Scherrer Institute, Switzerland.
[12] ETH Zürich, Switzerland.
[13] Ludwig-Maximilians-Universität, Munich, Germany

E-mail: `randolf.pohl@mpq.mpg.de`

Abstract. We have recently measured the $2S_{1/2}^{F=1} - 2P_{3/2}^{F=2}$ energy splitting in the muonic hydrogen atom μp to be 49881.88 (76) GHz. Using recent QED calculations of the fine-, hyperfine, QED and finite size contributions we obtain a root-mean-square proton charge radius of $r_p = 0.84184\,(67)$ fm. This value is ten times more precise, but 5 standard deviations smaller, than the 2006 CODATA value of $r_p = 0.8768\,(69)$ fm. The source of this discrepancy is unknown. Using the precise measurements of the 1S-2S transition in regular hydrogen and deuterium and our value of r_p we obtain improved values of the Rydberg constant, $R_\infty = 10973731.568160\,(16)\ \mathrm{m}^{-1}$, and the rms charge radius of the deuteron $r_d = 2.12809\,(31)$ fm.

1. Introduction

The hydrogen atom (H) is the simplest of all atoms, and this simplicity is beautiful. Quantum electrodynamics (QED), originally motivated by the discovery of the Lamb shift in hydrogen [1],

Journal of Physics: Conference Series **264** (2011) 012008 doi:10.1088/1742-6596/264/1/012008

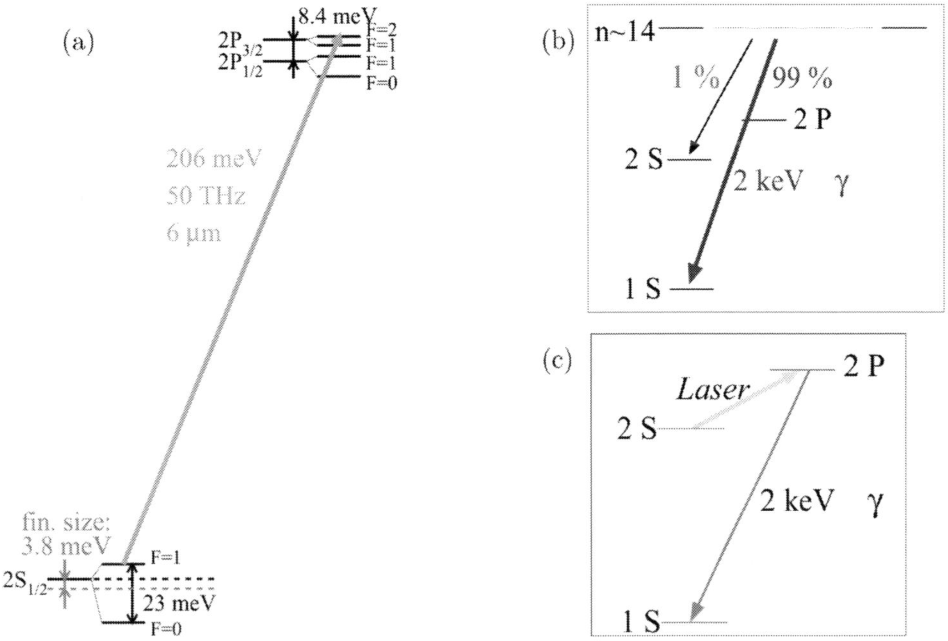

Figure 1. (a) The $n = 2$ levels in muonic hydrogen. Vacuum polarization is the dominant contribution of the 2S Lamb shift of 202 meV. The finite size effect is as large as 2% of the total Lamb shift. The measured $2S_{1/2}^{F=1} - 2P_{3/2}^{F=2}$ transition is indicated in green. The 1S ground state is 1.9 keV below the $n = 2$ states plotted here. (b,c) Experimental principle. (b) 99% of the muons stopped in H_2 gas at 1 mbar pressure proceed directly to the 1S ground state thereby emitting x-rays of the Lyman series around 2 keV. 1% of the muons form long-lived metastable 2S states with a lifetime $\tau_{2S} = 1\,\mu s$ at 1 mbar. (c) Laser light of suitable wavelength around $\lambda = 6\,\mu m$ drives the 2S-2P transition. The 2P state de-excites within 8 ps to the 1S ground state via emission of a Lyman-α x-ray at 1.9 keV.

can be used to accurately calculate the transition frequencies in the hydrogen atom [2, 3]. High-precision spectroscopy in hydrogen [4–9] can then be used to test the laws of physics, and to deduce accurate fundamental physical constants [10].

For more than a decade, the test of bound-state QED using hydrogen has been hampered by the lack of an accurate value of the proton rms charge radius, r_p. This quantity is required in the calculation of the nuclear size effects on the energy levels (in particular the S-states) in the hydrogen atom. For example, the $1S$ ground state of H is shifted by as much as 1 MHz.

Traditionally, the proton's rms charge radius has been determined by electron scattering [11–14]. A recent reevaluation of the world e-p scattering data has resulted in $r_p = 0.897(18)$ fm [15, 16], i.e., with a relative uncertainty $u_r = 1.8\,\%$. Very recently, a preliminary value $r_p = 0.879(8)$ fm from a new measurement in Mainz has become available [17].

Both spectroscopy and QED calculations of H have reached a truly astonishing accuracy. Although the finite size effect on the S states in H enters only at the 10^{-10} level, the proton charge radius deduced from the hydrogen measurements $r_p = 0.880(8)$ fm (see Tab. XLV of Ref. [10], adjustment 10), is twice more accurate than the e-p scattering world average [15, 16]. The 2006 CODATA value of $r_p = 0.8768(69)$ fm [10], is hence completely dominated by H spectroscopy.

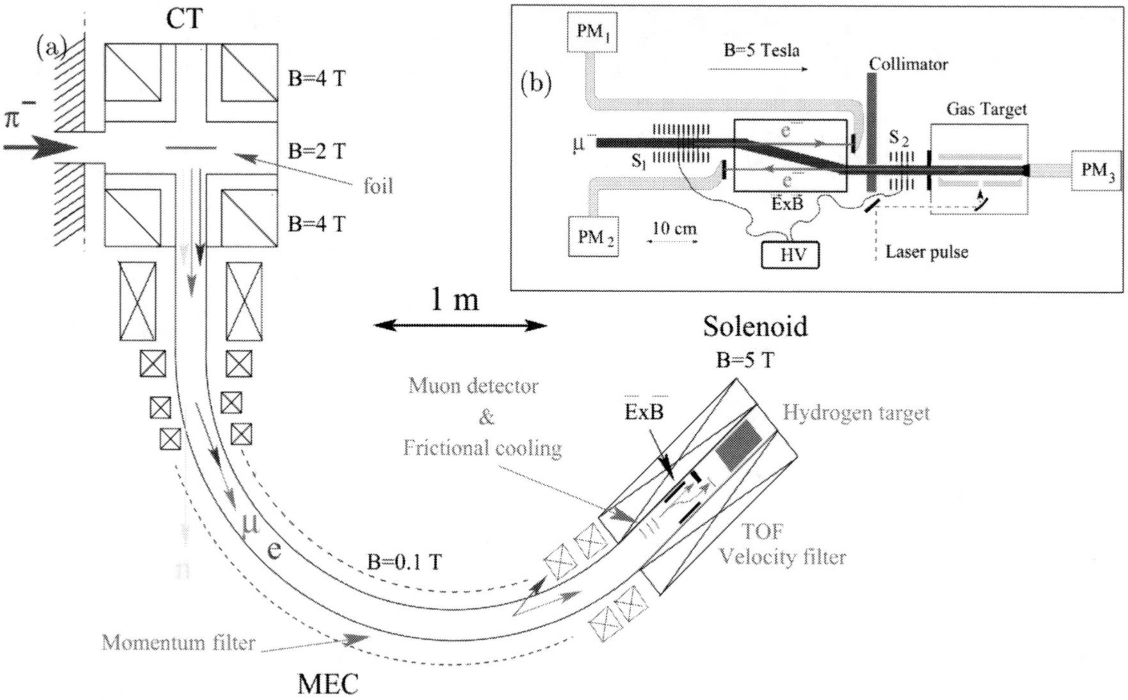

Figure 2. (a) Low-energy negative muon beam line: Negative pions are injected into the cyclotron trap (CT) and decay into muons μ^- which are decelerated in a foil. Slow μ^- leave the CT axially and are separated from background in the muon extraction channel (MEC). The experiment takes place in the 5 T solenoid shown in (b): Slow μ^- pass two stacks of ultra-thin carbon foils ($S_{1,2}$) where a few electrons are ejected. The e^- are detected in plastic scintillators read out by PMTs ($PM_{1..3}$), while the slower μ^- are separated from the e^- by an $\vec{E} \times \vec{B}$ drift in a capacitor. The delayed coincidence [$PM_1 \wedge (PM_2 \vee PM_3)$] triggers the laser system (Fig. 3). The μ^- stops inside the gas target filled with 1 mbar of H_2 gas. The stop volume is illuminated by the laser pulse injected into a multi-pass cavity, and viewed by 20 LAAPDs (not shown).

2. Muonic hydrogen

The leading order finite size effect on the S states in hydrogen-like systems with low Z is

$$E_{\rm fin.size} = \frac{2}{3}\left(\frac{m_r}{m_e}\right)^3 \frac{(Z\alpha)^2}{n^3} m_e c^2 \left(\frac{2\pi Z\alpha\, r_N}{\lambda_C}\right)^2, \tag{1}$$

where Z is the nuclear charge, α the fine structure constant, n the principal quantum number, r_N the nuclear radius, λ_C the Compton wavelength of the electron, m_e the electron mass, and

$$m_r = \frac{m_e\, m_N}{m_e + m_N} \tag{2}$$

is the reduced mass of the hydrogen-like system with nuclear mass m_N. One observes immediately the scaling of the finite size effect with the third power of the reduced mass m_r. This scaling reflects the overlap of the electron's wave function with the nucleus.

Muonic hydrogen μp is a hydrogen atom where the proton p is orbited by a negative muon μ^-, instead of an electron. The muon mass $m_\mu = 207\, m_e$, and so the reduced mass m_r is 186 times larger for μp than for H. This results in a 186 times smaller Bohr radius of muonic hydrogen, and hence a dramatic enhancement of the finite size effect on the S states of $186^3 \approx 6 \cdot 10^6$.

Journal of Physics: Conference Series **264** (2011) 012008 doi:10.1088/1742-6596/264/1/012008

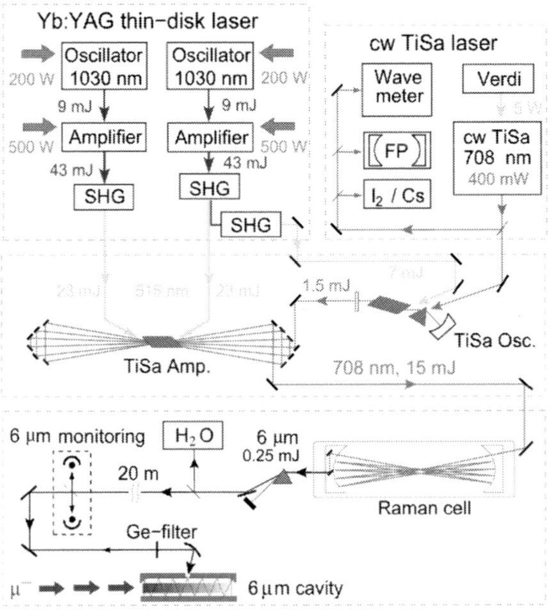

Figure 3. Laser system. A cw pumped Yb:YAG thin disk laser can deliver light within $\sim$ 200 ns of a randomly occurring trigger, given by a single detected muon. After frequency doubling (SHG) the green light is used to pump a pulsed MOPA TiSa laser system. The oscillator of the pulsed TiSa is injection-seeded with cw light from a frequency controlled TiSa ring laser. The TiSa pulses are converted to the required IR light at 6 μm via three sequential Raman Stokes shifts in a high-pressure (16 bar) H_2 Raman cell. Water vapour absorption spectroscopy is used to calibrate the IR light at 6 μm.

The classical Lamb shift (2S-2P energy difference) in electronic hydrogen H is dominated by the self energy (SE) of the electron, and the finite size effect contributes to the Lamb shift only at the $1 \cdot 10^{-4}$ level. In contrast, vacuum polarization (VP) dominates the Lamb shift in muonic hydrogen μp, and the finite size effect is as large as 2 % of the Lamb shift (see Fig. 1). This is the reason why we set out in 1998 to measure the $2S_{1/2}^{F=1} - 2P_{3/2}^{F=2}$ energy difference in μp by means of pulsed laser spectroscopy at $\lambda \approx 6\,\mu$m.

The energy difference $\Delta\widetilde{E}$ between the $2S_{1/2}^{F=1}$ and the $2P_{3/2}^{F=2}$ states in muonic hydrogen is the sum of radiative, recoil, and proton structure contributions, and the fine and hyperfine splittings for this particular transition. It is given by [2, 18–22]

$$\Delta\widetilde{E} = 209.9779\,(49) - 5.2262\,r_{\rm p}^2 + \; 0.0347\,r_{\rm p}^3 \text{ meV} \tag{3}$$

where $r_{\rm p} = \sqrt{< r_{\rm p}^2 >}$ is the rms charge radius of the proton, given in fm. The uncertainty of 0.0049 meV in $\Delta\widetilde{E}$ is dominated by the proton polarizability term [20] of 0.015(4) meV. A detailed derivation of Eq. (3) is given in the Supplementary Information of Ref [23].

3. Experiment

The details of the experiment have been given elsewhere [23–29]. In brief, single low-energy negative muons ($\sim$ 5 keV kinetic energy) from our novel beam line [26] (Fig. 2 (a)) at the Paul-Scherrer-Institute (PSI, Switzerland) enter a 5 T solenoid and are individually detected through secondary electrons emitted in 20 nm thin carbon foils (Fig. 2 (b)). The electrons are detected in plastic scintillators which are read out by photomultipliers ($PM_{1..3}$), a delayed coincidence of which provides the trigger signal for the laser [30, 31] (Fig. 3).

The muons are then stopped in a low-pressure hydrogen gas target filled with 1 mbar H_2 gas at room temperature. About 1 % of the muons form long-lived μp(2S) atoms with a lifetime of $\sim 1\,\mu$s at this gas pressure [32–35] (see Fig. 1(b)).

The pulsed laser system (Fig. 3) basically pulse-amplifies the cw light from a frequency controlled cw Ti:sapphire (TiSa) laser, creating 5 ns long pulses of 15 mJ energy at $\lambda \approx 708$ nm. After three sequential vibrational Stokes shifts in a Raman cell [36] filled with 16 bar of H_2 gas,

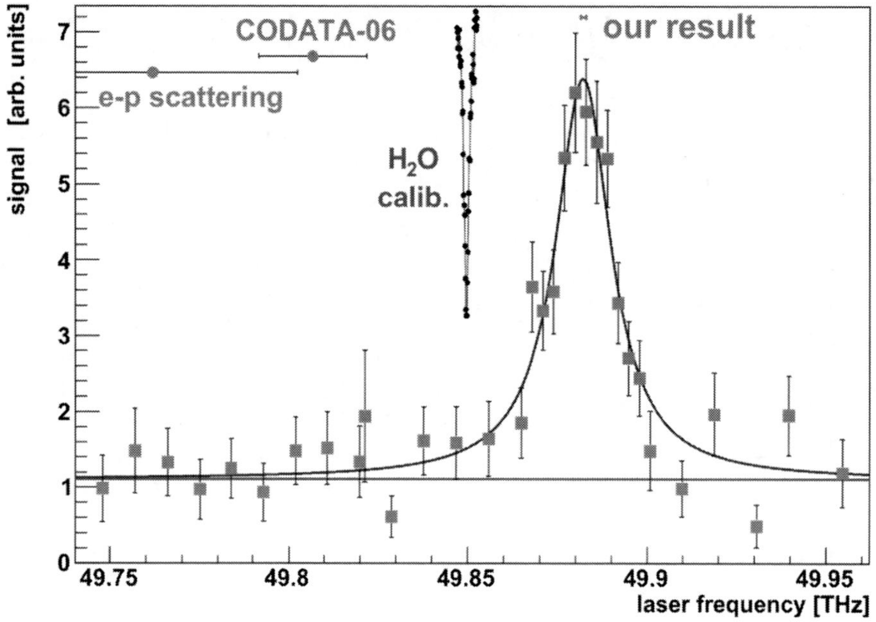

Figure 4. The measured resonance, a water absorption measurement used for calibration, and the predicted line positions using the proton rms charge radius from electron scattering [15, 16] and CODATA [10].

0.2 mJ pulses at $\lambda \approx 6\,\mu$m enter a multi-pass cavity [28] surrounding the muon stop volume. We detect x-rays using large-area avalanche photo diodes (LAPPDs). Our signal (Fig. 4) is the number of K_α x-rays detected at the time when the laser illuminates the muon stop volume (Fig. 1(c)), *versus* the laser frequency.

Tuning the cw TiSa by some frequency offset $\Delta\nu$ changes the IR light at $\lambda \approx 6\,\mu$m by the same $\Delta\nu$. Although the frequency of each laser in the step is well-known, we rely on well-known water vapour absorption lines [37, 38] for the final laser frequency calibration. The absolute water vapour line positions are known to 1 MHz [37, 38], but pulse to pulse instabilities of our laser system limit the frequency determination to 300 MHz.

3.1. Lamb shift in muonic hydrogen

The centre of the $2S_{1/2}^{F=1}$ to $2P_{3/2}^{F=2}$ transition in μp is at 49881.88(76) GHz [23]. The uncertainty of 0.76 GHz (15 ppm) contains 700 MHz statistical uncertainty from the free fit of a Lorentzian resonance line on top of a flat background, and the 300 MHz total systematic uncertainty which is exclusively due to our laser wavelength calibration procedure using H_2O vapour absorption.

Other systematic effects we have considered are Zeeman shift in the 5 T field (< 30 MHz), AC and DC Stark shifts (< 1 MHz), Doppler shift (< 1 MHz) and pressure shift (< 2 MHz). Our measured resonance position is not influenced by molecular effects, because the formed muonic molecules ppμ^+ are known to deexcite quickly [34, 39] and cannot contribute to our signal. Also, the width of our resonance line of 18.0(2.2) GHz agrees with the expected width of 20(1) GHz, whereas molecular lines would be wider.

The free fit gives $\chi^2 = 28.1$ for 28 degrees of freedom (dof). A fit of a flat line, assuming no resonance, gives $\chi^2 = 283$ for 31 dof, making this resonance line 16 σ significant. The frequency of the resonance centre corresponds to an energy of $\Delta\widetilde{E} = 206.2949(32)$ meV (see Eq. (3)).

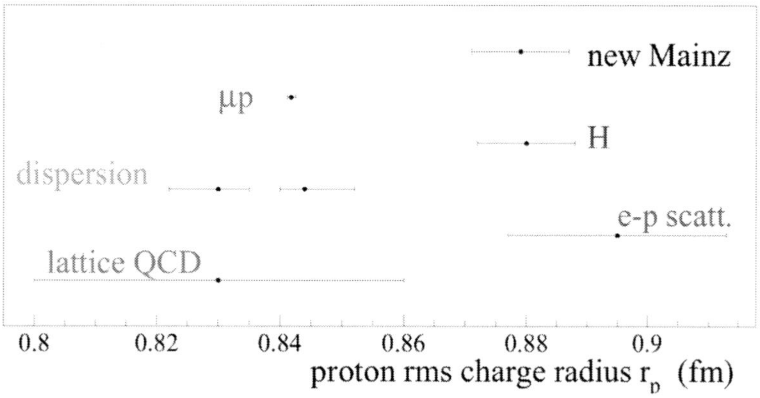

Figure 5. Proton charge radius. Our new value r_p= 0.84184(67) fm from muonic hydrogen (μp) spectroscopy [23] is in strong disagreement with the values extracted from hydrogen spectroscopy ("H", Ref. [10], see Tab. XLV, adjustment 10), the world average from electron scattering ("e-p scatt.") [15, 16], and the new electron scattering value from Mainz ("new Mainz") [17]. It agrees, however, nicely with "Lattice QCD" from Ref. [40], and "dispersion" (two model-dependent results from Ref. [41]).

3.2. The rms charge radius of the proton

The rms charge radius of the proton we find is

$$r_p = 0.84184(36)(56)\,\text{fm}. \tag{4}$$

The first uncertainty is due to the experimental uncertainty of 0.76 GHz (700 MHz statistical and 300 MHz systematic, added in quadrature), and the second uncertainty originates from the first term in Eq. (3). Theory, mainly the proton polarizability, gives the dominant contribution to our total relative uncertainty of 8×10^{-4}. Our experimental precision would allow us to deduce r_p two times better, with a relative uncertainty of 4×10^{-4}.

3.3. The proton radius puzzle

This new value of the proton radius r_p=0.84184(67) fm is 10 times more precise, but 5.0 σ smaller, than the 2006 CODATA value r_p = 0.8768(69) fm [10], which is dominated by spectroscopy in regular hydrogen (H). Our new r_p is 26 times more accurate, but 3.1 σ smaller, than the previously accepted, hydrogen-independent value extracted from electron proton scattering [15, 16] of r_p = 0.895(18) fm. Furthermore, new data from the Mainz MAMI electron accelerator [17] give r_p = 0.879(8) fm, in agreement with the hydrogen result as given by "adjustment 10" in Tab. XLV of Ref. [10].

Recent lattice QCD calculations [40], on the other hand, obtain r_p = 0.83(3) fm, favouring a lower radius than the one from H or electron scattering. Also, dispersion analysis of the nucleon form factors has recently [41] also produced smaller values of r_p: Their "SC approach" gives $r_p = 0.844^{+0.008}_{-0.004}$ fm, in agreement with our accurate value, whereas their "explicit pQCD approach" gives an even smaller value of $r_p = 0.830^{+0.005}_{-0.008}$ fm. The situation is summarized in Fig. 5.

3.4. A new value of the Rydberg constant

Assuming for now the correctness of the QED calculations in hydrogen [2, 3] and μp [18–22], we can use our precise value of r_p and the most accurately measured transition frequency in hydrogen (1S-2S) [6, 7] to deduce a new value of the Rydberg constant.

$$R_\infty = 10973731.568160(16)\,\text{m}^{-1}. \tag{5}$$

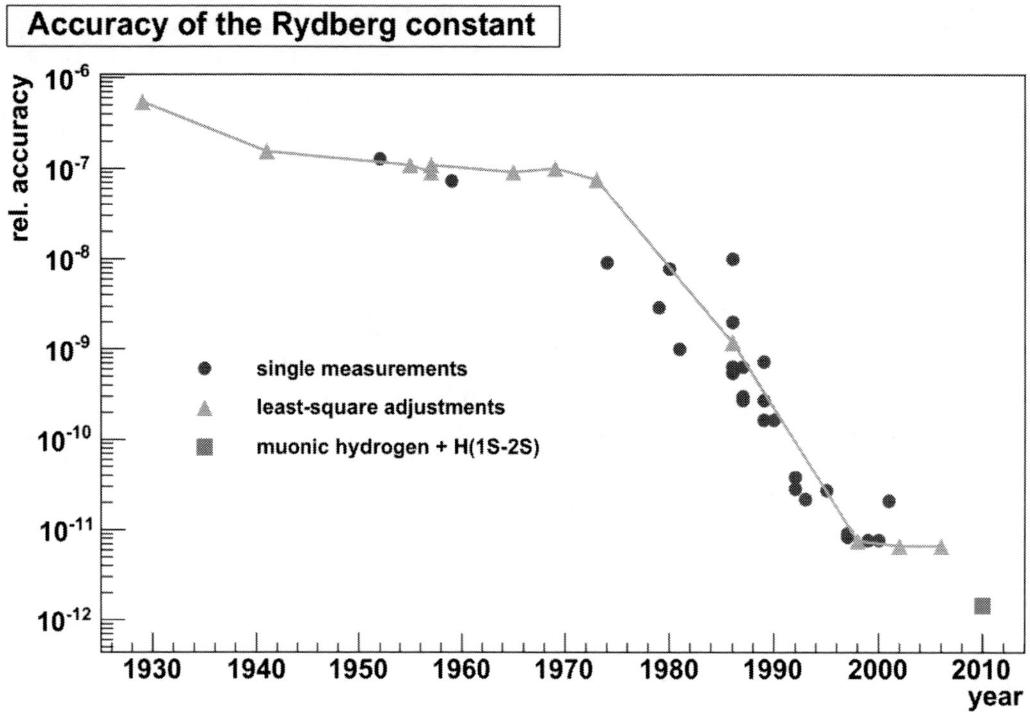

Figure 6. The Rydberg constant is a corner stone of the CODATA adjustment of fundamental constants [10]. Its accuracy is now 1.5 parts in 10^{12}, using our r_{p} and the super-accurate H(1S-2S) transition [6, 7].

This is $-110\,\mathrm{kHz/c}$ or $4.9\,\sigma$ away from the CODATA value [10], but 4.6 times more precise [1.5 parts in 10^{12}]. The new determination continues the astonishing improvement in the accuracy of the most accurately determined fundamental physical constant (Fig. 6).

3.5. The charge radius of the deuteron

The *difference* of the squared charge radii of the proton and the deuteron, $r_{\mathrm{d}}^2 - r_{\mathrm{p}}^2 = 3.82007(65)\,\mathrm{fm}^2$, is accurately determined from the precise measurement of the isotope shift of the 1S-2S transition in regular hydrogen and deuterium atoms [42]. This, together with our value of r_{p} , gives for the rms charge radius of the deuteron

$$r_{\mathrm{d}} = 2.12809(31)\,\mathrm{fm}. \quad (6)$$

Figure 7 compares this to recent results. The CODATA value of $r_{\mathrm{d}} = 2.1394(28)\,\mathrm{fm}$ is 4σ away. This can be understood because in the CODATA adjustment r_{d} is rigidly tied to r_{p} via the very precise isotope shift of the 1S-2S transition in regular H and D atoms. The proton charge radius deduced from H and μp disagree by $5\,\sigma$ for unknown reasons; hence the deuteron radius must also disagree by a large amount.

However, adjustment 11 of the 2006 CODATA adjustment (see Ref. [10] Tab. XLV), which uses only the deuterium data (scattering and spectroscopy), but ignores both electron scattering and spectroscopy on hydrogen, suggests a smaller value of r_{d}, in accord with our value. This is due to the fact that the value of r_{d} from electron-deuteron scattering [43] agrees with ours.

Neutron-proton scattering [44] gives a value which agrees both with the electron scattering value and with our new value, but not very well with the full CODATA result.

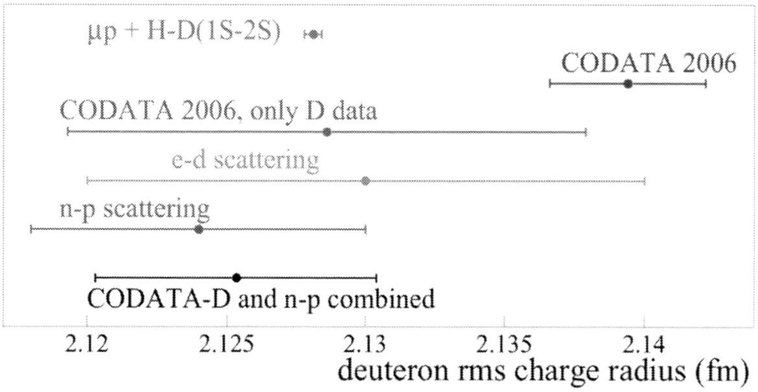

Figure 7. Our deuteron charge radius $r_d = 2.12809(31)$ fm deduced from our r_p together with the H-D (1S-2S) isotope shift [42] disagrees with the CODATA value of r_d, but agrees with the CODATA 2006 adjustment 11 which uses only deuterium data (Ref. [10], see Tab. XLV, adj. 11), the value from electron scattering [43], and the value from neutron-proton scattering [44].

The average of the independent values "CODATA, D only" [10] and neutron scattering [44] is r_d= 2.1254(50) fm, 2.4σ away from the final CODATA value $r_d = 2.1394(28)$ fm, but in good agreement with our result.

4. Conclusions and Outlook

The world's most precise value of the rms proton charge radius $r_p = 0.84184(67)$ fm that we have obtained from laser spectroscopy of the Lamb shift in muonic hydrogen μp has created a puzzle. The disagreement with the previous values from hydrogen spectroscopy and electron scattering is stunning.

Using this new value of r_p and the accurately measured hydrogen-deuterium isotope shift [42] we obtain $r_d = 2.12809(31)$ fm. This value agrees with several *hydrogen*-independent results.

Our new project, the measurement of the Lamb shift in muonic helium ions, will hopefully contribute to the solution of the "proton size puzzle".

References

[1] Lamb W E and Retherford R C 1947 *Phys. Rev.* **72** 241
[2] Eides M I, Grotch H and Shelyuto V A 2001 *Phys. Rep.* **342** 63
[3] Karshenboim S G 2005 *Physics Reports* **422** 1
[4] Lundeen S R and Pipkin F M 1981 *Phys. Rev. Lett.* **46** 232
[5] Hagley E W and Pipkin F M 1994 *Phys. Rev. Lett.* **72** 1172
[6] Niering M *et al.* 2000 *Phys. Rev. Lett.* **84** 5496
[7] Fischer M *et al.* 2004 *Phys. Rev. Lett.* **92** 230802
[8] de Beauvoir B *et al.* 2000 *Eur. Phys. J. D* **12** 61
[9] Schwob C *et al.* 1999 *Phys. Rev. Lett.* **82** 4960
[10] Mohr P J, Taylor B N and Newell D B 2008 *Rev. Mod. Phys.* **80** 633
[11] Lehmann P, Taylor R and Wilson R 1962 *Phys. Rev.* **126** 1183
[12] Hand L N, Miller D G and Wilson R 1963 *Rev. Mod. Phys.* **35** 335
[13] Murphy J J, Shin Y M and Skopik D M 1974 *Phys. Rev. C* **9** 2125, erratum *ibid.* **10** 2111
[14] Simon G G, Schmitt C, Borowski F and Walther V H 1990 *Nucl. Phys. A* **333** 381
[15] Sick I 2003 *Phys. Lett. B* **576** 62
[16] Blunden P G and Sick I 2005 *Phys. Rev. C* **72** 057601
[17] Bernauer J C *et al.* 2010 *ArXiv* (*Preprint* `nucl-ex 1007.5076`)
[18] Pachucki K 1996 *Phys. Rev. A* **53** 2092
[19] Pachucki K 1999 *Phys. Rev. A* **60** 3593

Journal of Physics: Conference Series **264** (2011) 012008 doi:10.1088/1742-6596/264/1/012008

[20] Borie E 2005 *Phys. Rev. A* **71** 032508
[21] Martynenko A P 2005 *Phys. Rev. A* **71** 022506
[22] Martynenko A P 2008 *Physics of Atomic Nuclei* **71** 125
[23] Pohl R, Antognini A, Nez F *et al.* 2010 *Nature* **466**(7303) 213
[24] Taqqu D *et al.* 1999 *Hyp. Interact.* **119** 311
[25] Pohl R *et al.* 2000 *Hyp. Interact.* **127** 161
[26] Kottmann F *et al.* 2001 *Hyp. Interact.* **138** 55
[27] Kottmann F *et al.* 2001 *AIP Conf. Proc.* **564** 13
[28] Pohl R *et al.* 2005 *Can. J. Phys.* **83** 339
[29] Nebel T *et al.* 2007 *Can. J. Phys.* **85** 469
[30] Antognini A *et al.* 2005 *Optics Communications* **253** 362
[31] Antognini A *et al.* 2009 *IEEE J. Quant. Electr.* **45** 993
[32] Pohl R 2001, Ph.D. thesis 14096, ETH Zurich, Switzerland
URL `http://e-collection.ethbib.ethz.ch/view/eth:23936`
[33] Pohl R *et al.* 2001 *Hyp. Interact.* **138** 35
[34] Pohl R *et al.* 2006 *Phys. Rev. Lett.* **97** 193402
[35] Pohl R 2009 *Hyp. Interact.* **193** 115
[36] Rabinowitz P, Perry B and Levinos N 1986 *IEEE J. Quant. Electr.* **22** 797
[37] Toth R A 1998 *J. Molec. Spectr.* **190** 379
[38] Rothman L S *et al.* 2009 *J. Quant. Spectr. and Rad. Transfer* **110** 533
[39] Kilic S, Karr J P and Hilico L 2004 *Phys. Rev. A* **70** 042506
[40] Wang P, Leineweber D B, Thomas A W and Young R D 2009 *Phys. Rev. D* **79** 094001
[41] Belushkin M A, Hammer H W and Meissner U G 2007 *Phys. Rev. C* **75** 035202
[42] Parthey C G *et al.* 2010 *Phys. Rev. Lett.* **104** 233001
[43] Sick I and Trautmann D 1998 *Nucl. Phys. A* **637** 559
[44] Babenko V A and Petrov N M 2008 *Phys. At. Nucl.* **71** 1730

Journal of Physics: Conference Series **264** (2011) 012009 doi:10.1088/1742-6596/264/1/012009

Gravitational redshift, equivalence principle, and matter waves

Michael A. Hohensee, Brian Estey, Francisco Monsalve, Geena Kim, Pei-Chen Kuan, Shau-Yu Lan, and Holger Müller

Department of Physics, University of California, Berkeley, CA 94720, USA

E-mail: hm@berkeley.edu

Abstract. We review matter wave and clock comparison tests of the gravitational redshift. To elucidate their relationship to tests of the universality of free fall (UFF), we define scenarios wherein redshift violations are coupled to ("type II"), or independent of ("type III"), violations of UFF, respectively. Clock comparisons and atom interferometers are sensitive to similar effects in both scenarios.

According to the Einstein Equivalence Principle, the laws of Special Relativity hold locally in any freely falling frame. As a result, lowering a clock into a gravitational potential by ΔU will slow it down by a factor of $1+\Delta U/c^2$, where c is the velocity of light. This gravitational redshift was the first consequence of General Relativity described by Einstein [1], and its measurement continues to be fundamental to our confidence in the theory. Clock comparison tests have measured the redshift with an accuracy of 7×10^{-5} [2], while matter wave tests, in which redshift anomalies modify material particles' Compton frequencies, have reached accuracies of 7×10^{-9} [3]. However, energy and momentum conservation intimately links the gravitational redshift to another consequence of the Einstein Equivalence Principle, the universality of free fall (UFF) [4, 5, 6]. The indirect bounds on redshift violations derived from such tests can be more stringent than the current direct bounds. Here, we review clock comparison and matter wave interference experiments, showing that modern redshift experiments can nevertheless bound anomalies that are presently poorly constrained. Using the Standard Model Extension (SME) [7], we present explicit lagrangians for redshift violations. In addition, we note that the most commonly used model for redshift violation does not produce physically measurable effects, although these can be recovered in more subtle scenarios. We identify the similarities and differences between matter wave and clock comparison tests.

The proper time interval $d\tau$ in General Relativity is given by [8]

$$c\,d\tau = \sqrt{-g_{\mu\nu}\dot{x}^\mu\dot{x}^\nu}\,d\lambda, \tag{1}$$

where $\dot{x}^\mu = dx^\mu/d\lambda$ and λ is the affine parameter that parameterizes the trajectory. This expression combines the gravitational redshift as a function of location r with special relativistic time dilation. The metric $g_{\mu\nu}$ (where $\mu,\nu = 0,1,2,3$) encodes the properties of space-time and the x^μ are the coordinates. In particular, the proper times of two clocks at rest at different locations x^μ and x'^μ have a ratio $d\tau/d\tau' = \sqrt{g_{00}/g'_{00}}$, which is the expression for the gravitational redshift in General Relativity. Since a freely falling particle seeks the path that

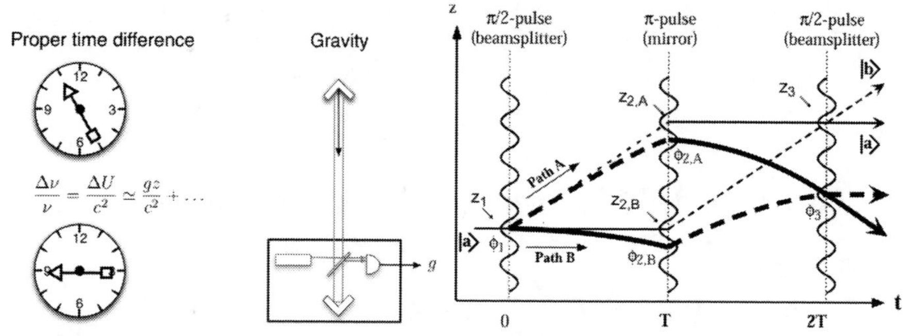

Figure 1. Left: Generic redshift experiment: Two clocks at different locations are compared; the gravitational potential difference ΔU between the two locations is measured by monitoring a geodesic trajectory. **Right:** Mach-Zehnder atom interferometer. A laser-cooled atom encounters three laser pulses that act as beam splitters and mirrors for the matter wave. Each of these pulses transfers the momentum $\hbar k$, where $k = k_1 + k_2$, to the atom The first laser pulse places the atom into a coherent superposition of two quantum states, which physically separate. The second pulse redirects the atom momentum so that the paths merge at the time of the third pulse.

maximizes its proper time, the gravitational redshift as given by g_{00} is directly responsible for the gravitational force. For a particle moving radially in the Schwarzschild metric in Schwarzschild coordinates t, r, θ, ϕ, this expression reduces to

$$\tau = \int \sqrt{-g_{00}(r) - g_{rr}v^2 dt} \approx \int \left(1 + \frac{U}{c^2} - \frac{1}{2}\frac{v^2}{c^2}\right) dt = \int \left(1 + \frac{gz}{c^2} - \frac{1}{2}\frac{v^2}{c^2}\right) dt \tag{2}$$

for a clock moving with velocity v in the radial direction. For taking the nonrelativistic limit, we dropped a constant term and denoted $z \ll r$ the vertical coordinate in the laboratory frame and g the acceleration of free fall.

Figure 1 (left) shows a generic redshift measurement between two stationary clocks at different locations. In past experiments, exchange of electromagnetic signals (*e.g.*, in Ref. [9]) or clock transport (in Ref. [10]) has been applied for making the comparison. Note that the redshift is $\Delta U/c^2$, independent of the method used for the comparison.

In order to compare the measured redshift with the theoretical prediction, the gravitational potential difference between the two clocks has to be known. This knowledge is obtained by monitoring the geodesic trajectory of an object. In satellite redshift tests [2, 11], this may be accomplished by monitoring the trajectory of the craft itself. For a laboratory experiment, the inhomogeneity of the gravitational field can usually be ignored to first order, so that the measurement of the potential difference $\Delta U = gz + \mathcal{O}(z^2)$ reduces to a measurement of the clock's separation z and the gravitational acceleration g. Since the geodesic trajectory maximizes the proper time, a change of the gravitational redshift might change the outcome of the measurement of the potential difference. If the outcome of both measurements change by the same amount, no anomaly will be detectable. It is therefore important to study the phenomenology of redshift violations, to see in which cases a meaningful (*i.e.*, measurable) anomaly is obtained.

Do matter wave interferometers really measure the redshift?
Peter Wolf and colleagues [12] have pointed out that the contribution of the gravitational redshift on the atom's proper time is canceled by the time dilation due to the atom's motion, as the motion is determined by the principle of least action from the same Lagrangian that

also determines the redshift. They also argue that redshift tests based on clock comparisons, in contrast to AI tests, measure different properties of the gravitational interaction, where the former probe the gravitational potential U, and the latter measure the gravitational acceleration $g = \nabla U$. Wolf *et al.* conclude that AIs are not measurements of the gravitational redshift, but merely tests of UFF, unless maybe if one calls into question the principle of least action.

In the context of tests of relativity, the distinction drawn by Wolf *et al.* between the ΔU in a clock comparison test and $g = \nabla U$ in an AI experiment does not exist. For the AI experiment, g is simply the dominant contributor to the potential difference ΔU between the interferometer arms over the length scale of the AI.

Let us now consider their main argument, that the total accumulated phase due to the redshift is canceled by the corresponding phase caused by time dilation signal due to the atom's motion around any closed loop ("the free evolution phase is always zero"): Assuming that General Relativity's prediction is valid, this is true. It means that we have to extract the redshift signal by subtracting the time dilation phase, which is known if the velocity of the atom as function of time is known. This is also true for a clock-comparison experiment: The most precise of these used a hydrogen maser clock in a sounding rocket that was compared to a ground-based one using a radio link [2]. During most of the flight, the maser is in free fall. The proper time interval $d\tau$ of the clock inside the rocket

$$d\tau = \left(1 + \frac{U(z)}{c^2} - \frac{1}{2}\frac{v^2}{c^2}\right) dt, \tag{3}$$

a sum of a gravitational redshift signal and a velocity dependent time dilation. Two clocks following paths like the matter wave packets in the interferometer (Fig. 1, right) would show the same proper time as the atomic wave packets. This includes the interaction phase, which for the clocks would result from an accelerational redshift during the times the clock's velocities are being changed. This will be explained below.

For any redshift measurement, the velocity-dependent time dilation is assumed to be known and the redshift is recovered. This happens in both AIs (see section 3.2) and clock comparisons. In particular, for atom interferometers it doesn't preclude us from measuring *deviations* from General Relativity's prediction, which is the purpose of a redshift test.

Thus, the two arguments presented by Wolf *et al.* apply to clock comparisons and AIs in a similar way; they do not point to any fundamental difference between the two. Rather, they may serve as useful *gedankenexperimente* that help elucidating which features of General Relativity are being tested by redshift experiments: A redshift test searches for deviations of the proper time from its value in General Relativity. Such deviations might offset the balance between the influences of gravity and motion on the proper time and lead to a detectable variation $\delta\sqrt{-g_{\mu\nu}\dot{x}^\mu\dot{x}^\nu}$ of the proper time [13]. As such, AIs are valid measurements of the gravitational redshift.

The question whether or not such variations can exist has been debated in the context of clock comparison experiments since the 1960s [14, 15, 16, 17]. As we will show in this paper, the answer does not depend on whether matter waves or clocks are used, but on the mechanism underlying the redshift violation. We will apply a consistent Lagrangian formalism and see which experiment measures which parameters. The Standard Model Extension (SME) [7] is such a formalism. It has the advantage of preserving energy and momentum conservation, renormalizability of the non-gravitational parts, as well as including all particles and interactions. We will use the SME for these reasons, noting that other formalisms like the parameterized post-Newtonian (PPN) theory also exist.

As an aside, we note that even the interaction phase (which Wolf *et al.* argue is the only observable phase) can, in the context of General Relativity, be interpreted as resulting from a redshift. This redshift is induced by the acceleration of the atom during the beam splitter as opposed to the Earth's gravitational field: Conventionally, the phase for each interaction is given by $\pm kz$, where the sign is for absorption and the minus sign for emission of a photon, and z is the vertical position. The atom's velocity before and after the interaction differs by $\pm v_r = \pm\hbar k/m$. For our General Relativistic picture, we can view that as the result of a constant acceleration $a = \pm v_r/(\delta t)$ during a short time δt. The Einstein Equivalence Principle states that the effects of acceleration are the same as those of gravity, so there should be an "accelerational" redshift $\pm\omega_C(v_r/\delta t)z/c^2$ of atom's Compton frequency relative to a reference atom held at $z = 0$. Inserting ω_C, v_r and integrating over the time interval yields

$$\pm\int_0^{\delta t} \omega_C \frac{az}{c^2} dt' = \int_0^{\delta t} \left(\frac{mc^2}{\hbar}\right) \frac{(\hbar k/m)z}{c^2 \delta t} dt' = \pm kz, \tag{4}$$

recovering the conventional result. (In this picture, there is no phase due to the propagation of the photon from 0 to z on a Null geodesic.) Thus, we see that the atom interferometer actually measures the sum of the gravitational and accelerational redshifts less the contribution of the time dilation. This gives it additional significance as a test of the Einstein Equivalence Principle. In this work, however, we focus on the gravitational redshift.

1. Phenomenology of redshift violations

1.1. *Type I*

At first glance, we might expect to obtain an anomalous redshift from attaching a factor to the redshift term in Eq. (2):

$$\mathcal{L} = mc^2\left[(1+\beta)\frac{gz}{c^2} - \frac{1}{2}\frac{\dot{z}}{c^2}\right]. \tag{5}$$

Here the dimensionless constant β vanishes if General Relativity is valid. This Lagrangian can be shown to yield no measurable anomalies: Rescaling the units of time and mass according to $t \to (1-\beta/2)t$ and $m \to (1+\beta)m$ absorbs the β into an unmeasurable redefinition of coordinates. In the generic experiment shown in Fig. 1 (left), for example, the redshift measurement will find $(1+\beta)gz/c^2$ but the measurement of the gravitational potential difference will now yield $(1+\beta)gz$. No anomaly will be evident in comparing the two.

1.2. *Type II*

If the parameter $\beta = \beta^{\mathrm{T}}$ varies with the composition of the clock or test particle T, measurable effects given by the difference $\beta^{\mathrm{T}} - \beta^{\mathrm{U}}$ of two particles T and U will be obtained. Tests of the gravitational redshift and of UFF will both lead to nonzero results. However, since UFF has been experimentally confirmed to an accuracy of 10^{-13} [18], redshift tests must be extremely accurate to yield new constraints.

The situation for clock comparisons is a little better if the parameter β^{T} is nonzero for electromagnetic masses only: Electromagnetism contributes all of the signal in atomic clocks such as hydrogen masers or cesium clocks, but only a fraction of a particle's rest mass. Thus, for a given violation, signals in redshift tests are larger than in UFF tests. However, given the current accuracy of UFF tests, redshift experiments with atomic clocks would need to reach an accuracy of 2×10^{-7} before they can become competitive even in this specialized situation, see p. 29 in Ref. [6] and [4, 5]. A famous, though unproven, conjecture by Schiff [14, 15] states that in any consistent theory of gravitation, all violations of the Einstein Equivalence Principle are of type II. Theories that may evade this conjecture will be discussed below.

1.3. *Type III*

We define type III violations as the case where the gravitational redshift is changed by a factor of $1 + \beta_{\mathrm{III}}$ while the acceleration of free fall is universal. In Sec.(4.2), it will turn out that the leading order type III effect depends on the velocity v of the experiment along the gradient of the gravitational potential. Hence, β_{III} is a function of that velocity. Specialized models for type III violations have been proposed, *e.g.*, based on nonminimal coupling [16]. Gravitoscalar and gravitovector fields [17] produce a model that is strictly speaking of type II, because electromagnetic masses fall at different rates than ordinary matter. These models are nevertheless close to type III, because electromagnetism contributes only a small part to particle rest masses. Unfortunately, the detailed experimental consequences of these models have not yet been worked out. Below, we will derive a detailed model for type III violations.

2. Clock comparison tests

Absolute redshift measurements compare clocks at locations whose gravitational potential differs by ΔU and test whether their relative frequency change $\Delta\nu/\nu$ satisfies $\Delta\nu/\nu = \Delta U/c^2$. For example, between 1959 and 1964, Pound, Rebka and Snider [9] studied Mössbauer sources vertically separated by $h = 22.5\,\text{m}$ in a tower. Vessot *et al.* [2] launched Gravity Probe A (GP-A), a hydrogen maser on board a Scout rocket, to a height of 10,000 km. The high elevation increases the redshift, but comes at the cost that now the clock is moving. Therefore, the special relativistic time dilation of the moving clock has to be corrected for.

3. Matter wave tests

The first experiments to detect the influence of gravity on the phase of matter waves were performed by Colella and others [19]. They reached an agreement with the expected value of better than 1%. The basis for matter wave tests of the gravitational redshift is the action of a point particle in General Relativity is given by $S = mc^2 \int d\tau$ [8], pp.315-324. In Feynman's path integral formulation [20] of quantum mechanics, the state $\psi(t_A, \vec{x}_A)$ is given by the path integral

$$\psi(t_A, \vec{x}_A) = \int d^3x_B \psi(t_B, \vec{x}_B) = \int_A^B Dq(t) e^{\frac{i}{\hbar}\int mc^2 d\tau}. \tag{6}$$

This shows that the evolution of a quantum state is given by the gravitational redshift and time dilation as encoded in the proper time τ. In the semiclassical limit, the path integral reduces to an ordinary integral over the classical, geodesic path. This leads to the expression [21]

$$\phi = \frac{i}{\hbar} \int mc^2 d\tau = \int \omega_C d\tau, \tag{7}$$

where ω_C is the Compton frequency.

3.1. "Pound-Rebka like" tests: stationary atoms in optical lattices

Bloch oscillations of cold atoms in optical lattices [22] are a well known effect in atomic physics. In the experiments of interest here, atoms are held in a coherent superposition of states located above one another with a separation of $\lambda/2$. In such a setup, a periodic modulation of the atom's quasimomentum is observed. If we evaluate Eq. (7) for atomic states located in adjacent lattice sites at heights $z = n\lambda/2$ and $(n+1)\lambda/2$, we obtain a difference

$$\omega_B = (1+\beta)\frac{\lambda}{2}\frac{g}{c^2}\frac{mc^2}{\hbar}, \tag{8}$$

in their Compton frequencies, known as the Bloch frequency. As usual, we have included a factor of $1+\beta$ to account for the redshift anomaly, which can be either β_{II} or β_{III}. This frequency can be observed as an oscillation of the atom's velocity expectation value over time.

Ivanov *et al.* [23] used strontium atoms (^{88}Sr) in an optical lattice of $\lambda/2$=266 nm. Comparing the measured Bloch frequency with the expected one led to [3]

$$\beta = (4.0 \pm 6.0) \times 10^{-5}. \tag{9}$$

A similar measurement [24] with rubidium (^{87}Rb) atoms and $\lambda/2 = 394$ nm leads to [3]

$$\beta = (3 \pm 1) \times 10^{-6}. \tag{10}$$

Pound-Rebka like atomic physics redshift measurements lead to limits of a few parts in 10^6 at present. This is already an improvement by a factor of more than 10 compared to the best classical tests.

3.2. "GP-A like" tests: atom interferometers

Just as GP-A used a large clock separation to obtain a higher sensitivity than Pound-Rebka tests, atom interferometers (Fig. 1, right) obtain higher sensitivity than the Bloch oscillation tests by increasing the distance separating the atoms. At the same time, however, the atoms are moving. Thus, corrections for the special relativistic time dilation will have to be applied in order to extract the gravitational redshift.

The atom interferometer contains a built-in mechanism for monitoring the trajectories and applying this correction. To see this, we derive the phase difference between the interferometer arms using the standard procedures [25]. Since we do not assume prior knowledge of the acceleration of free fall, we will label it g' and treat it as unknown. The specific value taken by g' depends upon which of the three anomalous redshift phenomenologies scenarios is under test. The free evolution phase is derived by integrating the Compton frequency along the two paths 1,2 (Fig. 1, right)

$$\Delta\phi_{\rm free} = -\omega_C \int \left[(1+\beta)g\frac{z_1-z_2}{c^2} - \frac{1}{2}\frac{\dot{z}_1^2-\dot{z}_2^2}{c^2}\right] dt = (1+\beta)kgT^2 - kg'T^2. \tag{11}$$

The first term, due to the redshift, contains g because it is given by the integral $\omega_C(g/c^2)\int(\Delta U/c^2)dt$, which depends on the vertical separation of the trajectories but not their acceleration of free fall. The second depends on g', because it is given by how fast the atoms actually move.

Whenever a photon is absorbed by the atom, the photon's phase is added to the phase of the matter wave. For stimulated emission of a photon, the opposite occurs. The photon's phase $kz + \phi_0$ is a function of the location z of the interaction occurs (where ϕ_0 is its value at $z = 0$). Summing up the photon's phases over all interactions, we obtain $\phi_i = +kg'T^2$. This, again, contains the acceleration of free fall g'. The final result for the atom interferometer phase can be written as

$$\Delta\phi = \underbrace{(1+\beta)kgT^2 - kg'T^2}_{\Delta\phi_{\rm free}} + kg'T^2. \tag{12}$$

Thus, the atom-light interaction phase cancels the one due to special relativistic time dilation no matter what the acceleration of free fall g'. This is analogous to measuring the trajectory and subtracting time dilation effects in classical redshift tests. This derivation can be made specific to each of the three classes of redshift anomalies described above by determining the expression for g':

- In type I scenarios, $g' = (1+\beta)g$ as derived by extremizing the action Eq. (5). Thus, $\Delta\phi_{\rm free} = 0$ and $\Delta\phi = (1+\beta)kgT^2$. The parameter β, however, is not measurable. This becomes obvious when we express $\Delta\phi = kg'T^2$ and note that g' is the acceleration of free fall as measured by the gravimeter.

- For type II scenarios, β is specific to the test particle T. We find $\Delta\phi_{\rm free} = 0$ and $\Delta\phi = (1+\beta^{\rm T} - \beta^{\rm gravimeter})kgT^2$, where $\beta^{\rm gravimeter}$ is the parameter specific to the gravimeter. For a specific experiment, we may set it to zero by definition of coordinates.
- For type III, $g = g'$ by definition, and $\Delta\phi = (1 + \beta_{\rm III})kgT^2$, where $\beta_{\rm III}$ is now universal for all particles but will turn out to be a fixed function of the particle positions, momenta, and the gravitational acceleration. See Sec. 4.2 for a detailed model.

The most accurate data to date comes from an interferometer using caesium atoms in an atomic fountain [26]. After correcting for a number of relatively small fundamental [26, 27] and systematic [26] effects, we obtain the redshift parameter $\beta = z_{\rm meas}/z_0 = (7 \pm 7) \times 10^{-9}$. In that experiment, the acceleration of gravity g was measured using an FG-5 falling corner cube gravimeter (Fig. 1, left). It basically measures the acceleration of free fall of a corner cube reflector which forms an optical interferometer. Note that, in contrast to the atom interferometer, this is insensitive to the proper time experienced by the mirror, as the photons are interfered, not the wave packet (or density matrix) describing the mirror.

4. Description of matter wave tests in some test theories

4.1. Type II

The standard model extension (SME) by Kostelecký and co-workers [28, 7] provides us with a model for type II violations if taken to post-Newtonian order PNO(3). The relevant quantities are the pure-gravity sector coefficients $\bar{s}_{\mu\nu}$ as well as the matter sector coefficients $\alpha(\bar{a}^w_{\rm eff})_\mu$ and $(\bar{c}^w)_\mu$. The superscript w can take the values e, n, p indicating the electron, neutron, and proton, respectively. The Lorentz-violating properties of an object B composed of these particles can be represented by effective coefficients $(c^B)_{\mu\nu}, (a^B_{\rm eff})_\mu$ [7]. By imposing the condition

$$\alpha(\bar{a}^w_{\rm eff})_0 = \frac{1}{3} m^w (\bar{c}^w)_{00}, \tag{13}$$

we obtain the "isotropic parachute model" [7] in which UFF is valid.

To determine the gravitational shift in the Compton frequency $mc^2/\hbar$ of matter waves, the dependence of the rest mass m on the metric $g_{\mu\nu} = \eta_{\mu\nu} + h_{\mu\nu}$ has to be found. This can be accomplished by collecting all terms that are independent of the particle momentum in the hamiltonian $H_{\rm NR}$ given in Ref. [7]. At PNO(3), we find

$$H_{\rm NR} = \frac{1}{2} m^{\rm T} v^2 \left(1 + \frac{5}{3}(c^{\rm T})_{00} + \frac{3}{2} h_{00}\right) - \frac{1}{2} m^{\rm T} h_{00} \left(1 + \frac{2\alpha}{m^{\rm T}}(\bar{a}^{\rm T}_{\rm eff})_0 + (c^{\rm T})_{00}\right) \tag{14}$$

If Eq. (13) and thus UFF, hold, both the kinetic and potential terms carry a constant factor $1 + \frac{5}{3}(c^{\rm T})_{00}$. This can be eliminated from the equations of motion by rescaling the particle masses to become $m^{\rm T\prime} = m^{\rm T}(1 + \frac{5}{3}(c^{\rm T})_{00})$, an unmeasurable scaling of the unit of mass. Thus, to PNO(3), there is no type III redshift violation in the SME.

Without Eq. (13), however, the kinetic term is scaled differently to the potential term. We can still redefine the particle mass to be $m^{\rm T\prime} = m^{\rm T}(1 + \frac{5}{3}(c^{\rm T})_{00})$, to put the kinetic term in the form expected by Eq. (11), but then we find that the potential term is scaled by $-\frac{1}{2} m^{\rm T\prime} h_{00}[1 + \frac{2\alpha}{m^{\rm T\prime}}(\bar{a}^{\rm T}_{\rm eff})_0 - \frac{2}{3}(\bar{c}^{\rm T})_{00}]$. This yields Eq. (11) with

$$\beta^{\rm T} = \frac{2}{3}\left[\frac{3\alpha}{m^{\rm T\prime}}(\bar{a}^{\rm T}_{\rm eff})_0 - (\bar{c}^{\rm T})_{00}\right]. \tag{15}$$

The superscript T at the redshift parameter β indicates the nature of the particle. From the matter-wave experiments described above, we obtain limits of

$$\beta^{\rm 88Sr} = (4 \pm 6) \times 10^{-5}, \quad \beta^{\rm 87Rb} = (3 \pm 1) \times 10^{-6}, \quad \beta^{\rm 133Cs} = (7 \pm 7) \times 10^{-9}. \tag{16}$$

4.2. Type III

The Standard Model Extension also provides a model for type III violations if it is taken to PNO(4). We will present such a model in a separate publication, because taking the SME to PNO(4) is quite challenging given that the nonlinearity of General Relativity plays a substantial role at this order. In this section, we offer a simple model [29] in which particle masses anomalously vary with the gravitational potential. The anomaly is represented by a single parameter $\beta^0_{\rm III}$ (which represents a combination of SME parameters). This generates gravitational redshift anomalies unconstrained by existing tests of UFF or Local Lorentz Invariance. The effects predicted by the simple model are qualitatively similar to those generated by the SME at PNO(4), although it should also be noted that similar effects exist in the fully Lorentz covariant theory, due to the nonlinearity of General Relativity.

We consider a model where the particles' masses anomalously vary in their local gravitational potential

$$L = mc^2 + mc^2\left(1+\beta^0_{\rm III}\frac{gz}{c^2}\right)\left(\frac{gz}{c^2}-\frac{\dot{z}^2}{2c^2}\right). \tag{17}$$

This model includes terms resembling those appearing at higher order in the expansion of the Schwarzschild metric. ($\beta^0_{\rm III}\neq 0$ may be understood as several anomalous terms in the expansion. Higher order gradients and GR must be considered when experiments are sensitive to $\beta^0_{\rm III}\sim 6$.) This Lagrangian modifies point-particle kinematics: the local acceleration of free-fall is, to first order in $\beta^0_{\rm III}$,

$$\ddot{z} = -g\left(1+\beta^0_{\rm III}\left[\frac{gz}{c^2}+\frac{\dot{z}^2}{2c^2}\right]\right), \tag{18}$$

and is the same for all objects. Expanding to $\mathcal{O}\left(T^2\right)$, where T is the pulse separation time, and assuming that photon momenta scale like the masses with height (dropping this assumption changes the second term given below by a factor of 3), the total AI phase is

$$\Delta\phi = kgT^2 - \beta_{\rm III}kgT^2 - \beta^0_{\rm III}\frac{v_0 v_r}{c^2}kgT^2, \quad \beta_{\rm III}\equiv\beta^0_{\rm III}\frac{1}{2}\frac{v_0^2}{c^2}. \tag{19}$$

General relativity, due to its nonlinearity, predicts an interferometer phase similar to the second term in (19) (with $\beta^0_{\rm III}=10$) [27]. Thus the type III violations are quantitative modifications of effects that already exist in General Relativity.

The most sensitive tests of the gravitational redshift in AIs are precise to 7 ppb in $\beta_{\rm III}$ [3], which in light of the small launch velocity (v_0 = 1.53 m/s), translates to a bound of $|\beta^0_{\rm III}|\leq 5\times 10^8$. This sensitivity is not unique to the AI. Moving Pound-Rebka-Snider tests with $\vec{v}\cdot\vec{g}/|\vec{g}| = v_0 \ll c$ observe

$$\nu_{2,1} = \nu_0\left(1+\frac{gh}{c^2}\left[1-\beta^0_{\rm III}\frac{v_0}{c}\left(1-\frac{v_0}{c}\right)\right]\right), \tag{20}$$

where we ignore higher order terms in the Schwarzschild expansion which become relevant when $\beta^0_{\rm III}$ becomes 1 or lower. Motion of the Earth in the Sun's gravitational potential would permit Earth-stationary Pound-Rebka-Snider experiments with 1% accuracy to bound $|\beta^0_{\rm III}| < 10^8$ by repeated measurements made while the Earth moves between orbital perihelions.

5. Clock-comparison versus matter wave tests

In a type II model, clock comparison redshift tests have been studied in [7]. The gravitational modifications in the clock's hamiltonian modify the redshift according to $\Delta\nu/\nu = (\tilde{g}z/c^2)(1+$

$\xi_{\rm clock}$), where $\xi_{\rm clock}$ is a clock-dependent parameter that can be calculated if the hamiltonian of the clock is known. For transitions in a Bohr-model hydrogen atom, for example [7]

$$\xi_{\rm H} = -\frac{2}{3}\left(\frac{m^p}{m^{\rm H}}(\bar{c}^e)_{00} + \frac{m^e}{m^{\rm H}}(\bar{c}^p)_{00}\right). \tag{21}$$

This shows that in type II models, atomic clock redshift tests are predominantly sensitive to the electron coefficients $\bar{c}^e$ and have a suppressed sensitivity towards proton and neutron terms. The reverse is true for matter wave tests. Both, however, compete with tests of UFF in this scenario. As mentioned before, classical redshift tests would need to reach 2×10^{-7} before they can become competitive.

In type III systems, the redshift violation parameter is independent of the nature of the clock by definition. Hence, there is no difference whatsoever between matter wave redshift tests and clock comparisons. Experimentally, the question as to the existence of type III effects as the much more interesting question, as it has not yet been answered by tests of UFF.

6. Summary

The phase in atom interferometers (AIs) is given by the integral of the proper time over the trajectories times the Compton frequency. This is a consequence of Feynman's relativistic path integral formulation of quantum mechanics and also part of Borde's "5D" formulation of atom optics [21]. The time shown by a moving clock is also the integral of the proper time. Complete or partial cancelations between the influences of gravity and the motion on the proper time happen in both moving clocks and moving atoms. To make a redshift test, one can (and has to) assume one part is known and subtract it. In both experiments, there is also a linear Doppler effect due to the clock's (or atom's) motion, called the interaction phase in AIs. These similarities suggest that the AI does indeed measure the redshift the way clocks do. To show this explicitly, we applied the Standard Model Extension (SME) as a consistent theoretical formalism [7].

The effects of the SME can be ordered by powers of $1/c$. To order $1/c^2$ ["PNO(2)"], AIs and clock comparisons (CCs) measure similar parameters: Atom interferometers are more sensitive to $(\bar{c}^p)_{00}$ and $(\bar{c}^n)_{00}$ coefficients for protons and neutrons and have a suppressed sensitivity to electron parameters $(\bar{c}^e)_{00}$. Comparisons of atomic clocks are more sensitive to electron coefficients, whereas the Mössbauer sources used in earlier redshift tests were predominantly sensitive to the proton and neutron coefficients measured by AIs. However, these parameter combinations overlap with tests of UFF. The only way to avoid this overlap at PNO(2) seems to be a violation of energy conservation. It is important that this is so for CCs just as for AIs.

This overlap of UFF and redshift experiments, however, need not hold for higher order effects: At PNO(4), UFF can be preserved locally while violations of local position invariance may still exist. In fact, there is a velocity-dependent force in General Relativity which can be measured in AIs. These violations cannot be ruled out by UFF experiments. Redshift experiments based on AIs and CCs will both be sensitive to this effect and measure the same parameter combinations. An advantage of the AI, however, is that the mc^2 in the Compton frequency cancels one $1/c^2$ in the PNO(4)-effect.

Acknowledgements We thank Steven Chu, Juna Kollmeier, Alan Kostelecký, Shau-Yu Lan, Silke Ospelkaus, Achim Peters, Jay Tasson, Guglielmo Tino and Nan Yu for discussions. Thanks also to Peter Hannaford and all the Organizers of the 22$^{\rm nd}$ ICAP held in 2010 at Cairns, Australia. Support by the David and Lucile Packard Foundation, the Alfred P. Sloan Foundation, and the National Institute of Standards and Technology is acknowledged.

References

[1] Einstein A 1911, *Ann. Phys.* **35** 898.
[2] Vessot R F C *et al.* 1980, *Phys. Rev. Lett.* **45** 2081.
[3] Müller H, Peters A and Chu S 2010, *Nature* **463** 926.
[4] Nordtvedt K 1975, *Phys. Rev. D* **11** 245.
[5] Haugan M P 1979, *Ann. Phys.* **118** 156.
[6] Will C M, *Theory and Experiment in Gravitational Physics* (Cambridge University Press, Cambridge, 1993).
[7] Kostelecký V A and Tasson J 2010, e-print: arXiv:1006.4106
[8] Misner C M, Thorne K S, and Wheeler J A, *Gravitation* (Freeman, 1970).
[9] Pound R V and Rebka G A Jr., 1960 *Phys. Rev. Lett.* **4** 337; Pound R V and Snider J L 1964, *Phys. Rev. Lett.* **13** 539; 1965 *Phys. Rev.* **140**, B788.
[10] Hafele J and Keating R 1972, *Science* **177** 166; *Science* **177** 168.
[11] Cacciapuoti L and Salomon Ch 2009, *Eur. Phys. J. Spec. Top.* **172** 57.
[12] P. Wolf *et al.*, Nature **467,** E1 (2010);
[13] H. Müller, A. Peters, and S. Chu, Nature **467,** E2 (2010).
[14] Schiff L I 1960, *Am. J. Phys.* **28** 340.
[15] Dicke R H 1960, *Am. J. Phys.* **28** 344.
[16] Coley A 1982, *Phys. Rev. Lett.* 49, 853; Ni W T 1977, *Phys. Rev. Lett.* **38** 301; Accioly A J, Wichoski U F, and Beratarello N 1993, *Brazilian J. Phys.* **23** 392.
[17] Nieto M M and Goldmann T 1991, Phys. Rep. **205** 221.
[18] Schlamminger S *et al.* 2008, *Phys. Rev. Lett.* **100** 041101.
[19] Colella R, Overhauser A W and Werner S A 1975, *Phys. Rev. Lett.* **34** 1472; Bonse U and Wroblewski T 1984, *Phys. Rev. D* **30** 1214; Horne M A 1986, Physica B 137; Werner S A, Kaiser H, Arif M, and Clothier R 1988, *Physica B* **151** 22; Littrell K C, Allman B E, and Werner S A 1997, *Phys. Rev. A* **56**, 1767.
[20] Feynman R P and Hibbs A R, *Quantum Mechanics and Path Integrals* (McGraw-Hill, 1965).
[21] Bordé Ch J 2008, *Eur. Phys. J. Spec. Top.,* **163** 315; Bordé Ch J 2002, *Metrologia* **39** 435.
[22] Peik E *et al.* 1997, *Phys. Rev. A* **55** 2989.
[23] Ivanov V V *et al.* 2006, *Phys. Rev. Lett.* **100** 043602.
[24] Cladé P *et al.* 2005, *Europhys. Lett.* **71** 730.
[25] Young B C, Kasevich M, and Chu S 1997, in Atom Interferometry (ed. Berman, P.) 363406 (Academic).
[26] Peters A, Chung K Y, and Chu S 1999, *Nature* **400** 849852 (1999); Peters A, Chung K Y, and Chu S 2001, *Metrologia* **38** 25.
[27] Dimopoulos S, Graham P W, Hogan J M, and Kasevich M A 2007, *Phys. Rev. Lett.* **98** 111102; Dimopoulos S, Graham P W, Hogan J M, and Kasevich M A 2008, *Phys. Rev. D* **78** 042003.
[28] Kostelecký V A 2004, *Phys. Rev. D* **69** 105009; Bailey Q and Kostelecký V A 2006, *ibid.* **74** 045001; Kostelecký V A and Tasson J 2009, *Phys. Rev. Lett.* **102** 010402.
[29] Hohensee M A and Müller H 2010, arXiv:1008.0611

Journal of Physics: Conference Series **264** (2011) 012010 doi:10.1088/1742-6596/264/1/012010

Astronomical and laboratory searches for space-time variation of fundamental constants

J. C. Berengut
School of Physics, University of New South Wales, Sydney, NSW 2052, Australia

V. V. Flambaum
School of Physics, University of New South Wales, Sydney, NSW 2052, Australia
European Centre for Theoretical Studies in Nuclear Physics and Related Areas (ECT*), Strada delle Tabarelle 286, I–38123, Villazzano (Trento), Italy

Abstract. We review recent activity searching for variations in the fundamental constants of nature in quasar absorption spectra and in the laboratory. While research in this direction has been ongoing for many decades, the topic has recently been stimulated by astronomical evidence for spatial variation in the fine-structure constant, α. This result could be confirmed using different quasar data and atomic clock measurements, but there are significant challenges to obtain the required accuracy. We review existing measurements and discuss some of the most promising systems where any variations would be strongly enhanced.

1. Introduction

Recently, a highly significant ($\sim 4\sigma$) indication that there is a spatial gradient in values of the fine-structure constant, α, was reported [1]. That is, in one direction on the sky α was larger in the past, while in the opposite direction it seems to have been smaller. This result has massive implications for the "fine-tuning" problem. It is well known that the constants of nature are finely tuned to allow life to exist. With the detection of spatial variation of coupling constants we begin to have a natural explanation of fine-tuning: we simply appeared in the region of the Universe where values of fundamental constants are suitable for our existence.

We can only detect the variation of dimensionless fundamental constants. In this review we will discuss systems that are sensitive to variation of the fine-structure constant, $\alpha = e^2/\hbar c$, the proton g-factor, g_p, and the dimensionless mass ratios $\mu = m_e/m_p$ and $X_q = m_q/\Lambda_{\rm QCD}$. Here m_e, m_p, and m_q are the electron, proton, and light-current quark masses, respectively, and $\Lambda_{\rm QCD}$ is the quantum chromodynamics scale, defined as the position of the Landau pole in the logarithm of the running strong coupling constant, $\alpha_s(r) \sim 1/\ln(\Lambda_{\rm QCD} r/\hbar c)$. The proton mass m_p is proportional to $\Lambda_{\rm QCD}$ (if we neglect the few percent contribution of quark masses to the proton mass); therefore the relative variation of $\mu = m_e/m_p$ is approximately equal to the relative variation of $X_e = m_e/\Lambda_{\rm QCD}$. In the Standard Model electron and quark masses are proportional to the vacuum expectation value of the Higgs field. In the same vein, g_p is not a fundamental constant, but we can express its variation in terms of variation of light quark mass using the relationship $\delta g_p/g_p \approx -0.1\,\delta X_q/X_q$ [2] (accurate calculations for nuclear g-factors can be found in [3]).

The evidence for a cosmological spatial gradient in α motivates us to interpret other data in terms of spatial variation. However, many of the limits on variations are best interpreted as limits on fundamental constants other than α, since the system being examined is more sensitive to these other constants. A minimal hypothesis is to expect all fundamental constants to vary along the same direction as the α-gradient. There are some good theoretical justifications for this postulation. If the constants vary because they are coupled to a scalar field Φ that varies over space-time, for example, the quintessence field Φ/c^2 or a dimensionless dilaton field, then a fundamental constant, X, will vary according to

$$\frac{\delta X}{X} = k_X \delta\Phi \tag{1}$$

where k_X is a dimensionless coupling coefficient. Then all constants will vary in the direction of the gradient of the scalar field $\Delta\Phi$, which in turn must be directed along the axis found by the α-variation data in [1].

Equation (1) implies that the relative variation of different constants can be related to each other using equations like

$$k_\mu = R_\mu^\alpha \, k_\alpha \tag{2}$$

where the constants $R_X^{X'}$ can be determined from observations and compared with theories of spatial variation. For example, applying Grand Unification of the interactions of the Standard Model to the relative variation of α and α_s (see, e.g., [4]) we find

$$\frac{\delta X_q}{X_q} \sim 35 \, \frac{\delta\alpha}{\alpha}$$

That is, the relative variation in X_q may be much larger than the relative variation in α. The coefficent here is model dependent, but large values are prevalent among models in which the variations come from high-energy scales. Therefore, variation of fundamental constants provides a unique probe of the predictions of Unification Theories (see, e.g., [5–9]).

In this review we discuss measurements of variation of fundamental constants in quasar absorption spectra and in atomic clocks. We interpret the astronomical measurements in the context of the observed spatial gradient in α, relating other constants to this variation using expressions such as (2). Atomic clocks may prove to be the most reliable way to corroborate the detection of spatial α-variation, and we discuss some of the proposed experiments where huge enhancements of the variation can be expected. There are many other systems which are not discussed here, including Big Bang Nucleosynthesis, where hints of variation have been reported [10, 11]. We refer the reader to recent reviews [4, 12].

2. Quasar absorption spectra

We start our discussion with quasar absorption systems, since the remarkable recent detection of spatial variation in α from optical absorption spectra [1], will provide us with a benchmark for all other studies.

2.1. Optical atomic spectra

It is natural to analyse fine-structure intervals in the search for variation of α. Measurements of α-variation by comparison of cosmic and laboratory optical spectra were first performed by Savedoff as early as 1956 [13]. There were numerous works successfully implementing this "alkali-doublet" method (see review [14]).

In 1999, a different approach was developed: the many-multiplet method [15, 16]. It exploits the fact that relativistic corrections to atomic transition frequencies can exceed the fine-structure

Journal of Physics: Conference Series **264** (2011) 012010 doi:10.1088/1742-6596/264/1/012010

interval between the excited levels by an order of magnitude (for example, an s-wave electron does not have spin-orbit splitting but it has the maximal relativistic correction to energy). We can express the α-dependence of a transition frequency ω as

$$\omega = \omega_0 + qx \tag{3}$$

where

$$x = (\alpha/\alpha_0)^2 - 1 \approx 2\frac{\alpha - \alpha_0}{\alpha_0}, \tag{4}$$

α_0 is the laboratory value of α and ω_0 is the laboratory frequency of a particular transition. The coefficients q can vary strongly from atom to atom and can have opposite signs in different transitions (for example, in s–p versus d–p transitions). Thus, any variation of α may be revealed by comparing different transitions in different atoms in cosmic and laboratory spectra. A statistical gain is also realised because many more spectral lines in different elements can be used. This method improves the sensitivity to any variation of α by more than an order of magnitude compared to the alkali-doublet method.

Relativistic many-body calculations are used to reveal the dependence of atomic frequencies on α^2 (the q coefficients). We have performed accurate many-body calculations of the q coefficient for all transitions of astrophysical interest [15–23]. These are strong E1 transitions from the ground state in Mg I, Mg II, Ti II, Fe I, Fe II, Cr II, Ni II, Al II, Al III, Si II, Zn II, Mn II, as well as many other atoms and ions which are seen in quasar absorption spectra, but have not yet been used in the quasar measurements because of the absence of accurate UV transition laboratory wavelengths. For a "shopping list" of needed measurements, see Ref. [24] (also [25] which is regularly updated).

From the very first analyses of quasar data that utilised the new method, hints of α-variation were reported [26, 27]. The largest analysis, with three independent samples of data containing 143 absorption systems spread over redshift range $0.2 < z < 4.2$, suggested that α was smaller in the past: $\delta\alpha/\alpha = (-0.543 \pm 0.116) \times 10^{-5}$ [28]. However, these studies all used spectra taken at the Keck telescope. Similar studies from another group, using our method and calculations but a much smaller sample of data taken at the Very Large Telescope (VLT), at first showed a stringent null constraints (Srianand *et al* [29]). More careful analysis of this sample suggested that the errors were underestimated by a large factor [30–32].

The latest results, combining the Keck data and a new sample of 153 measurements from the VLT, shows a spatial variation in α. This gradient, which we will refer to as the "Australian dipole", has a declination of around $-60°$. This explains why the Keck data, restricted mainly to the northern sky since the telescope is in Hawaii at a latitude of 20° N, originally suggested a time-varying α that was smaller in the past. The VLT is in Chile, at latitude 25° S, giving the new study much more complete sky coverage. The new results are entirely consistent with previous ones. The "Australian dipole" of α-variation found by [1] is

$$\frac{\delta\alpha}{\alpha_0} = (1.10 \pm 0.25) \times 10^{-6}\, r\cos\psi \tag{5}$$

where $\delta\alpha/\alpha_0 = (\alpha(\mathbf{r}) - \alpha_0)/\alpha_0$ is the relative variation of α at a particular place $\mathbf{r}$ in the Universe (relative to Earth at $\mathbf{r} = 0$). The function $r\cos\psi$ describes the geometry of the spatial variation: ψ is the angle between the direction of the measurement and the axis of the Australian dipole, (17.4 (0.9) h, $-58\,(9)°$) in equatorial coordinates. The distance function is the light-travel distance $r = ct$ measured in giga-lightyears. This is model dependent for large redshifts: the standard $\Lambda_{\rm CDM}$ cosmology parametrized by WMAP5 [33] is used to determine the light-travel time t. It is assumed here that $\delta\alpha/\alpha_0 = 0$ at zero redshift, which is supported by the data; however this assumption should be tested using the same absorption methods as are used at high redshift (e.g., by using absorbers within our own galaxy). Any monopole term (independent of ψ) in (5) detected at high redshift would be indicative of time-variation of α. The evidence for such a term is not strong [1].

The new results of α-variation are particularly striking in that if the VLT data and Keck data are considered separately there is a high level of agreement between their dipole fits. In their paper [1], the authors estimate the probability that the observed alignment between the independent Keck and VLT dipoles is due to chance at $\sim$ 4%. Similarly, if one breaks the data into subsamples consisting of absorbers at $z < 1.8$ and those at $z > 1.8$ (approximately half the data in each subsample), then there is again agreement between the directions of the independent dipole axes.

In addition to the previously mentioned many-multiplet method results of Srianand *et al* from the small VLT sample [29] (see also [34]), there are some high-accuracy single-absorption system α-variation results using alkali-doublets (Si IV) [35] and single-ion (Fe II) [36–39] measurements. The alkali-doublet result suffers from the same problems as the small-sample many-multiplet result of [29] (namely, sharp fluctuations in chi-squared vs. $\delta\alpha/\alpha$ graph which indicate failings in the chi-squared minimisation routine [30]).

The Fe II results give stringent constraints in two absorbing systems: at $z = 1.15$ towards HE0515–4414 [38] and at $z = 1.84$ towards Q1101–264 [39]. It is interesting to compare the results of these analyses with the variation expected in these systems if the Australian dipole result is correct. Using (5), we obtain an expected variation of $\delta\alpha/\alpha = (1.9 \pm 1.5) \times 10^{-6}$ for the former, which was measured to be $(-0.07 \pm 0.84) \times 10^{-6}$ [38]. The absorber towards Q1101–264 has a larger expected variation of $\delta\alpha/\alpha = (3.8 \pm 2.0) \times 10^{-6}$, and was measured to be $(5.4 \pm 2.5) \times 10^{-6}$ [39]. The single system results are therefore seen to be consistent with the Australian dipole.

2.2. Molecular rotational quasar spectra

Limits on variation of μ at high redshift can be obtained by comparison of different rotational-electronic excitations in molecular hydrogen. Taking advantage of newly available H_2 wavelengths, Ref. [40] reported a non-zero cosmological variation of μ in quasar spectra using molecular hydrogen transitions in the Ly-α forest. The authors obtained $\delta\mu/\mu = (2.4\pm0.6)\times10^{-5}$ at redshifts $z \approx 2.6 - 3.0$, i.e., a decrease of μ in the past 12 Gyr at the 3.5σ confidence level.

More recent determinations, using the laboratory wavelengths and sensitivity coefficients presented in [40–42] and carefully controlling systematics and accounting for known calibration errors in the VLT, obtain a more stringent constraint: $\delta\mu/\mu = (3.4 \pm 2.7) \times 10^{-6}$ [43, 44] from four quasar absorption systems.

In the context of an observed dipole in α-variation [1], it makes sense to check whether the (limited) data supports interpretation as a spatial variation. In Ref. [45] the data was shown to support a statistically significant spatial variation of μ aligned with the Australian dipole, although it was noted that the paucity of data prevents any firm conclusion.

2.3. Comparison of hydrogen hyperfine and UV transitions

A comparison of the 21-cm hyperfine transition in atomic hydrogen with UV metal lines was performed for 9 quasar spectra with redshifts $0.23 \le z \le 2.35$ [46, 47]. The ratio of the transition frequencies is proportional to the parameter $x = \alpha^2 \mu g_p$, and this was constrained to

$$\delta x/x = (6.3 \pm 9.9) \times 10^{-6} \ . \tag{6}$$

It was found that there was much more scatter in the data than would be expected from the statistical errors alone [47]. In principle this could mean that the model is wrong, e.g., a spatial variation should be considered rather than a time variation. However, even the best fit dipole

Table 1. Comparison of expected variation, given by (5), and measured variation of fundamental constants in different quasar absorption systems. $R^{\alpha}_{g_p}$ and R^{α}_{μ} are defined by equations like (2). As noted in Section 1, $R^{\alpha}_{g_p} \approx -0.1\, R^{\alpha}_q$ and $R^{\alpha}_{\mu} \approx R^{\alpha}_q$ where R^{α}_q is the variation of light quark mass X_q relative to α-variation. Errors in the expected α-variation (the prefactors in the third column) are of the order $\sim 1.5 \times 10^{-6}$.

System	Constant	Expected variation ($\times 10^{-6}$)	Measurement ($\times 10^{-6}$)	Ref.
HI + CI	$\alpha^2 \mu g_p$	$1.12\,(2 + R^{\alpha}_{\mu} + R^{\alpha}_{g_p})$	$6.64 \pm 0.84_{\text{stat}} \pm 6.7_{\text{sys}}$	[48]
		$-5.20\,(2 + R^{\alpha}_{\mu} + R^{\alpha}_{g_p})$	$7.0 \pm 1.8_{\text{stat}} \pm 6.7_{\text{sys}}$	[48]
HI + mol. rot.	$\alpha^2 g_p$	$0.50\,(2 + R^{\alpha}_{g_p})$	-2.0 ± 4.4	[49]
		$-5.47\,(2 + R^{\alpha}_{g_p})$	-1.6 ± 5.4	[49]
HI + OH	$(\alpha^2/\mu)^{1.57} g_p$	$-1.04\,(3.14 - 1.57 R^{\alpha}_{\mu} + R^{\alpha}_{g_p})$	$4.4 \pm 3.6_{\text{stat}} \pm 10_{\text{sys}}$	[50]
OH	$(\alpha^2/\mu)^{1.85} g_p$	$0.50\,(3.70 - 1.85 R^{\alpha}_{\mu} + R^{\alpha}_{g_p})$	-11.8 ± 4.6	[51]
NH_3	μ	$-5.47\, R^{\alpha}_{\mu}$	$< 1.8\ (2\sigma)$	[52]
		$1.34\, R^{\alpha}_{\mu}$	$< 1.4\ (3\sigma)$	[53]

model (which does not align with the Australian dipole) cannot explain the scatter [45]. It is probable that velocity offsets between the hydrogen and metal absorbers cause dominating systematics. Two new measurements comparing H I 21-cm with UV lines in neutral carbon in single absorption systems have reported slightly stronger constraints in x [48]; these are shown in the first two lines of table 1.

2.4. *Comparisons involving hyperfine and molecular rotational transitions*
Individual measurements of fundamental constant variation in quasar absorption spectra can be compared with the observed spatial α-variation [45]. That is, we can regard the α-variation, according to (5) with r and ψ calculated for the object, as the "expected" variation and compare other measurements with it (this is presented in table 1).

The frequency of the hydrogenic hyperfine line is proportional to $\alpha^2 \mu g_p$; molecular rotational frequencies are proportional to μ. Comparison of the two placed limits on variation of the parameter $\alpha^2 g_p$ [54] in two quasar absorption spectra. A similar analysis was repeated using more accurate data for the same objects [49], at $z = 0.247$ and at $z = 0.6847$, resulting in the limits shown in lines 3 and 4 of table 1, respectively. The object at $z = 0.6847$ is associated with the gravitational lens toward quasar B0218+357 and corresponds to lookback time ~ 6.2 Gyr.

Comparison of OH 18-cm with H I 21-cm, and different conjugate-satellite OH 18-cm lines, can be used to measure the combinations of fundamental constants $(\alpha^2/\mu)^{1.57} g_p$ and $(\alpha^2/\mu)^{1.85} g_p$, respectively. Resulting measurements are shown in lines 5 and 6 of table 1.

2.5. *Enhancement of variation of μ in the inversion spectrum of ammonia*
In 2004, van Veldhoven *et al* suggested using a decelerated molecular beam of ND_3 to search for the variation of μ in laboratory experiments[55]. The ammonia molecule has a pyramidal shape and the inversion frequency depends on the exponentially small tunneling of three hydrogen (or deuterium) nuclei through the potential barrier. Because of that, it is very sensitive to any changes of the parameters of the system, and particularly to the reduced mass for this vibrational mode.

High precision data on the redshifts of NH_3 inversion lines exist for the previously mentioned object B0218+357 at $z = 0.6847$ [56]. Comparing them with the redshifts of rotational lines of CO, HCO^+, and HCN molecules from Ref. [57] allows strong limits on the variation of μ, and hence X_e, to be obtained [58]. The most accurate measurements, utilising new rotational spectra, give very stringent limits on μ-variation in this quasar absorption system and the object PKS1830–211 at $z = 0.8858$ (last two lines of table 1).

3. Clocks

3.1. Optical atomic clocks

Atomic clocks can be used to measure time-variation of fundamental constants in the Earth frame. Different optical atomic clocks utilize transitions that have positive, negative or small contributions of the relativistic corrections to frequencies, so comparison of these clocks can be used to measure α-variation. The same methods of relativistic many-body calculations used in the quasar absorption studies can be used to calculate the dependence on α of different clocks [16, 17, 59–62]. A summary of results is presented in Ref. [63]. The relativistic effects are proportional to $(Z\alpha)^2$; therefore the q coefficients for optical clock transitions may be substantially larger than in cosmic transitions since the clock transitions are often in heavy atoms (Hg II, Yb II, Yb III, etc.) while cosmic spectra contain mostly light atoms ($Z \lesssim 33$).

The temporal variation of α measured in the laboratory can be compared with the variation expected from the observed spatial gradient in α [1] because the solar system has a component of its velocity (relative to the cosmic microwave background) along the axis of the dipole. The expected variation from the dipole was calculated to be [64]

$$\dot{\alpha}/\alpha = 1.35 \times 10^{-18} \cos\psi \ \mathrm{yr}^{-1} \tag{7}$$

where ψ is the angle between the motion of the Sun and the dipole. The best fit from [1] gives $\cos\psi \sim 0.07$ but this has an uncertainty of ~ 0.15. The α-variation is modulated by the annual motion of the Earth around the Sun, $\delta\alpha/\alpha = 1.4 \times 10^{-20} \cos\omega t$ [64] where ω refers to the angular frequency of the yearly orbit.

The current best constraint on time-variation of α was achieved by precisely measuring the frequency ratio of Hg II and Al II clocks several times over the course of a year [65]. Using our calculations, the rate of change of α is measured at $\dot{\alpha}/\alpha = (-1.6 \pm 2.3) \times 10^{-17}\ \mathrm{yr}^{-1}$. This limit will need to be improved by around two orders-of-magnitude in order to confirm or contradict the observed spatial gradient in α. On the other hand, the quasar observations do not exclude time-variation of α below the rate of $\sim 10^{-16}\ \mathrm{yr}^{-1}$, and here the current laboratory limits are already competitive.

3.2. Enhanced effect of α-variation in Dy atom

The sensitivity required by (7) may be obtained by finding systems where α-variation is strongly enhanced. Transitions between two almost degenerate levels in the Dy atom can give a very high relative enhancement because these levels move in opposite directions if α varies [15, 17, 62]. The relative variation may be presented as $\delta\omega/\omega = K\delta\alpha/\alpha$ where the coefficient K exceeds 10^8 ($q = 30\,000\ \mathrm{cm}^{-1}$, $\omega \sim 10^{-4}\ \mathrm{cm}^{-1}$). The values of $K = 2q/\omega$ are different for different hyperfine components and isotopes because ω changes. An experiment is currently underway to place limits on α variation using this transition [66, 67], however one of the levels has quite a large linewidth and this limits the accuracy. The current limit is $\dot{\alpha}/\alpha = (-2.7 \pm 2.6) \times 10^{-15}\ \mathrm{yr}^{-1}$. Several other enhanced effects of α variation in atoms have been calculated [68, 69].

3.3. Enhanced effects of α-variation in highly charged ions

Sensitivity to α-variation increases with ion charge as $(Z_i + 1)^2$ [70]. The most sensitive atomic systems will maximize the contributions from three factors: high nuclear charge Z, high ionization degree, and significant differences in the configuration composition of the states involved. Unfortunately, the interval between different energy levels in an ion also increases as $\sim (Z_i + 1)^2$, which can quickly take the transition frequency out of the range of lasers as Z_i increases. The phenomena of Coulomb degeneracy and configuration crossing can be used to combat this tendency. In a neutral atom, an electron orbital with a larger angular momentum is significantly higher than one with smaller angular momentum but with the same principal quantum number n. On the other hand, in the hydrogen-like limit orbitals with different angular momentum but the same principal quantum number are nearly degenerate. Therefore, somewhere in between there can be a crossing point where two levels with different angular momentum and principal quantum number can come close together: in such cases the excitation energy may be within laser range.

In Ref. [70] we showed, using the Ag isoelectronic sequence as an example, why high q-values can occur in highly charged ions, and how the tendency of such systems towards large transition frequencies could be overcome. A two-valence-electron ion, Sm^{14+}, was identified, which has optical transitions that are the most sensitive to potential variation of α ever found. While atomic spectroscopy in electron beam ion traps is currently not competitive with optical frequency standards (see, e.g., [71, 72] and review [73]) the technology continues to improve and, with the enhancements in sensitivity, highly-charged ions may prove to be a good system for detecting variation of α.

3.4. Enhanced effect of variation in UV transition of ^{229}Th nucleus

The ^{229}Th nucleus has the lowest known excited state, lying just 7.6 ± 0.5 eV above the ground state [74]. The position of this level was determined from the energy differences of many high-energy γ-transitions to the ground and first-excited states. The subtraction produces the large uncertainty in the position of the 7.6 eV excited state. The width of this level is estimated to be about 10^{-4} Hz [75], which explains why it is so hard to find the direct radiation in this very weak transition. Nevertheless, the search for the direct radiation continues.

Because the ^{229}Th transition is very narrow and can be investigated with laser spectroscopy, it is a possible reference for an optical clock of very high accuracy [76]. The near degeneracy of these isomers is a result of cancellation between very large energy contributions (order of MeV). Since these contributions would have different dependences on fundamental constants, this transition would be a very sensitive probe of possible variation of fundamental constants [77]. A rough estimate for the relative variation of the ^{229}Th transition frequency is

$$\frac{\delta\omega}{\omega} \approx 10^5 \left(0.1\frac{\delta\alpha}{\alpha} + \frac{\delta X_q}{X_q}\right) . \tag{8}$$

Therefore, the experiment would have the potential of improving the sensitivity to temporal variation of the fundamental constants by many orders of magnitude.

More accurate nuclear calculations give different values for the sensitivity of this transition to α. Refs. [78, 79] claim that both isomers have identical deformations and therefore there is no enhancement of α-variation. Other calculations give enhancement factors in the range 10^2 – 10^5, depending on particulars of the model used [80–83]. To resolve this, we have proposed a method of extracting sensitivity to α-variation using direct laboratory measurements of the change in nuclear mean-square charge radius between the isomers (isomeric shift) [84].

From (8), we obtain the following energy shift in the 7.6 eV ^{229}Th transition:

$$\delta\omega \approx \frac{\delta X_q}{X_q} \text{ MeV} . \tag{9}$$

This corresponds to the frequency shift $\delta\nu \approx 3 \cdot 10^{20}\, \delta X_q/X_q$ Hz. The width of this transition is 10^{-4} Hz, so one may hope to get the sensitivity to the variation of X_q about 10^{-24} per year. This is 10^{10} times better than the current atomic clock limit on the variation of X_q.

Note that there are other narrow low-energy levels in nuclei, for example, the 76 eV level in ^{235}U with lifetime 26.6 minutes is the second-lowest known. One may expect a similar enhancement there. Unfortunately, this level cannot be reached with usual lasers. In principle, it may be investigated using a free-electron laser or synchrotron radiation. However, the accuracy of the frequency measurements is much lower in this case.

Acknowledgments

This work is supported by the Australian Research Council, Marsden grant and ECT*.

References

[1] Webb J K, King J A, Murphy M T, Flambaum V V, Carswell R F and Bainbridge M B 2010 "Evidence for spatial variation of the fine structure constant" (*Preprint* `arXiv:1008.3907`)
[2] Flambaum V V, Leinweber D B, Thomas A W and Young R D 2004 *Phys. Rev. D* **69** 115006
[3] Flambaum V V and Tedesco A F 2006 *Phys. Rev. C* **73** 055501
[4] Flambaum V V and Berengut J C 2009 *Int. J. Mod. Phys. A* **24** 3342
[5] Marciano W J 1984 *Phys. Rev. Lett.* **52** 489
[6] Langacker P, Segrè G and Strassler M J 2002 *Phys. Lett. B* **528** 121
[7] Calmet X and Fritzsch H 2002 *Eur. Phys. J. C* **24** 639
[8] Wetterich C 2003 *J. Cosmol. Astropart. Phys.* **JCAP10** 002
[9] Dent T and Fairbairn M 2003 *Nucl. Phys. B* **653** 256
[10] Dmitriev V F, Flambaum V V and Webb J K 2004 *Phys. Rev. D* **69** 063506
[11] Berengut J C, Flambaum V V and Dmitriev V F 2010 *Phys. Lett. B* **683** 114
[12] Lea S N 2007 *Rep. Prog. Phys.* **70** 1473
[13] Savedoff M P 1956 *Nature* **178** 689
[14] Uzan J P 2003 *Rev. Mod. Phys.* **75** 403
[15] Dzuba V A, Flambaum V V and Webb J K 1999 *Phys. Rev. Lett.* **82** 888
[16] Dzuba V A, Flambaum V V and Webb J K 1999 *Phys. Rev. A* **59** 230
[17] Dzuba V A, Flambaum V V and Marchenko M V 2003 *Phys. Rev. A* **68** 022506
[18] Dzuba V A and Flambaum V V 2005 *Phys. Rev. A* **71** 052509
[19] Dzuba V A, Flambaum V V, Kozlov M G and Marchenko M 2002 *Phys. Rev. A* **66** 022501
[20] Berengut J C, Dzuba V A, Flambaum V V and Marchenko M V 2004 *Phys. Rev. A* **70** 064101
[21] Berengut J C, Flambaum V V and Kozlov M G 2005 *Phys. Rev. A* **72** 044501
[22] Berengut J C, Flambaum V V and Kozlov M G 2006 *Phys. Rev. A* **73** 012504
[23] Dzuba V A and Flambaum V V 2008 *Phys. Rev. A* **77** 012514
[24] Berengut J C, Dzuba V A, Flambaum V V, King J A, Kozlov M G, Murphy M T and Webb J K 2009 *Mem. Soc. Astron. It.* **80** 795
[25] Berengut J C, Dzuba V A, Flambaum V V, King J A, Kozlov M G, Murphy M T and Webb J K 2010 `arXiv:physics/0408017`
[26] Webb J K, Flambaum V V, Churchill C W, Drinkwater M J and Barrow J D 1999 *Phys. Rev. Lett.* **82** 884
[27] Webb J K, Murphy M T, Flambaum V V, Dzuba V A, Barrow J D, Churchill C W, Prochaska J X and Wolfe A M 2001 *Phys. Rev. Lett.* **87** 091301
[28] Murphy M T, Webb J K and Flambaum V V 2003 *Mon. Not. R. Astron. Soc.* **345** 609
[29] Srianand R, Chand H, Petitjean P and Aracil B 2004 *Phys. Rev. Lett.* **92** 121302
[30] Murphy M T, Webb J K and Flambaum V V 2007 *Phys. Rev. Lett.* **99** 239001
[31] Murphy M T, Webb J K and Flambaum V V 2008 *Mon. Not. R. Astron. Soc.* **384** 1053
[32] Srianand R, Chand H, Petitjean P and Aracil B 2007 *Phys. Rev. Lett.* **99** 239002
[33] Hinshaw G *et al* 2009 *Astrophys. J. Suppl. Ser.* **180** 225
[34] Chand H, Srianand R, Petitjean P and Aracil B 2004 *Astron. Astrophys.* **417** 853
[35] Chand H, Petitjean P, Srianand R and Aracil B 2005 *Astron. Astrophys.* **430** 47
[36] Quast R, Reimers D and Levshakov S A 2004 *Astron. Astrophys.* **415** L7
[37] Levshakov S A, Centurión M, Molaro P and D'Odorico S 2005 *Astron. Astrophys.* **434** 827
[38] Levshakov S A, Centurión M, Molaro P, D'Odorico S, Reimers D, Quast R and Pollmann M 2006 *Astron. Astrophys.* **449** 879
[39] Levshakov S A, Molaro P, Lopez S, D'Odorico S, Centurión M, Bonifacio P, Agafonova I I and Reimers D 2007 *Astron. Astrophys.* **466** 1077
[40] Reinhold E, Buning R, Hollenstein U, Ivanchik A, Petitjean P and Ubachs W 2006 *Phys. Rev. Lett.* **96** 151101

[41] Ubachs W, Buning R, Eikema K S E and Reinhold E 2007 *J. Mol. Spectrosc.* **241** 155
[42] Salumbides E J, Bailly D, Khramov A, Wolf A L, Eikema K S E, Vervloet M and Ubachs W 2008 *Phys. Rev. Lett.* **101** 223001
[43] King J A, Webb J K, Murphy M T and Carswell R F 2008 *Phys. Rev. Lett.* **101** 251304
[44] Malec A L, Buning R, Murphy M T, Milutinovic N, Ellison S L, Prochaska J X, Kaper L, Tumlinson J, Carswell R F and Ubachs W 2010 *Mon. Not. R. Astron. Soc.* **403** 1541
[45] Berengut J C, Flambaum V V, King J A, Curran S J and Webb J K 2010 "Is there further evidence for spatial variation of fundamental constants?" (*Preprint* arXiv:1009.0591)
[46] Tzanavaris P, Webb J K, Murphy M T, Flambaum V V and Curran S J 2005 *Phys. Rev. Lett.* **95** 041301
[47] Tzanavaris P, Webb J K, Murphy M T, Flambaum V V and Curran S J 2007 *Mon. Not. R. Astron. Soc.* **374** 634
[48] Kanekar N, Prochaska J X, Ellison S L and Chengalur J N 2010 *Astrophys. J. Lett.* **712** L148
[49] Murphy M T, Webb J K, Flambaum V V, Drinkwater M J, Combes F and Wiklind T 2001 *Mon. Not. R. Astron. Soc.* **327** 1244
[50] Kanekar N, Carilli C L, Langston G I, Rocha G, Combes F, Subrahmanyan R, Stocke J T, Menten K M, Briggs F H and Wiklind T 2005 *Phys. Rev. Lett.* **95** 261301
[51] Kanekar N, Chengalur J N and Ghosh T 2010 *Astrophys. J. Lett.* **716** L23
[52] Murphy M T, Flambaum V V, Muller S and Henkel C 2008 *Science* **320** 1611
[53] Henkel C, Menten K M, Murphy M T, Jethava N, Flambaum V V, Braatz J A, Muller S, Ott J and Mao R Q 2009 *Astron. Astrophys.* **500** 725
[54] Drinkwater M J, Webb J K, Barrow J D and Flambaum V V 1998 *Mon. Not. R. Astron. Soc.* **295** 457
[55] van Veldhoven J, Küpper J, Bethlem H L, Sartakov B, van Roij A J A and Meijer G 2004 *Eur. Phys. J. D* **31** 337
[56] Henkel C, Jethava N, Kraus A, Menten K M, Carilli C L, Grasshoff M, Lubowich D and Reid M J 2005 *Astron. Astrophys.* **440** 893
[57] Combes F and Wiklind T 1997 *Astrophys. J.* **486** L79
[58] Flambaum V V and Kozlov M G 2007 *Phys. Rev. Lett.* **98** 240801
[59] Dzuba V A and Flambaum V V 2000 *Phys. Rev. A* **61** 034502
[60] Angstmann E J, Flambaum V V and Karshenboim S G 2004 *Phys. Rev. A* **70** 044104
[61] Angstmann E J, Dzuba V A and Flambaum V V 2004 *Phys. Rev. A* **70** 014102
[62] Dzuba V A and Flambaum V V 2008 *Phys. Rev. A* **77** 012515
[63] Flambaum V V and Dzuba V A 2009 *Can. J. Phys.* **87** 25 (*Preprint* arXiv:0805.0462)
[64] Berengut J C and Flambaum V V 2010 "Manifestations of a spatial variation of fundamental constants on atomic clocks, Oklo, meteorites, and cosmological phenomena" (*Preprint* arXiv:1008.3957)
[65] Rosenband T, Hume D B, Schmidt P O, Chou C W, Brusch A, Lorini L, Oskay W H, Drullinger R E, Fortier T M, Stalnaker J E, Diddams S A, Swann W C, Newbury N R, Itano W M, Wineland D J and Bergquist J C 2008 *Science* **319** 1808
[66] Nguyen A T, Budker D, Lamoreaux S K and Torgerson J R 2004 *Phys. Rev. A* **69** 022105
[67] Cingöz A, Lapierre A, Nguyen A T, Leefer N, Budker D, Lamoreaux S K and Torgerson J R 2007 *Phys. Rev. Lett.* **98** 040801
[68] Dzuba V A and Flambaum V V 2005 *Phys. Rev. A* **72** 052514
[69] Angstmann E J, Dzuba V A, Flambaum V V, Karshenboim S G and Nevsky A Yu 2006 *J. Phys. B* **39** 1937
[70] Berengut J C, Dzuba V A and Flambaum V V 2010 *Phys. Rev. Lett.* **105** 120801
[71] Draganić I, Crespo López-Urrutia J R, DuBois R, Fritzsche S, Shabaev V M, Orts R S, Tupitsyn I I, Zou Y and Ullrich J 2003 *Phys. Rev. Lett.* **91** 183001
[72] Crespo López-Urrutia J R 2008 *Can. J. Phys.* **86** 111
[73] Beiersdorfer P 2009 *Phys. Scr.* **T134** 014010
[74] Beck B R, Becker J A, Beiersdorfer P, Brown G V, Moody K J, Wilhelmy J B, Porter F S, Kilbourne C A and Kelley R L 2007 *Phys. Rev. Lett.* **98** 142501
[75] Tkalya E V, Zherikhin A N and Zhudov V I 2000 *Phys. Rev. C* **61** 064308
[76] Peik E and Tamm Chr 2003 *Europhys. Lett.* **61** 181
[77] Flambaum V V 2006 *Phys. Rev. Lett.* **97** 092502
[78] Hayes A C and Friar J L 2007 *Phys. Lett. B* **650** 229
[79] Hayes A C, Friar J L and Möller P 2008 *Phys. Rev. C* **78** 024311
[80] He X-t and Ren Z-z 2007 *J. Phys. G* **34** 1611
[81] He X-t and Ren Z-z 2008 *J. Phys. G* **35** 035106
[82] Flambaum V V, Auerbach N and Dmitriev V F 2009 *Europhys. Lett.* **85** 50005
[83] Litvinova E, Feldmeier H, Dobaczewski J and Flambaum V V 2009 *Phys. Rev. C* **79** 064303
[84] Berengut J C, Dzuba V A, Flambaum V V and Porsev S G 2009 *Phys. Rev. Lett.* **102** 210801

Optical lattice clocks and frequency comparison

Hidetoshi Katori, Tetsushi Takano and Masao Takamoto

Department of Applied Physics, The University of Tokyo, Tokyo, Japan
CREST, Japan Science and Technology Agency, Saitama, Japan

E-mail: katori@amo.t.u-tokyo.ac.jp

Abstract. We consider designs of optical lattice clocks in view of the quantum statistics, relevant atomic spins, and atom-lattice interactions. The first two issues lead to two optimal constructions for the clock: a one-dimensional (1D) optical lattice loaded with spin-polarized fermions and a 3D optical lattice loaded with bosons. By taking atomic multipolar interactions with the lattice fields into account, an "atomic motion insensitive" wavelength is proposed to provide a precise definition of the "magic wavelength." We then present a frequency comparison of these two optical lattice clocks: spin-polarized fermionic ^{87}Sr and bosonic ^{88}Sr prepared in 1D and 3D optical lattices, respectively. Synchronous interrogations of these two optical lattice clocks by the same probe laser allowed canceling out its frequency noise as a common mode noise to achieve a relative stability of 3×10^{-17} for an averaging time of $\tau = 350$ s. The scheme, therefore, provides us with a powerful means to investigate intrinsic uncertainty of the clocks regardless of the probe laser stability. We discuss prospects of the synchronous operation of the clocks on the measurement of the geoid height difference and on the search of constancy of fundamental constants.

1. Introduction

To date, optical clocks based on singly trapped ions [1] and ultracold neutral atoms trapped in optical lattices that operate at the magic wavelength to eliminate light shift perturbation [2] are regarded as promising candidates for future atomic clocks. Since 2006, Sr-based "optical lattice clocks" have been evaluated close to the Cs clocks' uncertainty limit internationally [3, 4]. To further reduce their uncertainty and instability, there remain essential experimental challenges. One is to find out better lattice geometries as well as interrogated atom species that bring out the potential performance of the clock scheme, taking into account the collisional frequency shift, the black body radiation shift, and the lattice induced light shifts. The other is to establish experimental schemes that fully utilize the advantage of a large number N of atoms to improve the clock stability, which is currently hampered by the instability of the probe laser of typically 5×10^{-16} at 1 s that is due to the thermal noise of a reference cavity [5].

In this paper, we first describe designs of optical lattice clocks in view of the quantum statistics and relevant atomic spins. This leads to two optimal configurations for the clock: a one-dimensional (1D) optical lattice loaded with spin-polarized fermions and a 3D optical lattice loaded with bosons [6]. In particular, we refer to an "atomic motion insensitive" wavelength that provides a precise definition [7] of a "magic wavelength" including atomic multipolar interactions.

We then present a frequency comparison of these two optical lattice clocks using fermionic ^{87}Sr and bosonic ^{88}Sr. Synchronous interrogations of the two clocks by the same probe laser allow canceling out its frequency noise as a common mode noise in evaluating the relative

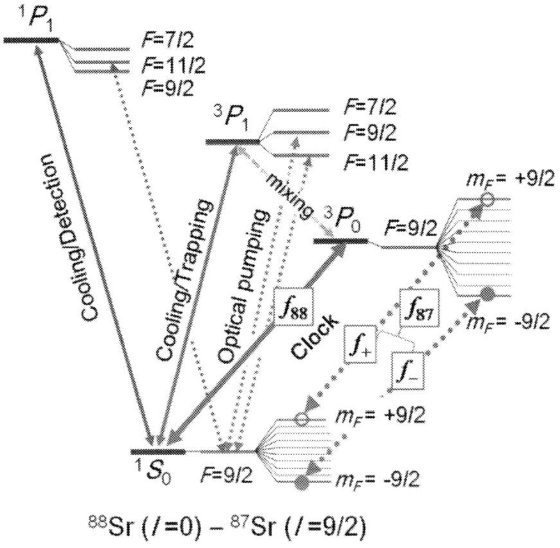

Figure 1. (Color online) Energy levels for ^{88}Sr and ^{87}Sr atoms. Spin-polarized ultracold ^{87}Sr atoms were prepared by optical pumping on the $^1S_0(F = 9/2) - {}^3P_1(F = 9/2)$ transition at $\lambda = 689$ nm with circularly polarized light. The first-order Zeeman shift and the vector light shift on the clock transition at $\lambda = 698$ nm were eliminated by averaging the transition frequencies $f_\pm$. The lack of nuclear spin makes the energy structure of ^{88}Sr simpler.

stability of the clocks; therefore the scheme enables us to explore the intrinsic uncertainty of the clock regardless of the probe laser stability. On the other hand, in order to overcome the clock instability that is predominantly determined by the "thermal noise limit" of the probe laser, we mention a novel frequency stabilization scheme that does not employ the "electron shelving" technique for the clock state detection. We discuss possible impacts of these approaches.

2. Designing optical lattice clocks: Quantum statistics and optical lattice geometry

In addition to the removal of light shift perturbations, the control and prevention of atomic interactions are a serious concern in designing optical lattice clocks. The collisional frequency shift of atomic clocks operated with ultracold atoms is related to the mean field energy shift $\mathcal{E} = 4\pi\hbar^2 a n g^{(2)}(0)/m$ of the relevant electronic state, with a the s-wave scattering length, n atomic density, and m atomic mass. Here, $g^{(2)}(0)$ is the two-particle correlation function at zero distance, which is zero for identical fermions and $1 \leq g^{(2)}(0) \leq 2$ for distinguishable or bosonic atoms. Hence, collisional shifts are suppressed for ultracold fermions, while they are intrinsically unavoidable for bosons. The quantum statistical nature of atoms is determined by their total spins; that is, bosons have zero or integer spins and fermions have half-integer spins. In particular, for atoms with an even number of electrons that have a $J = 0$ state suitable for optical lattice clocks, their nuclear spins I may be zero for bosons and $I \geq 1/2$ for fermions. Consequently, the total angular momentum $F = J + I$ of the clock states can be zero for bosonic atoms, but not for fermions, which causes coupling to the light polarization of the lattice field. Figure 1 shows the relevant energy levels for fermionic ^{87}Sr and bosonic ^{88}Sr isotopes of Sr that we employ in the following experiments.

We consider two lattice geometries. A one-dimensional (1D) (see Fig. 2(a)) or 2D lattice composed of a single electric field vector has spatially uniform light polarization. In contrast, a 3D lattice (see Fig. 2(b)) requires at least two electric field vectors; therefore, the synthesized

Journal of Physics: Conference Series **264** (2011) 012011 doi:10.1088/1742-6596/264/1/012011

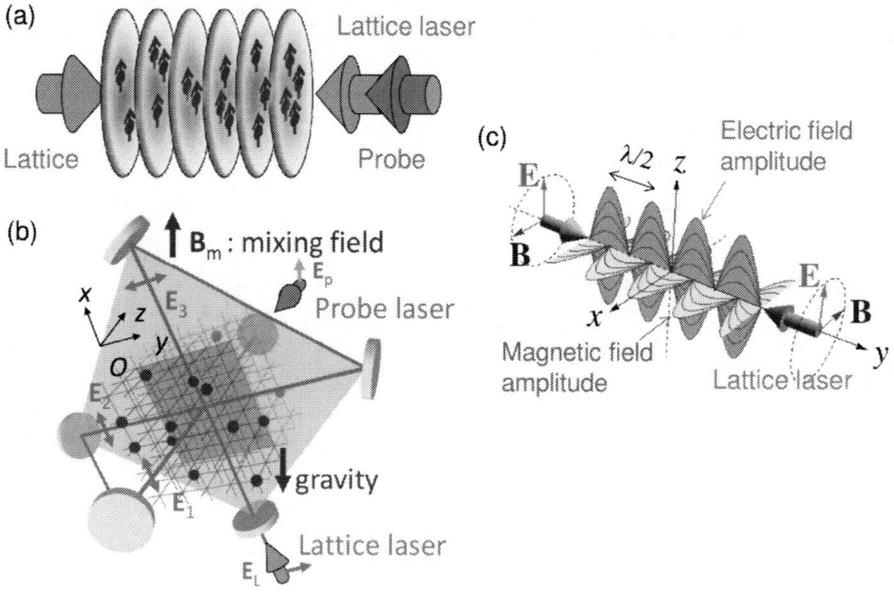

Figure 2. (Color online) (a) A one-dimensional optical clock is realized by a standing wave of light tuned to the magic wavelength. Multiply trapped spin-polarized fermions, which are indicated by allows, in a single pancake potential may be protected from collisions by the Pauli blocking. (b) Single occupancy 3D lattice. (c) Electric and magnetic field amplitudes in a standing wave.

field exhibits a polarization gradient that varies in space depending on the intensity profile and relative phases of the lattice lasers. In 1D optical lattice clocks with multiple atoms in each lattice site, the application of spin-polarized fermions [8, 9] may minimize the collisional frequency shift owing to their quantum-statistical properties. Figure 2(a) shows a schematic diagram for the "spin-polarized" 1D optical lattice clock [10], where the upward arrows correspond to spin-polarized fermionic atoms. The spatially uniform light field polarization of the 1D optical lattice allows us to cancel out the vector light shift [10] by alternately interrogating the transition frequencies $f_\pm$, corresponding to two outer Zeeman components ${}^1S_0(F, \pm m_F) - {}^3P_0(F, \pm m_F)$ of the clock transition (see Fig. 1). This vector light shift cancellation technique simultaneously cancels out the Zeeman shift, thereby realizing virtual spin-zero atoms.

The application of 3D optical lattices with a single (or less) atom in each lattice site is effective to suppress atomic collisions. However, light polarization inhomogeneity inevitable in 3D optical lattices makes a vector light shift for atoms with its angular momentum $F \neq 0$ problematic, as the "vector light shift cancellation" technique is no longer applicable. From this viewpoint, the 3D optical lattice clock is suitable for bosonic atoms with scalar clock states ($J = 0$) and has been demonstrated with bosonic ^{88}Sr atoms [6]. Its systematic uncertainties were investigated at 3×10^{-15} by referencing a spin-polarized 1D optical lattice clock with fermionic ^{87}Sr atoms.

Finally in order to moderate the hyperpolarizability effects that cannot be eliminated, in particular, at the red-detuned magic wavelength where atoms are trapped at lattice intensity maxima, we investigated a blue-detuned lattice that confines atoms in the intensity minima of the electric field. For Sr, such wavelengths are found on the blue side of the $5s^2\,{}^1S_0 - 5s5p\,{}^1P_1$ transition at 461 nm; one such wavelength is found at $\lambda_L \approx 390$ nm on the blue side of the $5s5p\,{}^3P_0 - 5s6d\,{}^3D_1$ transition at 394 nm. For this magic wavelength, a laser intensity of $I_L = 10$ kW/cm^2 gives a trap depth of about 200 kHz. As the atoms are trapped near the

Journal of Physics: Conference Series **264** (2011) 012011 doi:10.1088/1742-6596/264/1/012011

nodes of the standing wave, the effective light intensity that atoms experience is about one-tenth of the intensity at the anti-nodes. Assuming a trap depth of 10 μK, the 4th-order light shift is estimated to be 0.1 mHz, corresponding to a fractional uncertainty of 2×10^{-19} [11]. The blue magic wavelength was experimentally determined to be 389.889(9) nm [11] employing ^{87}Sr atoms trapped in a 1D optical lattice operated at the (red-detuned) magic wavelength of $\lambda_L = 813.4$ nm.

3. Multipolar interactions of atoms with lattice field and atomic-motion-insensitive wavelength

Optical lattice clocks operated at the blue magic wavelength seem to be ideal in view of the electric dipole (E1) interaction of atoms with the lattice laser field. However, when taking atomic multipolar interactions into account, things are not that simple. Consider, for example, the linearly polarized ($\|\mathbf{e}_z$) standing wave electric field $\mathbf{E} = \mathbf{e}_z E_0 \sin ky \cos \omega t$ with wave number k and frequency ω, as shown in Fig. 2(c). Following the Maxwell equation $\nabla \times \mathbf{E} = -\frac{1}{c}\frac{\partial \mathbf{B}}{\partial t}$, with c the speed of light, the corresponding magnetic field is given by $\mathbf{B} = -\mathbf{e}_x E_0 \cos ky \sin \omega t$. This indicates that the electric and magnetic field amplitudes are one quarter of the wavelength $\lambda/4 = \pi c/(2\omega)$ out of phase in space. Consequently, the magnetic dipole (M1) interaction is largest at the nodes of the electric field. Furthermore, as the electric quadrupole (E2) interaction is proportional to the gradient of the electric field, the E2 interaction is also largest at the nodes of the electric field. Hence, the blue magic wavelength is not necessarily free of light shift due to atomic multipolar interactions with the lattice field. The energy shift of atoms in optical lattices is obtained by the second-order perturbation in the E1, M1, and E2 interactions that vary as $V_{\rm E1} \sin^2 ky$, $V_{\rm M1} \cos^2 ky$, and $V_{\rm E2} \cos^2 ky$. Consequently, it is no longer possible to perfectly match the total light shift in two clock states, as the differential light shifts due to the M1 and E2 interactions introduce an atomic-motion-dependent light shift because of their spatial mismatch with the E1 interaction [12].

Although the contributions of the M1 and E2 interactions are 6-7 orders of magnitude smaller than that of the E1 interaction in optical lattice clocks in Sr [2], they have a non-negligible contribution in pursuing a 1×10^{-18} level uncertainty. Therefore, a more precise definition of the magic wavelength, including multipolar interactions, is necessary. Assuming the differential polarizabilities of the E1, M1, and E2 interactions in the clock transition to be $\Delta\alpha_{\rm E1}(\lambda_L)$, $\Delta\alpha_{\rm M1}(\lambda_L)$, and $\Delta\alpha_{\rm E2}(\lambda_L)$, and the corresponding spatial distributions to be $q_{\rm E1}(\mathbf{r})$, $q_{\rm M1}(\mathbf{r})$, and $q_{\rm E2}(\mathbf{r})$, the transition frequency of atoms in the optical lattices can be given by

$$\nu(\lambda_L) = \nu_0 - \frac{1}{2h}\left[\Delta\alpha_{\rm E1}(\lambda_L) q_{\rm E1}(\mathbf{r}) + \Delta\alpha_{\rm M1}(\lambda_L) q_{\rm M1}(\mathbf{r}) + \Delta\alpha_{\rm E2}(\lambda_L) q_{\rm E2}(\mathbf{r})\right] E^2, \quad (1)$$

where the 4th- and higher-order terms and light polarization dependences are omitted. We have shown that one can eliminate the atomic-motion-dependent light shift caused by the multipolar interactions by choosing particular 3D optical lattice geometries that make $q_{\rm M1}(\mathbf{r})$ or $q_{\rm E2}(\mathbf{r})$ terms in phase or out of phase with respect to the spatial dependence of $q_{\rm E1}(\mathbf{r})$ [7]. For example, in the case of a 1D lattice with the E1 spatial dependence $q_{\rm E1}(\mathbf{r}) = \sin^2 ky \, (= 1 - \cos^2 ky)$, the corresponding M1 and E2 interactions can be expressed as $q_{\rm M1}(\mathbf{r}) = q_{\rm E2}(\mathbf{r}) = \cos^2 ky = \Delta q - q_{\rm E1}(\mathbf{r})$ with $\Delta q = 1$. Therefore, by taking $\Delta\alpha_{\rm EM} \equiv \Delta\alpha_{\rm E1} - \Delta\alpha_{\rm M1} - \Delta\alpha_{\rm E2}$ and $\Delta\alpha_0 \equiv \Delta\alpha_{\rm M1} + \Delta\alpha_{\rm E2}$, Eq. (1) can be rewritten as

$$\nu(\lambda_L) = \nu_0 - \frac{1}{2h}\Delta\alpha_{\rm EM}(\lambda_L) q_{\rm E1}(\mathbf{r}) E^2 - \frac{1}{2h}\Delta\alpha_0(\lambda_L)\Delta q E^2, \quad (2)$$

where the second term on the right side varies in phase with the E1 interaction. This equation suggests the precise definition of the magic wavelength to be an "atomic-motion insensitive" wavelength that satisfies $\Delta\alpha_{\rm EM}(\lambda_L) = 0$. The last term provides a spatially constant offset of

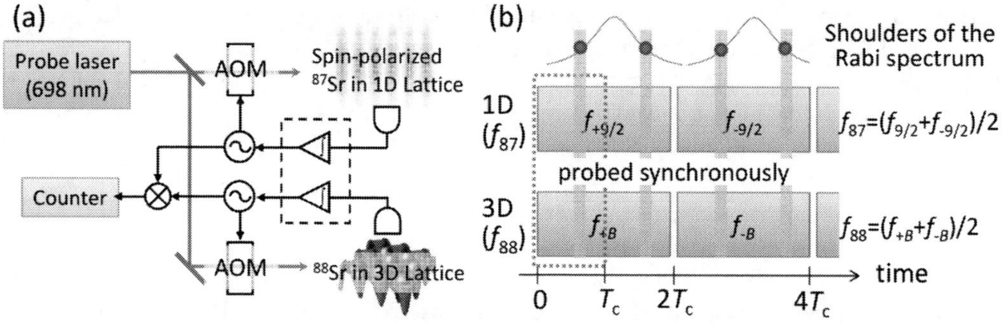

Figure 3. (Color online) (a) Experimental setup for frequency comparison of 1D and 3D optical lattice clocks. A single laser with an RF frequency offset of ≈ 62 MHz probes the clock transitions of the two isotopes simultaneously. (b) Timing chart for the frequency comparison. A cycle time of $T_c = 1.4$ s is used to probe a shoulder of the Rabi excitation. Cancellation of the slowly varying residual magnetic fields is completed in $4T_c$. Thick vertical lines indicate a clock probing period of 100 ms.

typically 10 mHz or below and is solely dependent on the total laser intensity $\propto \Delta q E^2$ used to form the lattice. This offset frequency can be accurately determined by measuring the atomic vibrational frequencies in the lattice [7].

4. Frequency comparison of 1D and 3D optical lattice clocks

Optical lattice clocks [2] are expected to be highly stable, e.g., $\sigma_y(\tau) \sim 10^{-18}/\sqrt{\tau}$, if they operate at the quantum projection noise (QPN) limit of $\sigma_y(\tau) \simeq (\Delta f/f_0)/\sqrt{N\tau}$, by taking advantage of the large number $N \sim 10^6$/s of atoms and moderately narrow linewidth $\Delta f = 1$ Hz. However, the frequency comparison of two optical lattice clocks [6] shows that their actual stabilities are essentially limited by those of the probe lasers due to the thermal noise of the reference cavity, e.g., $\sigma_y \sim 1 \times 10^{-15}$ for a 7.5-cm-long cavity made of ultralow expansion (ULE) glass [5] and by the Dick effect [13], which is attributed to the down-conversion of the high frequency (n/T_c with n an integer) components of the probe laser noise $\delta\omega(t)$ introduced by the discrete interrogation process with a cycle time T_c of the clock operation.

When exciting the clock transition, the population fluctuation δP of the excited state caused by the frequency noise $\delta\omega(t)$ is expressed using the sensitivity function $g(t)$ as in [13],

$$\delta P = \frac{1}{2}\int_0^{T_c} g(t)\delta\omega(t)dt, \tag{3}$$

where T_c is a clock cycle time. The frequency servo translates a population fluctuation δP into a frequency fluctuation of the probe laser, which degrades the stability of an atomic clock, known as the Dick effect.

We consider a frequency comparison of two optical lattice clocks that are simultaneously probed by a single laser, which allows the two clocks to be probed by the same frequency noise $\delta\omega(t)$. As long as the sensitivity functions $g(t)$ are the same for both clocks, their population fluctuations δP should be equal; therefore the stability degradation due to δP may be rejected as a common mode noise in evaluating the frequency difference of the two clocks. This allows an investigation of the relative stability of the two clocks that is only limited by the quantum projection noise of atoms regardless of the probe laser instability. Such a synchronous frequency comparison was previously demonstrated in the microwave frequency domain using Cs and Rb fountain clocks [14].

Figure 3(a) shows an experimental setup for frequency comparison of the 1D and 3D lattice clocks, loaded with ^{87}Sr and ^{88}Sr, respectively, as described in Sec. 2. After cooling and loading atoms in each optical lattice, we interrogated the two clock transitions synchronously with a 100 ms-long Rabi pulse. A cycle time of $T_c = 1.4$ s included the time spent for cooling and trapping of atoms in the lattices and for measuring the excitation probability by observing a shoulder of the Rabi spectrum, as depicted in Fig. 3(b). The clock laser frequencies f_{87} and f_{88} were stabilized to the clock transitions of ^{87}Sr and ^{88}Sr by steering the frequencies of acousto-optic modulators (AOMs) by respective digital servos. The beat note $\delta f = f_{88} - f_{87}$, which corresponds to their isotope shift $\delta f \simeq 62$ MHz, was investigated to evaluate the stabilities and uncertainties of the two optical lattice clocks.

In a 1D lattice, we trapped 3×10^3 fermionic ^{87}Sr atoms. They were spin-polarized in the $m_F = 9/2$ (or $-9/2$) stretched state in the $^1S_0$ ground state (see Fig. 1) to avoid atomic collisions via Pauli blocking [6, 8]. A bias magnetic field of 0.22 mT was applied to define the quantization axis. We excited the $^1S_0(F = 9/2, m_F = \pm 9/2) \rightarrow {}^3P_0(F = 9/2, m_F = \pm 9/2)$ transitions with frequencies of $f_\pm$ successively and calculated $f_{87} = (f_+ + f_-)/2$ to eliminate the first order Zeeman and vector light shifts. For a 3D optical lattice clock, 2×10^3 bosonic ^{88}Sr atoms were loaded into a single occupancy 3D lattice (see Fig. 2(b)). A bias magnetic field of $|\mathbf{B}_m| = 2.36$ mT was applied to magnetically induce the $^1S_0 \rightarrow {}^3P_0$ clock transition. A probe laser intensity of $I_p = 86$ mW/cm^2 was applied to compensate the relatively small magnetically-induced transition moment. These external fields introduced rather large frequency shifts; the second order Zeeman shift of $\Delta_B \simeq -130$ Hz and the probe laser light shift of $\Delta_p \simeq -1.5$ Hz, whose stabilities were evaluated with fractional uncertainties of 2×10^{-17} and 1×10^{-17}, respectively. To eliminate the influence of a slowly varying environmental magnetic field $\mathbf{B}_{\rm ext}$ that affected the second order Zeeman shift as $\propto \pm\mathbf{B}_m \cdot \mathbf{B}_{\rm ext}$, we reversed the direction of the bias magnetic field as $\pm\mathbf{B}_m$ in successive measurements and calculated $f_{88} = (f_{+B} + f_{-B})/2$, where we note that the contribution of $\propto |\mathbf{B}_{\rm ext}|^2$ was negligibly small. Moreover, we installed fiber noise cancellation systems referencing the end mirrors of the 1D and 3D lattices to reduce the Doppler shifts arising from the relative motion between the probe lasers and atoms in the lattices.

Figure 4(a) shows the temporal response of the frequency difference $\delta f = f_{88} - f_{87}$ of the 1D-3D lattice clocks in the synchronous measurement. Thanks to the common mode noise rejection of the laser frequency noise, atomic frequency differences of a few times 10^{-16} are visible in tens of seconds. This will find useful applications in investigating possible perturbations on probed atomic transitions nearly in real time.

Figure 4(b) shows the relative stabilities of the 1D and 3D optical lattice clocks. The triangles show the relative stability of the two clocks measured for an asynchronous interrogation, where the 1D and 3D clocks operated alternately every 1 s [6]. The short-term stability is limited by that of the probe laser operated at the thermal noise limit of a reference cavity made of ULE glass. After an averaging time of a few tens of seconds, the Allan deviation starts to decrease due to the feedback control with a Nyquist frequency of $f_N = (4\text{ s})^{-1}$. The Allan deviation decreased with $\sigma_y(\tau) = 6 \times 10^{-15}/\sqrt{\tau}$ (green dashed line) and reached 1×10^{-16} at $\tau = 2 \times 10^3$ s. In contrast, the relative stability measured for the synchronous interrogations is shown by the circles. The Allan deviation decreased with $\sigma_y(\tau) = 6 \times 10^{-16}/\sqrt{\tau}$ and reached 3×10^{-17} for an averaging time of $\tau = 350$ s, which shows 10 times better stability than that of the asynchronous interrogation. It is noted that the stability in the synchronous measurement is approaching the QPN limit for $N = 1 \times 10^3$ and $\Delta f = 8$ Hz (blue dashed line). These parameters are similar to those used in the measurements.

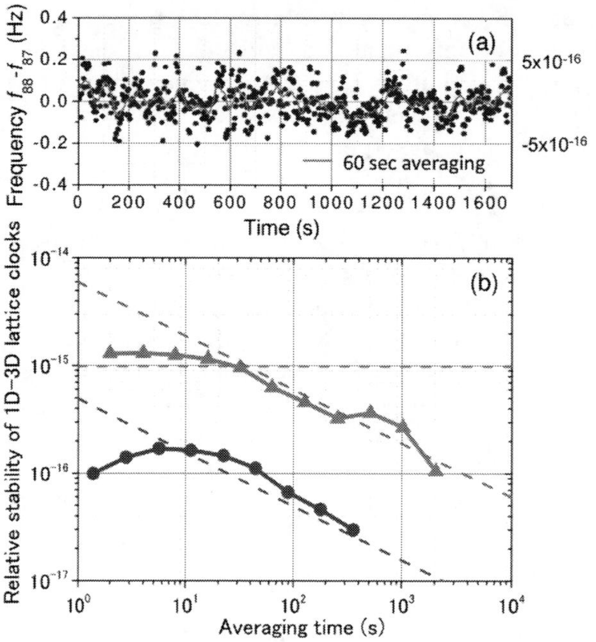

Figure 4. (Color online) (a) Frequency difference $\delta f = f_{88} - f_{87}$ of the 1D-3D lattice clocks measured for the synchronous clock operation, where the isotope shift of ≈ 62 MHz is subtracted. (b) Allan standard deviation calculated for $\delta f = f_{88} - f_{87}$ for sequential interrogations (triangles) and for synchronous interrogations (circles). In the synchronous interrogations, the Allan standard deviation decreased with $\sigma_y(\tau) = 6 \times 10^{-16}\tau^{-1/2}$ and reached 3×10^{-17} for an averaging time of $\tau = 350$ s. The blue dashed line shows the QPN limit for $N = 1 \times 10^3$ and $\Delta f = 8$ Hz.

5. Summary and outlook

We have discussed possible realizations of optical lattice clocks taking quantum statistics, the higher order and the multipolar atom-field interactions into account. Two configurations, 1D optical lattices loaded with spin-polarized fermions and 3D lattices loaded with bosons with a single atom in each lattice site, will allow an investigation of the clock scheme free from the collision shifts and the vector light shift. In addition, the concept of "atomic motion insensitive" wavelength is introduced to redefine the "magic wavelength" including the multipolar atom-field interactions. The "blue atomic-motion-insensitive" wavelength may provide neutral atoms with ideal traps for spectroscopy, as it closely simulates Paul traps for a single ion. We have demonstrated the advantage of employing a "large number of atoms" in optical lattice clocks by the synchronous operation of the 1D and 3D optical lattice clocks, which allowed us to explore a hitherto uninvestigated clock stability of 3×10^{-17} with an averaging time as short as $\tau = 350$ s.

The synchronous frequency comparison scheme provides a convenient means to investigate systematic uncertainties of optical lattice clocks, such as the collision shift, the blackbody radiation shift, the higher order and the multipolar light shifts at fractional uncertainties of 1×10^{-17} and beyond. The scheme is readily applicable to the frequency comparison of two clocks located at remote sites by sharing a local oscillator with the help of precise optical frequency link technologies [4, 15]. Assuming the achieved frequency stability shown in Fig. 4(b), the remote frequency comparison will allow an investigation of the geoid height difference of two sites with an uncertainty of 30 cm in minutes' averaging time. On the other hand, the synchronous interrogation may significantly increase the stability in the frequency

Journal of Physics: Conference Series **264** (2011) 012011 doi:10.1088/1742-6596/264/1/012011

comparison of optical lattice clocks made of different atomic species to investigate the constancy of fundamental constants [16] and the local position invariance [3]. In the case of optical lattice clocks employing Sr (f_0 = 429 THz), Yb (f_0 = 518 THz) and Hg (f_0 = 1129 THz) atoms that are synchronously interrogated by probe lasers frequency-synthesized by an optical frequency comb driven by a single thermal-noise-limited local oscillator operated at 5×10^{-16}, relative stabilities of $2.4 \times 10^{-17}/\sqrt{\tau}$ and $1.5 \times 10^{-16}/\sqrt{\tau}$ are expected for Sr/Yb and Sr/Hg frequency comparisons, where we assumed one million atoms and the excitation linewidth of $\Delta f = 8$ Hz.

In pursuit of a stability goal for the optical lattice clocks [2], e.g., $1 \times 10^{-18}/\sqrt{\tau}$, a novel frequency stabilization scheme needs to be developed. Considering the instability of the thermal noise (at 300 K) limited probe lasers that show a minimum of $(0.5 - 1) \times 10^{-15}$ independent of the averaging time in the range of ms to a few second [5], a servo-loop time constant much less than a second is beneficial to achieve better stability. In this direction, we are considering to make use of the transmitted probe laser through millions of lattice-trapped atoms, since the transmitted light carries full information of the atomic dipoles through the phase shift and absorption/emission that can be extracted as quadrature components using the FM spectroscopy technique. A detailed discussion will be given elsewhere.

This research was partially supported by the Photon Frontier Network Program of the Ministry of Education, Culture, Sports, Science, and Technology, Japan.

References

[1] Rosenband T, Hume D B, Schmidt P O, Chou C W, Brusch A, Lorini L, Oskay W H, Drullinger R E, Fortier T M, Stalnaker J E, Diddams S A, Swann W C, Newbury N R, Itano W M, Wineland D J and Bergquist J C 2008 *Science* **319** 1808–1812

[2] Katori H, Takamoto M, Pal'chikov V G and Ovsiannikov V D 2003 *Phys. Rev. Lett.* **91** 173005

[3] Blatt S, Ludlow A D, Campbell G K, Thomsen J W, Zelevinsky T, Boyd M M, Ye J, Baillard X, Fouche M, Targat R L, Brusch A, Lemonde P, Takamoto M, Hong F L, Katori H and Flambaum V V 2008 *Phys. Rev. Lett.* **100** 140801

[4] Hong F L, Musha M, Takamoto M, Inaba H, Yanagimachi S, Takamizawa A, Watabe K, Ikegami T, Imae M, Fujii Y, Amemiya M, Nakagawa K, Ueda K and Katori H 2009 *Opt. Lett.* **34** 692–694

[5] Numata K, Kemery A and Camp J 2004 *Phys. Rev. Lett.* **93** 250602

[6] Akatsuka T, Takamoto M and Katori H 2008 *Nat. Phys.* **4** 954–959

[7] Katori H, Hashiguchi K, Il'inova E Y and Ovsiannikov V D 2009 *Phys. Rev. Lett.* **103** 153004

[8] Takamoto M and Katori H 2009 *J. Phys. Soc. Jpn.* **78** 013301

[9] Gibble K 2009 *Phys. Rev. Lett.* **103** 113202

[10] Takamoto M, Hong F L, Higashi R, Fujii Y, Imae M and Katori H 2006 *J. Phys. Soc. Jpn.* **75** 104302

[11] Takamoto M, Katori H, Marmo S I, Ovsiannikov V D and Pal'chikov V G 2009 *Phys. Rev. Lett.* **102** 063002

[12] Taichenachev A V, Yudin V I, Ovsiannikov V D, Pal'chikov V G and Oates C W 2008 *Phys. Rev. Lett.* **101** 193601

[13] Santarelli G, Audoin C, Makdissi A, Laurent P, Dick G J and Clairon A 1998 *IEEE Trans. Ultrason. Ferroelectr. Freq. Control* **45** 887–894

[14] Bize S, Sortais Y, Lemonde P, Zhang S, Laurent P, Santarelli G, Salomon C and Clairon A 2000 *IEEE Trans. Ultrason. Ferroelectr. Freq. Control* **47** 1253–1255

[15] Newbury N R, Williams P A and Swann W C 2007 *Opt. Lett.* **32** 3056–3058

[16] Uzan J P 2003 *Rev. Mod. Phys.* **75** 403

Thermodynamics of the unitary Fermi gas

F. Chevy[1], S. Nascimbène[1,2], N. Navon[1], K. Jiang[1,3], C. Lobo[4] and C. Salomon[1]

[1] Laboratoire Kastler Brossel, École Normale Supérieure, CNRS, UPMC, 75505 Paris, France
[4] Cavendish Laboratory, Cambridge, UK

E-mail: chevy@lkb.ens.fr

Abstract. The understanding of quantum many-body systems is one of the most daunting challenges of modern physics. Thanks to recent progress in cooling and trapping techniques, it is now possible to investigate their properties in the well controlled environment of ultra-cold gas systems. In this article, we present experimental results on the thermodynamics of strongly correlated Fermi gases and we provide a reinterpretation of the equation of state of a strongly polarized Fermi gas in terms of Fermi liquid parameters

1. Introduction

Combined with the high level of control in atomic systems, the achievement of a strongly correlated regime in ultra-cold gases offers the opportunity to reproduce accurately model hamiltonians developed in condensed matter physics, and to provide insight on complex systems such as high critical temperature superconductors, heavy fermion compounds or even neutron stars [1, 2]. Among the recent breakthroughs, let us cite the observation of the Mott transition with bosons [3] and fermions [4, 5], or the observation of fermionic superfluidity and the exploration of the crossover interpolating between weakly and strongly interacting regimes [6, 7, 8, 9].

To achieve the degree of accuracy required by the fulfilment of the quantum simulation program, new experimental tools are being develop to further cool the gases to reach regimes where new phases are expected. In this article, we demonstrate a method giving access to the equation of state of a strongly interacting system. This scheme is based on a proposal by E.A Mueller, T.L. Ho and Q. Zhou [10] and was recently implemented at ENS [11, 12, 13]. We illustrate the method in the case of a unitary Fermi gas, both at finite temperature and spin population imbalance, and we exhibit some of the Fermi liquid features of the normal phase.

2. Experimental setup

Details on our experimental apparatus are presented in [11, 14]. We start from a double species magneto-optical trap of bosonic ^{7}Li and fermionic ^{6}Li that we transfer into a magnetic trap where we perform radio-frequency (rf) sympathetic cooling of the mixture. When the temperature of the sample reaches $\sim 100\,\mu$K, we transfer the mixture in a single beam dipole trap of waist $w = 35\,\mu$m

[2] Present address: Ludwig-Maximilians Univertät Schellingstr. 4/II, 89799 München, Germany.
[3] Present address: Wuhan Institute of Physics and Mathematics (WIPM), Chinese Academy of Sciences (CAS), Wuhan, 430071, China.

and power 60 W, supplemented by a magnetic field curvature to increase axial confinement. We transfer the atoms in the hyperfine states $|F=1, m_F=1\rangle$ (^{7}Li) and $|F=1/2, m_F=\pm 1/2\rangle$ (^{6}Li) using a rf sweep. In the case of the fermionic isotope, we control the amount of spin imbalance between the $m_F=\pm 1/2$ states by changing the speed of the sweep. We ramp up the magnetic field close to the broad ^{6}Li Feshbach resonance located at 834 G where we perform further evaporative cooling as well as our measurements. At the final stage of evaporation, the fermionic cloud contains typically 5 to 10×10^4 atoms, and the depth of the dipole trap is 3.5 μK, corresponding to typical radial and axial trapping frequencies $\omega_r = 2\pi \times 830$ Hz and $\omega_z = 2\pi \times 37$ Hz for the coldest samples.

We keep ^{7}Li all along during the cooling stage in order to use it as a temperature probe for the fermions. Indeed, thanks to small inter- and intra-species scattering lengths $a_{67}=2$ nm and $a_{77}=-3$ nm, the presence of the bosonic atoms does not significatively perturb the fermion density profile and time of flight expansion of ^{7}Li can be used to measure the temperature of the cloud. This thermometric scheme was also used in the Li/K mixture in Innsbruck [15], and in our case can be applied down to $T=150$ nK. At this temperature, we reach Bose-Einstein condensation of the bosons which collapse as a consequence of attractive interactions associated with negative scattering lengths [16, 17, 18, 19]. Lower temperatures can be achieved by removing the bosonic atoms before transferring the fermions into the dipole trap. Based on comparison with other experiments, and in particular the determination of the tricritical point at MIT [20], we estimate in this case that we cool our sample to temperatures as low as $T/T_F=0.03$.

Comparison between experimental results and theories usually developed for homogeneous systems is made difficult by the inhomogeneity of the atomic density profile induced by the trap used to contain the atoms, as well as the absorption imaging scheme giving access only to the column-density of the cloud in three dimensional systems. As a consequence, the first studies of the properties of ultra-cold gases were restricted to the measurement of thermodynamic quantities averaged over the trap [21, 22] or the characterization of the low lying excitation spectrum [23].

Recently two schemes were proposed to connect directly experiments in trap to properties of homogeneous systems. Both of them rely on the Local Density Approximation (LDA) which assumes the existence of a local thermodynamic equilibrium allowing one to define local intensive quantities such as temperature, pressure or chemical potentials. As a consequence, the density profile can be calculated using the equilibrium condition

$$\mu_\sigma^0 = \mu_{\sigma,\mathrm{hom}}(T, n_{\sigma_1}(\boldsymbol{r}), n_{\sigma_2}(\boldsymbol{r}), ...) + V_\sigma(\boldsymbol{r}), \tag{1}$$

where μ_σ^0 is the global chemical potential of species σ, T is the temperature, $n_\sigma(\boldsymbol{r})$ the density profile of σ, $\mu_{\sigma,\mathrm{hom}}$ is the chemical potential of a *homogeneous* gas, which depends in general on the densities of all species, and V_σ is the trapping potential of species σ. Note that, in the following, indices σ can refer both to different spin states or different atomic species.

The first scheme is inspired from medical imaging techniques [24] and takes advantage of the cylindrical invariance of single beam optical traps to deconvolve the line of sight integration using a mathematical operation known as the inverse Abel transform [25, 20, 26]. This transformation is however very sensitive to noise and requires high signal to noise ratio to be effective. By contrast, following a suggestion by E.A. Mueller, Ho and Zhou noticed that in the presence of harmonic trapping, the pressure of the gas can be related to the doubly integrated densities [10] by the equation

$$P(T, \mu_{z\sigma_1}, \mu_{z\sigma_2}, ...) = \frac{1}{2\pi} \sum_\sigma m_\sigma \bar{\omega}_{r\sigma}^2 \bar{n}_\sigma(z). \tag{2}$$

In this equation, P is the pressure, $\mu_{z,\sigma} = \mu_\sigma^0 - m_\sigma \omega_{z\sigma}^2 z^2/2$ is the local chemical potential along the z axis, m_σ is the mass of the species σ, $\omega_{r\omega}^2 = \omega_{x\sigma}\omega_{y\sigma}$ is the geometric-mean trapping frequency in the transverse plane and $\bar{n}_\sigma(\boldsymbol{r})$ is the doubly integrated density defined as

$$\bar{n}_\sigma(z) = \int \mathrm{d}x\mathrm{d}y n_\sigma(x, y, z) \tag{3}$$

In other words, by measuring the doubly integrated density profile as a function of position, it is possible to reconstruct directly the equation of state $P(T, \mu_{\sigma_1}, \mu_{\sigma_2}...)$. The advantage of this scheme is threefold. First, taking a single picture gives access to a full piece of the equation of state, since the different values of z accessible along the cloud density profile correspond to different values of the chemical potential. Second, the calculation of the doubly integrated density profile requires the integration of the column density accessible by absorption imaging and thus improves the signal to noise ratio of the data. Third, the equation of state relating the pressure to the temperature and chemical potentials contains all the information necessary to reconstruct thermodynamic quantities such as the entropy or the atom density since according to Gibbs-Duhem relation we have $\mathrm{d}P = s\mathrm{d}T + \sum_\sigma n_\sigma \mathrm{d}\mu_\sigma$, where s is the entropy density.

In this article we illustrate this scheme for the case of a two component Fermi gas at unitarity where the scattering length is infinite. By a simple dimensional argument, the macroscopic state of a two component system can be characterized by two dimensionless numbers $\eta = \mu_2/\mu_1$ and $t = k_B T/\mu$, which allows one to recast the grand canonical equation of state as

$$P(\mu_1, \mu_2, T) = P_0(\mu_1) h(\eta, t), \tag{4}$$

where $P_0(\mu)$ is the pressure of a single component Fermi gas at zero temperature given by

$$P_0(\mu) = \frac{1}{15\pi^2}\left(\frac{2m}{\hbar^2}\right)^{3/2} \mu^{5/2} \tag{5}$$

and h is a universal dimensionless function characterizing the thermodynamic properties of the system. In the following we will discuss first the low temperature limit and we will show we can experimentally reconstruct with high accuracy the function $h(\eta, 0)$ corresponding to the low temperature limit of the equation of state. Second, we will consider the finite temperature, spin balanced case that will allow us to measure the critical temperature of the system.

3. The zero temperature spin-imbalanced Fermi gas

The fate of a superfluid against spin polarization was addressed very early after the publication of Bardeen-Cooper and Schrieffer's paper unveiling the microscopic mechanisms of superconductivity. Clogston [27] and Chandrasekhar [28] showed that the pairing gap was protecting the fully paired BCS state against chemical potential imbalances $h = \mu_1 - \mu_2$ between the two spin species. In the weakly interacting regime, they uncovered the existence of a first order transition located at $h = h_c = \Delta/\sqrt{2}$. Below h_c the the system is a superfluid with equal spin densities, while above h_c the system is spin polarized. The exact nature of the phase above h_c is still debated. At zero temperature, general arguments by Kohn and Luttinger suggest that the gas should remain superfluid [29], and several exotic pairing mechanisms were proposed to describe this polarized superfluid, from spatially inhomogeneous superfluids proposed by Fulde, Ferrell [30], Larkin and Ovchinnikov [31] (the FFLO phases) to breached pair superfluids (or Sarma phase) [32, 33] and p-wave superfluids [34]. However, Kohn and Luttinger's theorem does not specify the critical temperature of these new superfluids which may be too low to be realized experimentally in cold Fermi gases.

Although superconductors appear to be natural testing grounds for these theories, the Meissner effect hinders spin polarization of electrons using a magnetic field. As a consequence, the Clogston-Chandrasekhar limit could only be observed in exotic materials such as heavy fermion compounds [35] or aluminium films of thickness smaller than the size of the Cooper pairs

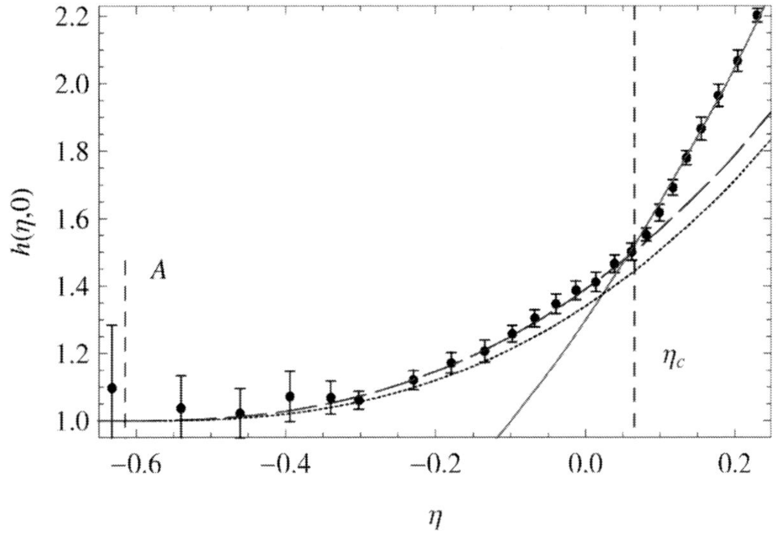

Figure 1. Equation of state of a low temperature, spin imbalanced Fermi gas. Here $h = P/P_0$ is the pressure of the system normalized to that of a single component Fermi gas and $\eta = \mu_2/\mu_1$ is the ratio of the chemical potentials. Full line: equation of state of a spin balanced superfluid (Eq. (7)); Dashed line: ideal mixture of polarons and majority atoms (Eq. (9)); dotted line Monte-Carlo simulations ref. [41]; data from ref. [11].

[36]. In ultra-cold atomic systems, spin imbalanced gases can be easily created using radio-frequency pumping and maintained for very long times thanks to the low inelastic collision rates characterizing these systems. The Clogston-Chandrasekhar limit was observed independently at Rice [37], and MIT [38], and later at ENS [14] (for reviews, see [39, 40]). In these experiments, the robustness of the superfluid against chemical potentials imbalance is revealed by the existence of a fully paired superfluid core surrounded by a spin imbalanced rim for spin population imbalances as high as $P = (N_1 - N_2)/(N_1 + N_2) = 0.76$ at unitarity [38]. As mentioned above, at zero temperature, the spin polarized region of the gas should also be superfluid. However, stirring experiments indicate that the border of the cloud does not sustain quantized vortices and thus suggest that experiments are performed at a temperature too high to sustain imbalanced superfluidity.

We present in Fig. 1 the equation of state of the spin imbalanced system obtained experimentally using Eq. (2). The graph displays three parts corresponding to three different regions of the low-temperature phase diagram. For $\eta < A \sim -0.6$, the gas is fully polarized and contains only majority atoms; hence $h \equiv 1$. For $\eta > \eta_c \sim 0.06$, the gas is a fully paired superfluid as we show below.

Let us first discuss the superfluid part of the equation of state. For equal chemical potentials, dimensional analysis shows that at $T = 0$ the pressure of the gas is proportional to that of an ideal gas and we thus have

$$P(\mu) = \frac{1}{15\pi^2}\left(\frac{2m}{\xi_s \hbar^2}\right)^{3/2} \mu^{5/2}, \tag{6}$$

where $\mu_1 = \mu_2 = \mu$, and $\xi_s \sim 0.41(1)$ is a universal numerical coefficient the determination of which has attracted the interest of both theoreticians [42, 43, 44, 45] and experimentalists [9, 12, 46, 47, 48, 37]. To deduce the equation of state in the presence of a chemical potential imbalance, we make use of the condition $n_1 = n_2$ characterizing the spin balanced superfluid.

Using the Gibbs-Duhem relation we have $\partial_{\mu_1} P = \partial_{\mu_2} P$, which is solved readily and indicates that P is a function of $(\mu_1 + \mu_2)/2$ only. Extrapolating to the value $\mu_1 = \mu_2$, we recover Eq. (6), and hence

$$P(\mu_1, \mu_2) = P(\mu) = \frac{1}{15\pi^2} \left(\frac{m}{\xi_s \hbar^2} \right)^{3/2} (\mu_1 + \mu_2)^{5/2}, \tag{7}$$

which agrees remarkably well with experiments as observed in Fig. 1.

We now turn to to the partially polarized phase and we focus first on the strongly polarized regime $\eta \sim A$ where the last minority atoms vanish from the cloud. In this extreme case, the last impurity atom is dressed by excitations of the surrounding Fermi sea and forms a quasi-particle called a Fermi polaron, the properties of which have been studied using Monte-Carlo [41, 49] as well as analytical methods [50, 51, 52]. These studies show that in the low momentum limit the dispersion relation of the Fermi polaron is similar to that of a free particle and can be written as

$$E(\boldsymbol{p}) = AE_{F1} + \frac{p^2}{2m^*} + ... \tag{8}$$

where E_{F1} is the Fermi energy of the majority atoms, $A = -0.61$ is a numerical parameter and $m^* = 1.20\, m$ is the effective mass. The values of A and m^* were determined experimentally using rf spectroscopy [53] and collective mode oscillations [14]. As indicated in Fig. (1) A sets the limit between the partially and fully polarized regions of the equation of state. Indeed, AE_{F1} is the minimal energy required to add an impurity to the system and, thus is equal to μ_2 by definition of the chemical potential. Moreover, in the limit of a weak impurity number, the majority Fermi sea is not modified by the minority atoms and we thus expect $\mu_1 = E_{F1}$, and hence $\mu_2/\mu_1 = A$.

Remarkably, the properties of the Fermi polaron provide all the necessary information to understand the thermodynamic of the partially polarized phase of Fig. 1. Indeed, at unitarity, $1/k_{F1}$ is the only length scale of the problem and thus sets the size of the cloud of excitations shrouding the Fermi polaron. As a consequence, in the dilute impurity limit $n_2/n_1 \ll 1$, we can consider the minority atoms as independent of each other, which suggests they can be described as an ensemble of non-interacting fermionic quasi-particles obeying the equation of state [41, 54]

$$P = \frac{1}{15\pi^2} \left(\frac{2m}{\hbar^2} \right)^{3/2} \mu_1^{5/2} + \frac{1}{15\pi^2} \left(\frac{2m^*}{\hbar^2} \right)^{3/2} (\mu_2 - A\mu_1)^{5/2}, \tag{9}$$

where the first term corresponds to the contribution of the majority atoms and the second to that of the gas of polarons. As shown in Fig. 1, this equation of state (dashed line) is in very good agreement with experiments as well as Monte-Carlo simulations [41] (dotted line).

This equation of state can be recast in the grand-canonical ensemble where we express the total energy as a function of atom number. After some straightforward algebra, we show that it satisfies the Landau-Pomeranchuck equation of state

$$E(N_1, N_2) = \frac{3}{5} N_1 E_{F1} \left(1 + \frac{5A}{3} x + \frac{m}{m^*} x^{5/3} + F x^2 + ... \right), \tag{10}$$

with $x = N_2/N_1$ and $F = 5A^2/9 \sim 0.2$. This formulation is reminiscent from Landau's Fermi liquid theory. The main assumption of this model is the existence of long-lived particle-hole excitations dictating the thermodynamic properties of the gas. In this case, the energy of the system is expanded as [55]

$$E = E_0 + \sum_{\boldsymbol{k},\sigma} \varepsilon_{\boldsymbol{k},\sigma} n_{\boldsymbol{k},\sigma} + \frac{1}{2V} \sum_{\boldsymbol{k},\boldsymbol{k}',\sigma\sigma'} f_{\sigma,\sigma'}(\boldsymbol{k}, \boldsymbol{k}') n_{\boldsymbol{k},\sigma} n_{\boldsymbol{k}',\sigma'}, \tag{11}$$

where E_0 is the energy of the non-interacting majority Fermi sea, $n_{\boldsymbol{k},\sigma}$ is the number of excitations of momentum $\boldsymbol{k}$ and spin σ, $\varepsilon_{\boldsymbol{k},\sigma}$ is the dispersion relation of the quasi-particles and $f_{\sigma,\sigma'}$ characterizes interactions between quasi-particles. In the absence of impurities ($n_{\boldsymbol{k},2}=0$), we recover the majority ideal gas, and hence $\varepsilon_{\boldsymbol{k},1}=\hbar^2k^2/2m$ and $f_{11}=0$. Let us now consider a weakly excited system where $n_{\boldsymbol{k},\sigma}$ is small. In the low minority density limit, the Fermi wavevector is vanishingly small and leads to third order terms in the interaction term of Eq. (11). Similarly, the momentum of majority atoms stays close to the Fermi surface and one can therefore assume $k\sim k_F$. Since the system is isotropic, there is no dependence on the angle either, which allows one to assume that $f_{\sigma\sigma'}$ does not depend on the momentum at all, and we can then rewrite (11) as

$$E=E_0+\sum_{\boldsymbol{k},\sigma}\varepsilon_{\boldsymbol{k},\sigma}n_{\boldsymbol{k},\sigma}+\frac{1}{2V}\sum_{\sigma\sigma'}f^0_{\sigma\sigma'}\delta N_\sigma\delta N_\sigma, \tag{12}$$

where $\delta N_\sigma=\sum_{\boldsymbol{k}}n_{\boldsymbol{k},\sigma}$ is the variation of atom number of species σ. In the case in which we do not excite the majority Fermi sea, we recover exactly the Landau-Pomeranchuk functional (10) if we take $f^0_{22}=A^2E_{F1}/3n_1$. Finally, we obtain the crossed term f^0_{12} by considering a situation with $N_1+\delta N_1$ majority atoms in the presence of an arbitrary distribution of polarons. We can describe this system either by starting with a non excited majority Fermi sea containing $N_1+\delta N_1$, or conversely as a Fermi sea of N_1 particles excited by adding δN_1 majority atoms. In the first case, the energy of the system is

$$E=E_0(N_1+\delta N_1)+\sum_{\boldsymbol{k}}\varepsilon_{\boldsymbol{k},2}(N_1+\delta N_1)n_{\boldsymbol{k},2}+\frac{f_{22}\delta N_2^2}{2V}, \tag{13}$$

while in the second case we expect

$$E=E_0(N_1)+\sum_{\boldsymbol{k}\sigma}\varepsilon_{\boldsymbol{k},\sigma}(N_1)n_{\boldsymbol{k},\sigma}+\frac{1}{2V}\sum_{\sigma\sigma'}f^0_{\sigma,\sigma'}\delta N_\sigma\delta N_{\sigma'}, \tag{14}$$

with $\sum_{\boldsymbol{k}}n_{\boldsymbol{k},1}=\delta N_1$. Identifying the crossed terms proportional to $\delta N_1\delta N_2$ in the two expressions, we finally get

$$f^0_{12}=A\frac{\partial E_{F1}}{\partial n_1}. \tag{15}$$

The knowledge of the Fermi-liquid parameters give access to some of the dynamical properties of the system [56], from the damping of excitations [57] to transport properties such as viscosity which was recently measured at high temperature in cold atoms [58].

4. Finite temperature equation of state

The Fermi liquid picture can be extended to the finite temperature regime as illustrated in Fig. 2 in the case of a spin balanced gas ($N_1=N_2$). Here, we observe a quadratic dependence of the pressure with temperature. Using thermodynamic argument, one can show that this corresponds to a specific heat C_V varying linearly with temperature, a characteristic feature of a Fermi liquid. In this model C_V/C_V^0, where C_V^0 is the specific heat of an ideal gas, is equal to the ratio m^*/m. Comparison with our data shows that $m^*=1.13m$, which is relatively close to the polaron's effective mass and demonstrate that the effective mass of the quasi particle is weakly sensitive to the spin population imbalance.

In addition, we observe that the experimental equation of state departs from the Fermi liquid behavior at $k_BT/\mu=0.32$ (Fig. 2b). We interpret this departure as the signature of the entrance in the superfluid regime. Expressed in the canonical ensemble, the critical condition can be recast as $t_c=k_BT_c/E_{F1}=0.15$, in agreement with the most advanced Monte-Carlo predictions $t_c=0.15$ [59, 60] and $t_c=0.18$ [61].

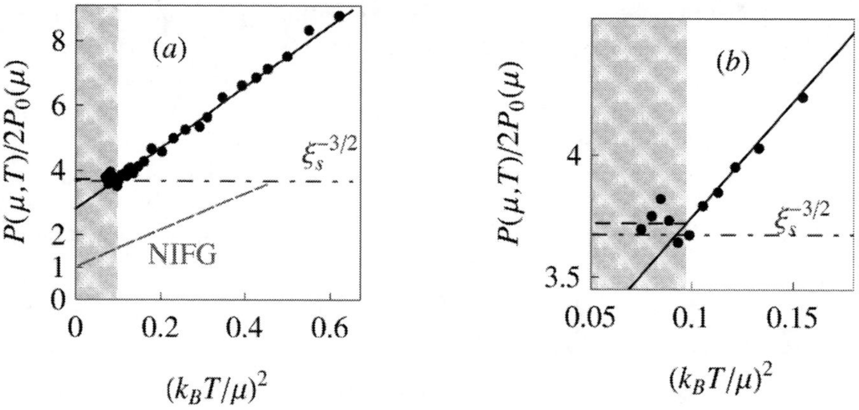

Figure 2. Left: Low temperature equation of state of a spin balanced unitary Fermi gas. The linear behavior of pressure with T^2 is a characteristic feature of a Fermi liquid associated with a specific heat varying linearly with temperature. The dashed line corresponds to the non interacting Fermi gas (NIFG). Right: Close up on the low temperature regime demonstrating the departure from the Fermi liquid behavior.

5. Conclusion and outlook

The experimental scheme presented here is very general and is limited only by the validity of the LDA and the harmonicity of the trap. The study presented here was extended to the BEC-BCS crossover in [12] which indicates that the polaronic phase vanishes in the BEC limit, as observed also in rf measurements [53]. In this region, the onset of bosonic degrees of freedom associated with the formation of tightly bound dimers leads to new physics and in particular the breakdown of the Fermi liquid description [49, 62, 63, 64, 65]. It can be further extended to many different situations, as illustrated in [13] in the case of a Bose gas in free space or in a lattice.

Acknowledgments

The authors thank C. Mora, X. Leyronas and R. Combescot for fruitful discussions. The authors acknowledge support from ERC (Ferlodim), ESF (Euroquam Fermix), ANR FABIOLA, Région Ile de France (IFRAF), and IUF. C.L. acknowledges support from the EPSRC (Contract No. EP/E053033/1).

References

[1] Inguscio M, Ketterle W and Salomon C 2006 *Proceedings of the International School of Physics Enrico Fermi, Course CLXIV, Varenna*
[2] Bloch I, Dalibard J and Zwerger W 2008 *Rev. Mod. Phys.* **80** 885–964
[3] Greiner M, Mandel O, Esslinger T, Hänsch T and Bloch I 2002 *Nature* **415** 39–44
[4] Schneider U, Hackermuller L, Will S, Best T, Bloch I, Costi T A, Helmes R W, Rasch D and Rosch A 2008 *Science* **322** 1520
[5] Jördens R, Strohmaier N, Günter K, Moritz H and Esslinger T 2008 *Nature* **455** 204–207
[6] Jochim S, Bartenstein M, Altmeyer A, Hendl G, Riedl S, Chin C, Hecker Denschlag J and Grimm R 2003 *Science* **302** 2101
[7] Greiner M, Regal C and Jin D 2003 *Nature* **426** 537–540
[8] Zwierlein M, Stan C, Schunck C, Raupach S, Gupta S, Hadzibabic Z and Ketterle W 2003 *Phys. Rev. Lett.* **91** 250401
[9] Bourdel T, Khaykovich L, Cubizolles J, Zhang J, Chevy F, Teichmann M, Tarruell L, Kokkelmans S and Salomon C 2004 *Phys. Rev. Lett.* **93** 050401
[10] Ho T and Zhou Q 2009 *Nature Physics* **6** 131
[11] Nascimbène S, Navon N, Jiang K, Chevy F and Salomon C 2010 *Nature* **463** 1057–1060
[12] Navon N, Nascimbène S, Chevy F and Salomon C 2010 *Science* **328** 729

[13] Nascimbène S, Navon N, Chevy F and Salomon C 2010 *New J. Phys.* **12** 103206
[14] Nascimbène S, Navon N, Jiang K, Tarruell L, Teichmann M, McKeever J, Chevy F and Salomon C 2009 *Phys. Rev. Lett.* **103** 170402
[15] Spiegelhalder F M, Trenkwalder A, Naik D, Hendl G, Schreck F and Grimm R 2009 *Phys. Rev. Lett.* **103** 223203
[16] Dodd R J, Edwards M, Williams C J, Clark C W, Holland M J, Ruprecht P A and Burnett K 1996 *Phys. Rev. A* **54** 661–664
[17] Kagan Y, Shlyapnikov G V and Walraven J T M 1996 *Phys. Rev. Lett.* **76** 2670–2673
[18] Dalfovo F and Stringari S 1996 *Phys. Rev. A* **53** 2477–2485
[19] Bradley C C, Sackett C A and Hulet R G 1997 *Phys. Rev. Lett.* **78** 985–989
[20] Shin Y, Schunck C, Schirotzek A and Ketterle W 2008 *Nature* **451** 689–693
[21] Stewart J, Gaebler J, Regal C and Jin D 2006 *Phys. Rev. Lett.* **97** 220406
[22] Luo L, Clancy B, Joseph J, Kinast J and Thomas J 2007 *Phys. Rev. Lett.* **98** 80402
[23] Altmeyer A, Riedl S, Kohstall C, Wright M, Geursen R, Bartenstein M, Chin C, Denschlag J and Grimm R 2007 *Phys. Rev. Lett.* **98** 40401
[24] Epstein C 2007 *Introduction to the Mathematics of Medical Imaging* (Cambridge University Press)
[25] Shin Y 2008 *Phys. Rev. A* **77** 041603
[26] Horikoshi M, Nakajima S, Ueda M and Mukaiyama T 2010 *Science* **327** 442
[27] Clogston A 1962 *Phys. Rev. Lett.* **9** 266–267
[28] Chandrasekhar B 1962 *App. Phys. Lett.* **1** 7–8
[29] Kohn W and Luttinger J M 1965 *Phys. Rev. Lett.* **15** 524–526
[30] Fulde P and Ferrell R 1964 *Phys. Rev* **135** A550
[31] Larkin A and Ovchinnikov Y 1964 *Zh. Eksp. Teor. Fiz* **47** 1136
[32] Sarma G 1963 *J. Phys. Chem. Solids* **24** 1029–1032
[33] Gubankova E, Liu W and Wilczek F 2003 *Phys. Rev. Lett.* **91** 032001
[34] Bulgac A, Forbes M and Schwenk A 2006 *Phys. Rev. Lett.* **97** 020402
[35] Heffner R H and Norman M R 1996 *Comments in Cond. Matt. Phys.* **17** 361
[36] Tedrow P M, Meservey R and Schwartz B B 1970 *Phys. Rev. Lett.* **24** 1004–1007
[37] Partridge G, Li W, Kamar R, Liao Y and Hulet R 2006 *Science* **311** 503–505
[38] Zwierlein M, Schirotzek A, Schunck C and Ketterle W 2006 *Science* **311** 492–496
[39] Radzihovsky L and Sheehy D 2010 *Rep. Prog. Phys.* **73** 076501
[40] Chevy F and Mora C 2010 *Rep. Prog. Phys.* **73** 112401
[41] Lobo C, Recati A, Giorgini S and Stringari S 2006 *Phys. Rev. Lett.* **97** 200403
[42] Carlson J, Chang S Y, Pandharipande V and Schmidt K 2003 *Phys. Rev. Lett.* **91** 050401
[43] Perali A, Pieri P and Strinati G 2004 *Phys. Rev. Lett.* **93** 100404
[44] Astrakharchik G, Boronat J, Casulleras J and Giorgini S 2004 *Phys. Rev. Lett.* **93** 200404
[45] Carlson J and Reddy S 2005 *Phys. Rev. Lett.* **95** 060401
[46] O'Hara K M, Hemmer S L, Gehm M E, Granade S R and Thomas J E 2002 *Science* **298** 2179
[47] Bartenstein M, Altmeyer A, Riedl S, Jochim S, Chin C, Hecker Denschlag J and Grimm R 2004 *Phys. Rev. Lett.* **92** 120401
[48] Kinast J, Turlapov A, Thomas J, Chen Q, Stajic J and Levin K 2005 *Science* **307** 1296
[49] Prokof'ev N and Svistunov B 2008 *Phys. Rev. B* **77** 020408
[50] Chevy F 2006 *Phys. Rev. A* **74** 063628
[51] Combescot R, Recati A, Lobo C and Chevy F 2007 *Phys. Rev. Lett.* **98** 180402
[52] Combescot R and Giraud S 2008 *Phys. Rev. Lett.* **101** 050404
[53] Schirotzek A, Wu C H, Sommer A and Zwierlein M 2009 *Phys. Rev. Lett.* **102** 230402
[54] Mora C and Chevy F 2010 *Phys. Rev. Lett.* **104** 230402
[55] Pines D and Nozières P 1966 *The Theory of Quantum Liquids, vol. I: Normal Fermi Liquids* (WA Benjamin, New York)
[56] Abrikosov A A and Khalatnikov I M 1959 *Rep. Prog. Phys.* **22** 329
[57] Bruun G, Recati A, Pethick C, Smith H and Stringari S 2008 *Phys. Rev. Lett.* **100** 240406
[58] Cao C, Elliott E, Joseph J, Wu H, Petricka J, Schaefer T and Thomas J 2010 *Arxiv preprint arXiv:1007.2625*
[59] Burovski E, Prokofev N, Svistunov B and Troyer M 2006 *Phys. Rev. Lett.* **96** 160402
[60] Bulgac A, Drut J and Magierski P 2008 *Phys. Rev. A* **78** 23625
[61] Goulko O and Wingate M 2010 *eprint* arXiv:1008.3348
[62] Combescot R, Giraud S and Leyronas X 2009 *Europhys. Lett.* **88** 6007
[63] Mora C and Chevy F 2009 *Phys. Rev. A* **80** 033607–033617
[64] Punk M, Dumitrescu P and Zwerger W 2009 *Phys. Rev. A* **80** 053605–053615
[65] Bruun G and Massignan P 2010 *Phys. Rev. Lett.* **105**

Universal structure of a strongly interacting Fermi gas

Eva Kuhnle, Paul Dyke, Sascha Hoinka, Michael Mark, Hui Hu, Xia-Ji Liu, Peter Drummond, Peter Hannaford and Chris Vale

ARC Centre of Excellence for Quantum Atom Optics, Swinburne University of Technology, Hawthorn 3122, Australia

E-mail: cvale@swin.edu.au

Abstract. This paper presents studies of the universal properties of strongly interacting Fermi gases using Bragg spectroscopy. We focus on pair-correlations, their relationship to the contact $\mathcal{C}$ introduced by Tan, and their dependence on both the momentum and temperature. We show that short-range pair correlations obey a universal law, first derived by Tan through measurements of the static structure factor, which displays a universal scaling with the ratio of the contact to the momentum $\mathcal{C}/q$. Bragg spectroscopy of ultracold ^{6}Li atoms is employed to measure the structure factor for a wide range of momenta and interaction strengths, providing broad confirmation of this universal law. We show that calibrating our Bragg spectra using the f-sum rule leads to a dramatic improvement in the accuracy of the structure factor measurement. We also measure the temperature dependence of the contact in a unitary gas and compare our results to calculations based on a virial expansion.

1. Introduction

Universality is a striking feature of certain classes of interacting Fermi systems. When the range of the interparticle scattering potential r_0 is much smaller than the mean interparticle spacing $n^{-1/3}$ and the s-wave scattering length a, which characterises elastic collisions is large compared to $n^{-1/3}$, the system may display universal properties. Here, universality means that all dilute Fermi systems with sufficiently strong interactions behave identically on a scale given by the average particle separation, independent of the details of the short range interaction. Two component gases of fermionic atoms near a Feshbach resonance are an excellent system in which to achieve this universal regime in a precisely controlled manner [1–3]. With the discovery of universality in the Bose-Einstein condensate (BEC) to Bardeen-Cooper-Schrieffer (BCS) crossover, ultracold Fermi gases near Feshbach resonances have become a central topic in atomic physics [4–10].

Understanding these strongly interacting Fermi gases, however, can present significant challenges for theorists [11]. As the scattering length becomes large it can no longer be treated as a small parameter and perturbative theories, which have had wide success for weakly interacting Bose gases, cannot be applied. In 2005 Shina Tan made dramatic progress by deriving several exact relations valid through the BEC-BCS crossover, which relate the microscopic properties of the constituent particles to bulk thermodynamic quantities [12–14]. These exact relations are applicable in broad circumstances: zero or finite temperatures, superfluid or normal phases, homogeneous or trapped systems, and in few or many-body systems. Tan invoked a new quantity in his theory, called the contact $\mathcal{C}$. The contact is defined as

$$\mathcal{C} = \lim_{k \to \infty} n_k k^4 \tag{1}$$

where n_k is the momentum density as a function of the wavevector k. At large k, this density profile has a $1/k^4$ dependence so the contact is a constant which depends on the strength of the two-body interactions. A more physically intuitive picture of the contact was also given by Tan as the number of closely spaced spin-up/spin-down pairs, i.e., $\mathcal{C}$ quantifies the likelihood of finding two fermions with opposite spin close enough to interact with each other. In systems with a large scattering length where the range of the interaction potential r_0 is negligible, this single parameter encapsulates all of the information required to determine the many-body properties [15, 16]. $\mathcal{C}$ depends upon two quantities, the dimensionless interaction parameter $1/(k_F a)$, where k_F is the Fermi wavevector and the relative temperature of the system T/T_F, where T_F is the Fermi energy divided by Boltzmann's constant. The contact $\mathcal{C}$ was first extracted [17] from the number of closed-channel molecules determined through photo-association [8] and Tan's adiabatic and virial relations were very recently verified experimentally [18]. Both of these studies investigated the interaction dependence of the $\mathcal{C}$. We will generally refer to the dimensionless contact $\mathcal{I}$ given by $\mathcal{C}/(Nk_F)$, where N is the number of particles.

2. Structure factor in the unitary limit - a universal quantity

Tan's exact analytic relations link various many-body quantities such as the total energy and momentum distribution to the short-range parameter $\mathcal{I}$. Short-range structure in a quantum fluid depends upon the relative wave-function of the interacting particles, in this case fermions in different spin states. In a two-component (spin-up/spin-down) Fermi gas this is given by $\psi_{\uparrow\downarrow}(r) \propto 1/r - 1/a$, where a is the s-wave scattering length. Tan used this to show that, for length scales $r_0 < r < 1/k_F$, the spin-antiparallel pair correlation function is given exactly by Eq. (2) which includes the contact as a pre-factor [12]

$$g^{(2)}_{\uparrow\downarrow}(r) = \frac{\mathcal{C}}{16\pi^2}\left(\frac{1}{r^2} - \frac{2}{ar}\right). \tag{2}$$

Such pair correlation functions are difficult to measure directly in ultracold gases; however, it is possible to measure macroscopic quantities which depend on correlation functions in a well defined way. A prime example is the static structure factor, $S(k)$, which is given by the Fourier transform of $g^{(2)}(r)$ ($q = \hbar k$ is the probe momentum). In a two component Fermi gas with an equal number of particles in each state N, the structure factor consists of two components, corresponding to correlations between particles in the same state and particles in different states $S(k) = S_{\uparrow\uparrow}(k) + S_{\uparrow\downarrow}(k)$. When the momentum is much larger than the Fermi momentum, particles in the same state will be uncorrelated and $S_{\uparrow\uparrow}(k \gg k_F) \simeq 1$, so all variation in $S(k)$ will then be due to changes in $S_{\uparrow\downarrow}(k)$ [19, 20]. The Fourier transform of Eq. (2) yields the following for $S_{\uparrow\downarrow}(k)$

$$S_{\uparrow\downarrow}(k \gg k_F) = \frac{\mathcal{I}}{4}\frac{k_F}{k}\left[1 - \frac{4}{\pi k_F a}\left(\frac{k_F}{k}\right)\right], \tag{3}$$

which has a straightforward dependence upon $\mathcal{I}$, the relative probe momentum k/k_F, and the interaction strength $1/(k_F a)$. At unitarity, $a \to \infty$, the second term vanishes and $S(k)$ varies linearly with k_F/k. Rewriting Tan's relation in this way points to a method to experimentally verify Tan's universal relation for pair correlations between spin-up/spin-down fermions since the static structure factor can readily be measured using Bragg spectroscopy.

3. Verification of Tan's universal pairing law

Inelastic scattering experiments are a well established technique to probe both the dynamic and static structure factors of many-body quantum systems [21]. Ultracold atoms are highly amenable to inelastic scattering through Bragg spectroscopy which has previously been used to measure both the dynamic [22, 23] and static structure factors [24, 25] of atomic Bose-Einstein condensates. In ultracold Fermi gases Bragg spectroscopy was used to measure both the dynamic and static structure factors over the BEC-BCS crossover, albeit at a single momentum [20]. The ability to vary the momentum differentiates Bragg spectroscopy from rf spectroscopy [26, 27], providing access to a broad range of the excitation spectrum and avoiding final state interaction effects as no third atomic state is involved.

Conceptually, one can understand how Bragg spectroscopy probes the contact (the number of closely spaced spin-up/spin-down pairs). The spatial period of the Bragg lattice ($\lambda_{Br} = k^{-1}$) sets the size scale being probed. Pairs, whose size is small compared to λ_{Br} can be scattered as single particles with twice the atomic mass ($2m$). Varying the ratio k/k_F we can map the fraction of pairs within a particular size range $0 < r < \lambda_{Br}$. At unitarity, these short-range pair correlations grow linearly with λ_{Br}, Eq. (3).

To demonstrate the validity of Eq. (3) we perform measurements of $S(k)$ for a range of momenta (k/k_F) and at different values of the interaction parameter $1/(k_F a)$. We start by preparing clouds containing $N \approx 3 \times 10^5$ ^{6}Li atoms in an equal mixture of the two lowest lying ground states $|F = 1/2, m_F = \pm 1/2\rangle$ evaporatively cooled in a single beam optical dipole trap ($\lambda = 1075$ nm) at a magnetic field of 834 G [31]. Next, we adiabatically ramp up a second far detuned laser ($\lambda = 1064$ nm) which intersects the first trapping beam at an angle of 74° forming a crossed beam dipole trap. By appropriately selecting the intensities of each of the two trap lasers we tune the mean harmonic confinement frequency of the crossed trap $\bar{\omega}$ over the range $\bar{\omega} = 2\pi \times (38 - 252)$ s^{-1} while the aspect ratio, $\omega_{x,y}/\omega_z$, varies from 3.4 to 16. Controlling the trap frequencies in this way allows us to tune the atom density and hence the Fermi wavevector over a broad range. Once in the final crossed beam trap, we ramp the magnetic field to select the scattering length a which gives the desired value of $1/(k_F a)$. This allows us to tune k_F and $1/(k_F a)$ independently. The cloud is then held at the final magnetic field for a time $\tau \gg 10\,\bar{\omega}^{-1}$ to reach equilibrium before applying the Bragg pulse and imaging at that magnetic field.

Bragg scattering is achieved by illuminating these ^{6}Li clouds with two counter propagating laser beams with a small frequency difference δ. This creates a standing wave which moves with velocity δ/k_{Br} where $k_{Br} = 2\pi/\lambda_{Br}$ and $\lambda_{Br} = 671$ nm is the wavelength of the light used for Bragg scattering. Resonant Bragg scattering occurs when the energy difference between the two beams is equal to the energy required to eject a particle from the cloud with momentum $2\hbar k_{Br}$. The lasers used for Bragg scattering are approximately 2 GHz red-detuned from the two ground hyperfine states which is large compared to the $\sim$ 80 MHz splitting between the $|1/2, \pm 1/2\rangle$ states. This means that both states are coupled approximately equally to the Bragg beams (to within 4%).

To measure the static structure factor $S(k)$ we begin by recording a Bragg spectrum. This consists of measuring the momentum transferred to the cloud by the Bragg lasers as a function of δ. The transferred momentum is directly proportional to the resultant centre of mass displacement $\Delta X(k, \delta)$ measured after 2 ms time of flight [20]. A Bragg spectrum for a unitary gas is shown in the inset of Fig. 1 for $k/k_F = 8.5$ as a function of the Bragg frequency δ. At this momentum, the pair and free-atom excitations are clearly distinguished at frequencies of $\sim$ 150 kHz and $\sim$ 300 kHz, respectively. The Bragg pulse duration, 50 μs, is short compared to the two-photon Rabi cycling period, ensuring that spectra are obtained in the linear response regime.

The centre of mass displacement, $\Delta X(k, \delta)$, is proportional to the convolution of the spectral content of the Bragg pulse and the dynamic structure factor of the gas $S(k, \delta)$ [28]. The proportionality constant depends upon the two-photon Rabi frequency Ω_{Br} which can be difficult to measure accurately. Integrating $\Delta X(q, \delta)$ over δ yields a number proportional to the static structure factor, $S(k) = \hbar N^{-1} \int S(k, \delta) d\delta$, but Ω_{Br} remains unknown. We overcome the need to find Ω_{Br} by invoking the f-sum rule [21]: $NE_r = \int S(k, \delta)\delta d\delta$, where $E_r = 2\hbar^2 k_{Br}^2/m$ is the two-photon recoil energy. As both $S(k)$ and the f-sum rule involve N, we normalise the area under each measured spectra using

$$S(k) = \frac{2\hbar k_{Br}^2}{m} \frac{\int \Delta X(k, \delta) d\delta}{\int \Delta X(k, \delta)\delta d\delta}, \tag{4}$$

where the constant involving Ω_{Br} appears before each integral and therefore cancels. Equation (3) is absolute, requiring only knowledge of the recoil energy, which can be determined with high accuracy, leading to an accurate measure of $S(k)$.

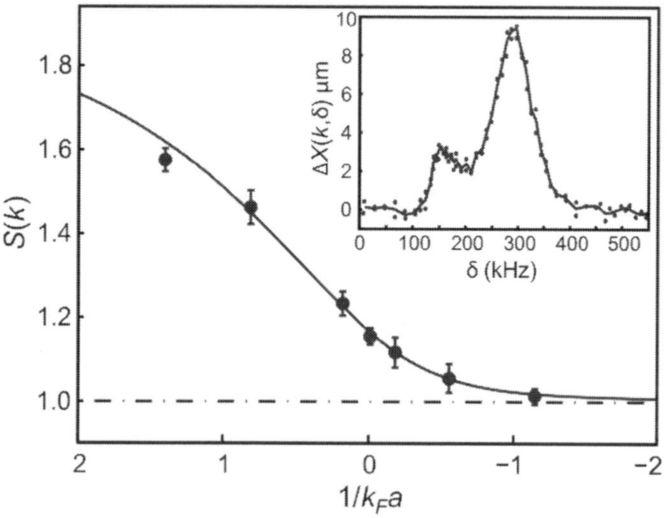

Figure 1. Static structure factor in the BEC-BCS crossover. Experimental points were obtained by integrating Bragg spectra normalized via the f-sum rule for $k/k_F = 4.8$. Error bars are due to shot to shot atom number fluctuations and uncertainties in measuring centre of mass. The theoretical line is a zero temperature result, calculated by interpolating the near-resonance structure factor Eq. (2) with the asymptotic results in the BCS and BEC limits. Inset: A Bragg spectrum obtained at $1/(k_F a) = 0$ and $k/k_F = 8.5$ showing centre of mass displacement $\Delta X(k, \delta)$ versus Bragg frequency δ. Points are single shots and the line is a guide to the eye. The pair ($\delta \sim 150$ kHz) and free atom peaks ($\delta \sim 300$ kHz) are clearly distinguished.

3.1. Interaction dependence

Bragg spectra reflect the composition of the gas, being dominated by bosonic molecules below the Feshbach resonance, pairs and free fermionic atoms near unitarity, and free fermions above resonance. To reveal the effects of interactions, scattering of trapped gases has to be considered [20]. This can be complicated by elastic collisions between scattered and unscattered particles which are hard to separate in the BEC-BCS crossover. Here, elastic collisions help as they preserve centre of mass motion, so $\Delta X(k, \delta)$ is a good measure of the Bragg signal throughout the BEC-BCS crossover, independent of cloud shape and scattering length. We have previously measured Bragg spectra across the BEC-BCS crossover at a momentum of $k/k_F = 4.8$ and observed a smooth transition in $S(k, \delta)$ from being dominated by molecular excitations below the Feshbach resonance to atomic excitations above resonance. From such spectra, we can obtain $S(k)$ using Eq. 4 as plotted in Fig. 1 which shows excellent agreement with the zero temperature theoretical calculation. This shows that we can quantitatively measure pair correlations at a given momentum using Bragg spectroscopy.

3.2. Momentum dependence

Next we come to the measurement of $S(k)$ as a function of k/k_F [29]. At unitarity we expect $S(k)$ to vary linearly with k_F/k according to Eq. (2). Rather than changing k directly by varying the angle between the two Bragg beams, we change $k_F = (48N)^{1/6}\sqrt{m\bar{\omega}/\hbar}$ by varying $\bar{\omega} = (\omega_x\omega_y\omega_z)^{1/3}$, the geometric mean frequency of the optical dipole trap. Universality allows us to change the relative length scale being probed simply by changing the density of the gas. Using the crossed beam optical trap described above we can tune $\bar{\omega}$ such that the Fermi wavevector can be tuned anywhere over the range $k_F = 2.1\ \mu\text{m}^{-1}$ - $5.3\ \mu\text{m}^{-1}$, or $k/k_F = 3.5$-9.1. A sequence of Bragg spectra were obtained for gases at three values of $1/(k_Fa) = +0.3$, 0.0 and -0.2, while varying k_F. From these, $S(k)$ was extracted and the results are plotted in Fig. 2. The solid lines are the prediction from Eq. (2) using the zero temperature contact with no free parameters. The experimental points closely follow the theory, confirming the exact analytic dependence of $S_{\uparrow\downarrow}(k) \cong S(k) - 1$ on q, Eq. (2). At unitarity $1/(k_Fa) = 0$, the dependence on a vanishes and a straight line fit (dashed line) shows the simple universal behavior of fermionic pairing. The fitted slope of 0.75 ± 0.03 is slightly below the zero temperature prediction of 0.81 which may be due to reduced pairing at the finite temperature ($T/T_F = 0.10 \pm 0.02$ at unitarity) discussed in the next section. This result is in good agreement with the contact evaluated from the equation of state of a homogeneous gas measured by the ENS group [10]. At $1/(k_Fa) = +0.3$ the data depart from a straight line displaying the downward curvature consistent with the first order term in Eq. (2). A similar upward curvature is seen at $1/(k_Fa) = $ -0.2. Our simple relation Eq. (2) is seen to accurately describe $S(k)$ on both sides of the Feshbach resonance demonstrating the wide applicability of the Tan relations.

4. Temperature dependence of $\mathcal{I}$

As discussed in section 1, the contact $\mathcal{I}$ depends on both the interaction parameter and the relative temperature. Here we look at the latter dependence in a unitary Fermi gas where the spin-up/spin-down static structure factor is given by $S_{\uparrow\downarrow}(k) = \frac{\mathcal{I}}{4}\frac{k_F}{k}$. Any variation in the measured structure factor with temperature is then linearly related to the contact. For this series of measurements we use a lower relative momentum of $k/k_F = 2.7$ to enhance the contribution of $S_{\uparrow\downarrow}(k)$ to $S(k)$. This was achieved by reducing the angle between the two Bragg beams to 49° which reduces the atomic recoil frequency to 51 kHz. The lower momentum also means that all Bragg scattered atoms now remain trapped after the Bragg pulse as the energy imparted to the atoms/pairs by the Bragg pulse is lower than the trap depth. This makes possible a second means of determining $S(k)$ from measurements of the increase in the mean square cloud width to determine the energy imparted by the Bragg pulse. The details of this measurement scheme will be published elsewhere [30], but here we note that this method provides an alternative means to determine $S(k)$ that we can compare to our measurements of the momentum imparted by the Bragg pulse through centre of mass displacement. Figure 3 shows the measured structure factors (left axis) and the corresponding contact (right axis) measured as a function of temperature. The different temperatures were obtained by evaporating down to the lowest temperature and then releasing and recapturing atoms in the optical trap and holding for 800 ms to allow rethermalisation. In this way we could repeatably vary the temperature over the range $0.1 \rightarrow 1T_F$ while holding N and the trapping parameters

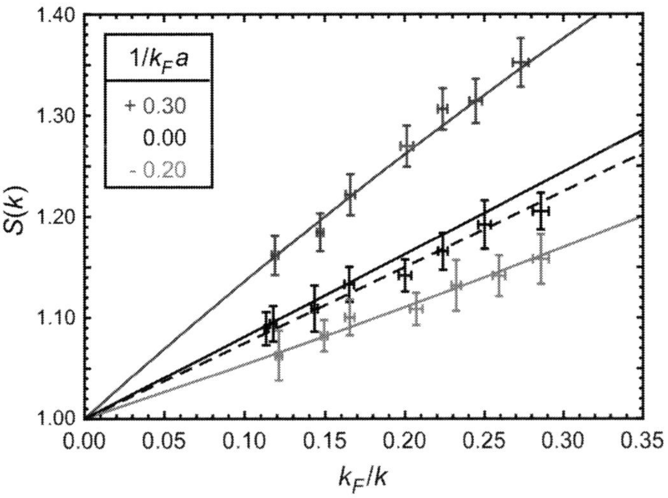

Figure 2. Universal dependence of the static structure factor of a strongly interacting Fermi superfluid. Measured and calculated static structure factor versus k_F/k for $1/(k_Fa) = +0.3$, 0.0, and -0.2. Bragg momentum k is fixed while k_F is varied by changing the mean trapping frequency $\bar{\omega}$. Vertical error bars are due to atom number fluctuations and uncertainties in measuring the centre of mass and horizontal error bars are due to atom number fluctuations and uncertainties in $\bar{\omega}$. Solid lines are the zero temperature theory and the dashed line is a straight line fit to the $1/(k_Fa) = 0$ data yielding a slope of 0.75 ± 0.03.

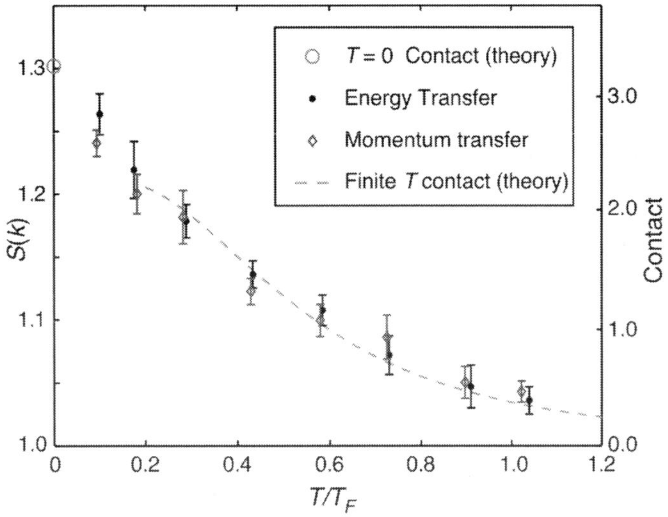

Figure 3. Temperature dependence of the static structure factor and contact for unitary Fermi gas. Open circles are the structure factors measured using centre of mass displacement as in section 3 and filled circles are the structure factors measured via the energy transferred to the cloud. The large open circle represents the zero temperature calculated contact and the dashed red line is a theoretical calculation based on the quantum virial expansion [32].

constant, by varying the time between release and recapture. Also shown in the figure are the calculated zero temperature contact $\mathcal{I} = 3.26$ (large red open circle) and finite temperature contact determined via a virial expansion calculation (red dashed line) [32]. The contact decays monotonically with increasing temperature reflecting the decay of pair correlations and agrees well with the theoretical prediction. We note that the critical temperature for superfluidity occurs around $0.2\,T_F$ and, while the data points near the critical temperature are relatively sparse, we do not observe any striking features near this point. It is also clear that pair correlations at this momentum persist well above the critical temperature, up to nearly the Fermi temperature. While this measurement only quantifies pair correlations at a particular momentum it does indicate that the build up of such correlations is relatively gradual and begins at high temperature.

5. Conclusions

In summary, we have shown that the structure factor of a strongly interacting ultracold Fermi gas follows a universal law which is a direct consequence of Tan's relation for the pair correlation functions. Our measurements provide one of the first demonstrations of a broadly applicable exact result for Fermi gases in the BEC-BCS crossover. We also provide the first measurements of the temperature dependence of the contact in a unitary Fermi gas and find good agreement with theoretical calculations based on the virial expansion. Bragg spectroscopic measurements provide a useful means to quantify pairing and to quantitatively map the phase diagram of the BEC-BCS crossover. In future work we will utilise low momentum Bragg spectroscopy to probe pairing at momenta nearer the Fermi momentum which may offer insight into pseudogap pairing.

This work is supported by the Australian Research Council Centre of Excellence for Quantum-Atom Optics and Discovery Projects DP0984522 and DP0984637.

6. References

[1] Heiselberg H 2001 *Phys. Rev. A* **63** 043606
[2] Ho T-L 2004 *Phys. Rev. Lett.* **92** 090402
[3] Hu H, Drummond P D and Liu X-J 2007 *Nat. Phys.* **3** 469
[4] Jochim S, Bartenstein M, Altmeyer A, Hendl G, Riedl S, Chin C, Hecker-Denschlag J and Grimm R 2003 *Science* **302** 2101
[5] Greiner M, Regal C A and Jin D S 2003 *Nature* **426** 537
[6] Zwierlein M W, Abo-Shaeer J R, Schirotzek A, Schunck C H and Ketterle W 2005 *Nature* **435** 1047
[7] Kinast J, Turlapov A, Thomas J E, Chen Q, Stajic J and Levin K 2005 *Science* **307** 1296
[8] Partridge G B, Strecker K E, Kamar R I, Jack M W and Hulet R G 2005 *Phys. Rev. Lett.* **95** 020404
[9] Nascimbene S, Navon N, Jiang K J, Chevy F and C. Salomon 2010 *Nature* **463** 1067 (2010)
[10] Navon N, Nascimbene S, Chevy F and Salomon C 2010 *Science* **328** 729
[11] Giorgini A, Pitaevskii L P and Stringari A 2008 *Rev. Mod. Phys.* **80** 1215
[12] S. Tan 2008 *Ann. Phys.* **323** 2952
[13] S. Tan 2008 *Ann. Phys.* **323** 2971
[14] S. Tan 2008 *Ann. Phys.* **323** 2987
[15] Braaten E and Platter L 2008 *Phys. Rev. Lett.* **100** 205301
[16] Zhang S and Leggett A J 2009 *Phys. Rev. A* **79** 023601
[17] Werner F, Tarruell L and Castin Y 2009 *Eur. Phys. J. B* **68** 401
[18] Stewart J T, Gaebler J P, Drake T E and Jin D S 2010 *Phys. Rev. Lett.* **104** 235301
[19] Combescot R, Giorgini S and Stringari S 2006 *Europhys. Lett.* **75** 695
[20] Veeravalli G, Kuhnle E, Dyke P and Vale C J 2008 *Phys. Rev. Lett.* **101** 250403
[21] Pines D and Nozières P 1966 *The Theory of Quantum Liquids, Vol. I* (Benjamin, New York)
[22] Stenger J, Inouye S, Chikkatur A P, Stamper-Kurn D M, Pritchard D E and Ketterle W 1999 *Phys. Rev. Lett.* **82** 4569
[23] Zambelli F, Pitaevskii L P, Stamper-Kurn D M and Stringari S 2000 *Phys. Rev. A* **61** 063608
[24] Stamper-Kurn D M, Chikkatur A P, Grlitz A, Inouye S, Gupta S, Pritchard D E and Ketterle E 1999 *Phys. Rev. Lett.* **83** 2876

[25] Steinhauer J, Ozeri R, Katz N and Davidson N 2002 *Phys. Rev. Lett.* **88** 120407
[26] Schunck C H, Shin Y, Schirotzek A and Ketterle W 2008 *Nature* **454** 739
[27] Stewart J T, Gaebler J P and Jin D S 2008 *Nature* **454** 744
[28] Brunello A, Dalfovo F, Pitaevskii L, Stringari S and Zambelli F 2001 *Phys. Rev. A* **64** 063614
[29] Kuhnle E, Hu H, Liu X-J, Dyke P, Mark M, Drummond P D, Hannaford P and Vale C J 2010 *Phys. Rev. Lett.* **105** 070402
[30] Kuhnle E, Hu H, Dyke P, Hoinka S, Hannaford P and Vale C J 2010 in preparation
[31] Fuchs J, Duffy G J, Veeravalli G, Dyke P, Bartenstein M, Vale C J, Hannaford P and Rowlands W J 2007 *J. Phys. B: At. Mol. Opt. Phys.* **40** 4109
[32] Liu X-J, Hu H and Drummond P D 2009 *Phys. Rev. Lett.* **102** 160401

22nd International Conference on Atomic Physics IOP Publishing
Journal of Physics: Conference Series **264** (2011) 012014 doi:10.1088/1742-6596/264/1/012014

Dipolar effects and collisions in an ultracold gas of LiCs molecules

J. Deiglmayr[1], M. Repp, A. Grochola[2], O. Dulieu[3], R. Wester[4] and M. Weidemüller

Physikalisches Institut, Universität Heidelberg, 69120 Heidelberg, Germany

E-mail: weidemueller@uni-heidelberg.de

Abstract. We review our recent experiments on the formation and optical trapping of LiCs molecules, which are photoassociated from a mixed ultracold gas of Li and Cs atoms. The electric dipole moment of the atoms in the lowest vibrational states is measured, yielding the predicted value of 5.5 Debye, and vibrational redistribution due to spontaneous emission and blackbody radiation is observed. We also show preliminary results on inelastic collision between optically trapped LiCs molecules and Cs atoms.

1. Introduction

In the past two years, the quest for ultracold molecules in deeply bound internal states, in particular the rovibrational ground state, has progressed with breath-taking speed. Previous work demonstrated the conceptual possibility to populate the vibrational ground state of ultracold homonuclear molecules (K_2) [1] and heteronuclear molecules (RbCs) [2] by applying photoassociation (PA). In 2008, several groups achieved major breakthroughs to increase the number of molecules in the vibrational ground state and to selectively populate the rotational ground state. The molecules were either optically pumped into the vibrational ground state by shaped broadband laser pulses following PA [3], or directly formed by PA of ultracold atomic gases [4]. In an alternative approach, the molecules were created by magnetoassociation at high phase-space density in the vicinity of a Feshbach resonance from a quantum-degenerate gas, and subsequently adiabatically transferred into the rovibrational ground state by one or two consecutive stimulated Raman adiabatic passage (STIRAP) steps [5, 6, 7] (see also Christoph Nägerl's contribution to these Proceedings).

In two of these experiments, one in the group of Deborah Jin and Jun Ye at Boulder and one in our group, formerly in Freiburg, now in Heidelberg, the molecules were formed from a mixture of ultracold atomic gases, thus creating molecules with a permanent body-fixed electric dipole moment (EDM) (KRb in Boulder, LiCs in Freiburg/Heidelberg). Ultracold gases of polar molecules (see the recent reviews [8, 9]) offer intriguing perspectives to the exploration of complex many-body quantum systems [10, 11], the development of quantum computation

[1] Permanent address: Physics Institute, University of Freiburg, 79104 Freiburg, Germany
[2] Permanent address: University of Warsaw, Poland
[3] Permanent address: Laboratoire Aimé Cotton, CNRS, Universit'e Paris-Sud, Orsay, France
[4] Permanent address: Physics Institute, University of Freiburg, 79104 Freiburg, Germany

techniques [12], precision measurements of fundamental constants [13], and the investigation and control of ultracold chemical reactions [14]. Current experimental efforts are therefore aiming at the investigation of effects related to the polar nature of the ultracold molecules. Here, we will concentrate on progress being made with a gas of ultracold LiCs molecules in low rovibrational states. LiCs is of particular interest as its dipole moment is as large as 5.5 Debye [15, 16], thus featuring the largest electric dipole moment of all bialkali molecules. We will briefly review our experimental approach to the formation of an ultracold gas of LiCs molecules and present a measurement of the LiCs dipole moment. Then, we will show evidence for vibrational relaxation and blackbody redistribution as a direct consequence of the large dipole moment. In the last part, we show first humble steps towards Ultracold Chemistry by presenting inelastic collisions between atoms and molecules trapped in a quasi-electrostatic optical dipole trap.

2. Photoassociation and detection of an ultracold gas of LiCs

The experimental sequence for formation, detection, and high-resolution spectroscopy of ground state molecules is schematically shown in Fig. 1. Ultracold LiCs molecules are formed by photoassociation (PA) [17, 18] of laser cooled ^{133}Cs and ^{7}Li atoms. The atoms are cooled and trapped in two overlapped magneto-optical traps (MOTs) which are loaded via a single Zeeman slower from a double species oven. For the Cs atoms the MOT is operated in the configuration of a dark spontaneous force optical trap (dark SPOT) [19] to reduce inelastic inter-species collisions. In this configuration 4×10^7 ^{133}Cs and 10^8 ^{7}Li atoms are trapped simultaneously at densities of $3 \times 10^9\,\mathrm{cm}^{-3}$ and $10^{10}\,\mathrm{cm}^{-3}$, and temperatures of 250(50) μK and 600(150) μK, respectively. Details of the experimental setup for the formation and detection of ultracold LiCs molecules have been given elsewhere [4, 20, 21]. For PA of LiCs molecules the two overlapped atom clouds are continuously illuminated by light from a tunable cw TiSa laser. The excited LiCs molecules decay spontaneously within few tens of nanoseconds into different electronic ground state levels. Molecules in given vibrational ground state levels are then ionized by resonant-enhanced multi-photon ionization (REMPI) and the resulting ions are detected in a time-of-flight (TOF) mass spectrometer [21].

Rotationally resolved spectroscopy of ground state molecules is achieved by depletion spectroscopy [22]: a narrow band laser pumps population out of a selected rovibrational ground state level leading to a reduction in the detected ion signal from this level. Further details have been given previously [20]. Electric fields between 0 and 500 V/cm can be applied to the atomic and molecular samples by two round electrodes inside the vacuum with an outer diameter of 80 mm and a distance of 40 mm, centered around the trapped atom clouds. The electrodes have a central hole (ø 30 mm) to allow optical access and extraction of produced ions (these electrodes are also part of the mass spectrometer used for the detection of the ground state molecules [21]). The electric field created at the trap position is derived from the measured dimensions and voltages using the simulation package SIMION (v8.0) with an uncertainty of 3% due to estimated inaccuracies in the electrode geometry and voltages. For the detection of molecules the fields are switched 0.6 ms before ionization to a fixed value of 38 V/cm.

In order to obtain trapped samples of LiCs molecules, PA of atoms stored simultaneously in a quasi-electrostatic trap (QUEST) is performed. The QUEST is formed by the single focus of a CO_2 laser (power 64(4) W, waist w_0=89(4) μm) emitting at 10.6 μm. Details of the QUEST setup can be found in [23] and references therein. The resulting trap depths for the different species are calculated from measured values for beam power and waist, experimental values for atomic polarizabilities [24], and *ab-initio* calculated values for the molecular polarizabilities [25] as 194(21) μK for Li and 479(53) μK for Cs. Due to the large detuning from all electronic transitions the trap depth for LiCs molecules varies only slowly with the vibrational index v'', from a minimum value of 435(48) μK for $X^1\Sigma^+, v''=0$ to a maximal trap depth of 700(77) μK for $v''=42$. Despite its significant detuning, the optical trapping field could drive Raman transitions

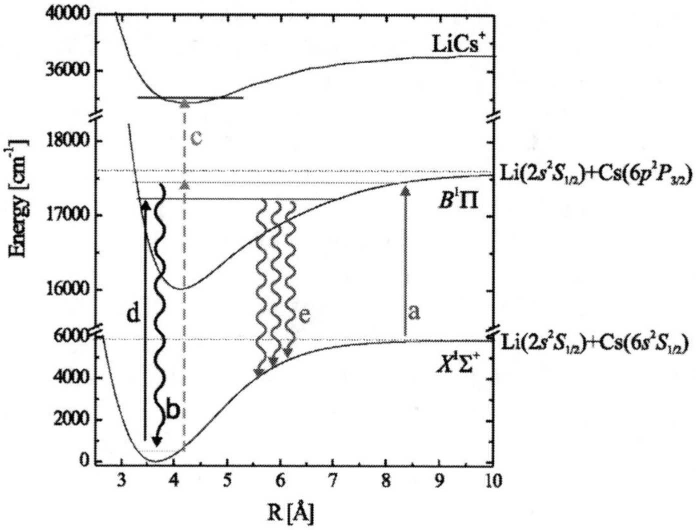

Figure 1. Sketch of the excitation and detection scheme. a) photo-association, b) spontaneous decay into deeply bound ground state molecules, c) two photon ionization with resonant intermediate state; for depletion spectroscopy: d) excitation of ground state molecules, e) redistribution of ground state population

between neighboring levels due to its high intensity. The rate for such transitions can be estimated from a simple classical model of the Raman process [26]. Using the *ab-initio* calculated polarizability tensor for the $X^1\Sigma^+$ state of LiCs [25] and the experimental laser intensity we derive transition rates which are several orders of magnitude below the transition rates for spontaneous decay and black-body radiation (BBR) transitions. Thus vibrational or rotational heating by the trapping CO_2 laser can be neglected even on longer timescales.

After a variable hold time in the dipole trap, molecules in given vibrational ground state levels are ionized by a REMPI pulse (typical intensities 2 J/cm^2) and the resulting ions are detected in the TOF spectrometer. The slowest measured decay of a molecular signal from the trap corresponds to a time constant of 24(4) s. This value is comparable to the lifetime of pure atomic samples in the trap and is thus attributed to trap loss due to collisions with thermal background gas at a pressure of 2×10^{-11} mbar. Further details of the experimental setup for the formation and detection of ultracold LiCs molecules can be found elsewhere [4, 20, 21].

3. Measurement of the electric dipole moment

In order to determine the permanent EDM of a given vibrational level of the $X^1\Sigma^+$ state we measure the electric-field induced shift in the transition frequency between the $J''=0$ component of this level and a rovibrational level in the $B^1\Pi$ state [16]. A shift in the transition frequency originates in the combined Stark shifts of the $X^1\Sigma^+$ level and the excited $B^1\Pi$ level. A rotationless level v'' ($J''=0$) of the $X^1\Sigma^+$ state experiences an electric-field induced shift $\triangle W = -\mu_v^2 E^2/(6hB_v)$, where μ_v is the permanent EDM, E the external field, and B_v the rotational constant [27]. Rotational constants are calculated from an experimental $X^1\Sigma^+$ potential energy curve [28] using Level 8.0 [29]. Due to selection rules the upper level in the depletion transition has to be rotationally excited ($J'=1$). The $B^1\Pi$ state is predicted to have a smaller EDM than the ground state: a pseudopotential configuration interaction calculation as described in ref. [15] yields for the $B^1\Pi$ state a permanent EDM of 0.47 Debye at the equilibrium distance. One can thus expect a significantly smaller Stark effect for the upper level than for the ground state level, as verified experimentally.

For the formation of deeply bound molecules we use the PA resonance $B^1\Pi,v'=18,J'=1$. After spontaneous decay the rotational states $J''=0$ of $X^1\Sigma^+,v''=2$ and higher are significantly populated. Then depletion spectroscopy of $v''=2,J''=0$ and $v''=3,J''=0$ is performed in electric fields varying from 0 to 488 V/cm. We did not obtain depletion spectra with sufficient signal to noise ratio for Stark spectroscopy from the $J''=0$ component of the $v''=0$ and $v''=1$ levels. Molecules in the lowest vibrational levels $v''=0$ and $v''=1$ are also formed by PA, but mostly in the rotationally excited $J''=2$ components [4, 30] for which the Stark effect at feasible electric fields could not be experimentally observed, yet. The change in the frequency of the transition $X^1\Sigma^+,v''=2,J''=0\rightarrow B^1\Pi,v'=18,J'=1$ is determined via depletion spectroscopy. We find a permanent EDM of $\mu=5.5(2)$ Debye for $v''=2$, where the reported error stems from the uncertainty of the fit and the estimated uncertainty in the electric field determination. The same measurement and analysis was performed for molecules in $X^1\Sigma^+,v''=3,J''=0$. Using $B_{v''=3}=0.1836\,\mathrm{cm}^{-1}$ the value for the EDM is determined as $\mu=5.3(2)$ Debye [16].

4. Dipolar relaxation of LiCs molecules

The presence of a permanent electric dipole moment in a molecule leads to a coupling of the internal molecular state to the thermal environment via black-body radiation (BBR). Thus if the system is prepared in a certain internal state, this state will be altered after a finite time. In particular this yields a limited lifetime even for molecules prepared in the absolute rovibrational ground state. In the case of LiCs one can expect significant dipole transition rates between vibrational ground state levels, enhanced by the large slope of the dipole moment function associated with its large average value. In the following the redistribution of population due to spontaneous decay and BBR will be modeled.

A transition between levels i and j with wave functions Ψ_i and Ψ_j is proportional to the dipole matrix element $\mu_{i,j} = \langle\Psi_i|\mu(R)|\Psi_j\rangle$, where $\mu(R)$ is the EDM function. For vibrational transitions only the variation of μ with the internuclear distance R leads to a finite transition dipole moment. The EDM function was calculated *ab-initio* and published previously [15]. Rovibrational wave functions and overlap integrals are calculated with Level 8.0 (`http://leroy.uwaterloo.ca/programs/`) using a precise experimental potential energy curve [28]. The resulting permanent EDMs of low lying vibrational levels are in very good agreement with measured values [16], as presented in Sect. 3. For modeling the spontaneous decay we calculate the Einstein A coefficients, and the Hönl-London factor $S_{J',J''}$ as defined in Ref. [31]. The coupling to BBR from the surrounding walls of the vacuum chamber is given by the Einstein B coefficients for stimulated absorption and emission multiplied by the spectral density of the BBR from an environment at a given temperature T.

For the rovibrational ground state $X^1\Sigma^+,v''=0,J''=0$ we calculate a depopulation time constant of 65.4 s, in good agreement with previously calculated values [32, 33]. It is noteworthy that at cryogenic temperatures (77 K) this depopulation time constant is lengthened to 1.3×10^3 s. Already at room temperature rotational heating due to BBR is greatly suppressed by the low spectral density at the relevant transition frequencies. Only for $J''>12$ rotational heating rates become comparable to molecular trap loss rates, which are far below $0.1\,\mathrm{s}^{-1}$.

The evolution of an initial distribution of populated ground state levels is finally given by the following rate equation for the population n_i in a level i (the single index i enumerates rovibrational levels v'',J'' in energetic ordering)

$$\dot{n}_i(t) = \sum_{j>i} A_{ji}n_j(t) - \sum_{j<i} A_{ij}n_i(t) + \sum_{j\neq i} \tilde{B}_{ji}n_j(t) - \sum_{j\neq i} \tilde{B}_{ij}n_i(t) - \Gamma n_i(t) \qquad (1)$$

where $\tilde{B}_{ij} = A/(\exp(h\tilde{\nu}c/kT)-1)$ is the black-body induced redistribution rate and Γ is a decay

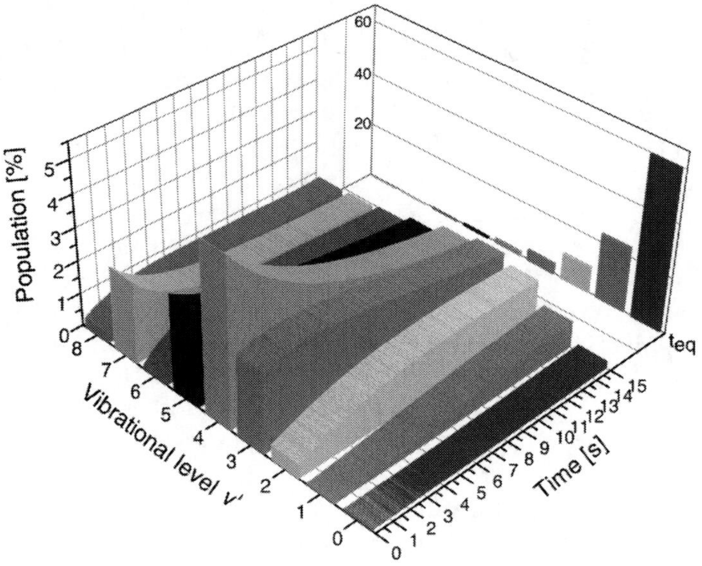

Figure 2. Time evolution of the population in the lowest nine vibrational levels of the $X^1\Sigma^+$ after spontaneous decay from $B^1\Pi, v'=26, J'=1$. The population is given as a percentage of the initial total ground state population. Also shown is the population (rescaled to unity) after $t_{eq} \equiv 5 \times A_{0,1}^{-1} \sim 200$ s.

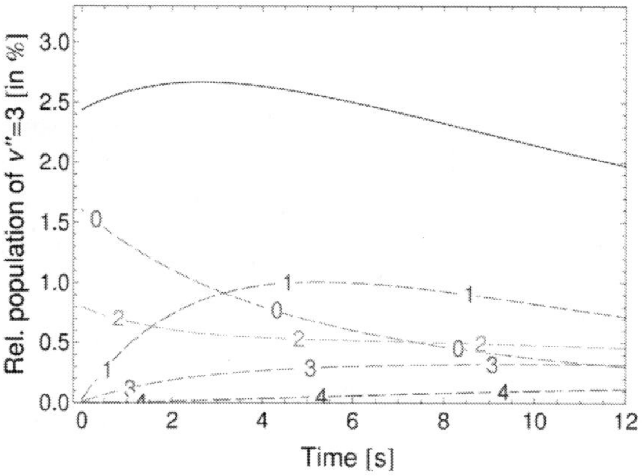

Figure 3. Simulated time evolution of the population in $X^1\Sigma^+, v''=3$. Shown are the total signal (solid line) and the population of the rotational sublevels (dashed lines, labeled by J'').

constant accounting for losses from the trap. The value of Γ is taken from measurements (as discussed in section 2) and is assumed to be independent of v''.

In the formation of ultracold molecules by PA (PA) [17, 18], the initial population of ground state levels is determined by the Einstein A coefficients for the spontaneous decay from the excited PA state. For the $B^1\Pi$ state of LiCs we have calculated this initial population for different PA levels and found agreement with experimental observations [20]. Starting from this initial distribution the rate equation Eq. (1) is integrated numerically in order to derive the time evolution of the population, where convergence of the results is reached for time steps smaller than 50 ms. As can be seen in Fig. 2 the initial population distribution in the lowest vibrational

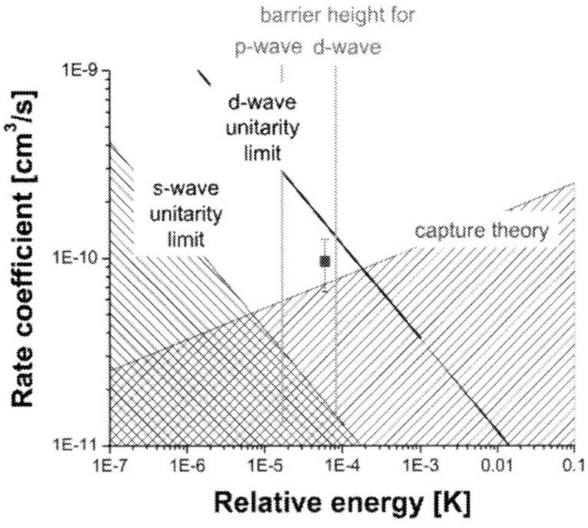

Figure 4. Measured rate coefficient for inelastic Cs-Cs_2 collisions in relation to the p- and d-wave barriers and the expected rate coefficient from capture theory and in the unitarity limit. The measured rate coefficient is found to be independent of the vibrational and rotational states.

levels of the $X^1\Sigma^+$ state is strongly modified. Most levels show either a steady increase or decrease essentially due to vibrational redistribution with adjacent states. As an exception the population in the v''=3 level exhibits a clear maximum after roughly 3 s, presenting a concise feature for testing the validity of the model. The time evolution of the population in this level is shown in more detail in Fig. 3 together with the individual contributions from the rotational sublevels. The significant increase of the population in v''=3,J''=1, being caused by the decay of molecules from v''=4 and the cycling of population between neighboring levels due to BBR with comparable rates, indicates stronger dynamics of the process than is visible from the vibrational populations only.

Our model shows that for LiCs population redistribution caused by BBR and spontaneous decay leads to features on timescales accessible to the experiment. We have indeed found experimental evidence for population redistribution in ground state LiCs molecules, which is in good qualitative agreement with the simulation presented here.

5. Cold atom-molecule collisions

As a first step towards chemistry at ultralow temperatures, we have studied inelastic collisions between atoms and molecules simultaneously trapped in the CO_2 dipole trap [23]. Ultracold collisions between weakly bound, magneto-associated molecules and atoms have been studied by several groups [34, 35]. In our case, the PA process prepares the molecules in deeply bound vibrational states. Since the energy released by inelastic collisions transforming either rotation or vibrational energy into kinetic energy is much larger than the trap depth, the reaction partners are expelled from the trap with near-unity probability. There are several predictions on the rate coefficients for specific inelastic atom-molecule collisions involving alkali atoms [36, 37], not including Cs. However, as Julienne and coworkers recently pointed out [38], the rate coefficient can be derived from a universal relation depending only on the long-range properties of the atom-biatom potential, if the inelastic process takes place with unity probability at short range. This assumption is safely fulfilled in the cases studied here, and typical rate coefficients predicted by this approach are in the range 10^{-10} cm^3/s. Similar results have been found by the group of Pierre Pillet under equivalent conditions [39].

A measurement of the rate coefficient for these inelastic collisions proceeds in the following

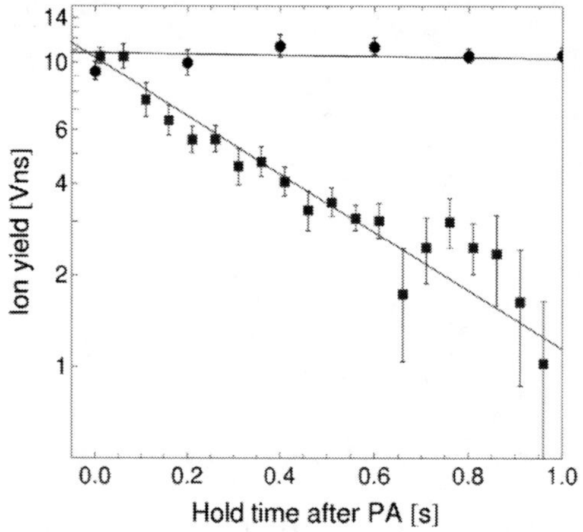

Figure 5. Molecular lifetime measurements in the dipole trap for a pure LiCs ensemble (dots, lifetime τ=24(3) s), and for LiCs molecules stored together with cesium atoms in F=3, produced via $B^1\Pi, v'=26, J'=1$ (squares, lifetime τ=0.45(3) s). Each point is an average of 16 measurements, where the given error bars are the standard deviations of the data sets. Solid lines are fits to the data.

way. The atoms are loaded into the optical trap and associated to molecules by illuminating the atomic cloud with the PA laser. Typical PA times are of the order of 100 ms to a second. Typically 5000 molecules are formed in a distribution of internal states which is determined by the Franck-Condon overlap of the excited rovibrational state addressed by the PA laser and the ground state. The molecules then interact with the atomic cloud for an adjustable time interval, until the number of remaining molecules is probed by REMPI with a short laser pulse. From the temporal decay of the number of trapped molecules one deduces a loss rate. By measuring the dependence of the molecular loss rate on the atomic density, one can directly determine the rate coefficient which is $\beta = 1.5 \times 10^{-10}\,\mathrm{cm}^3/\mathrm{s}$ as the slope of the curve. As expected from the prediction by Julienne *et al.*, the measured rate coefficients show no dependence on the Cs_2 vibrational level, even though they cover many vibrational quantum numbers. It should, however, be noted that, due to the temperature of the gas around 100 μK, there is a small contribution of p-wave scattering to the rate coefficient. Figure 4 shows the measured rate coefficient in relation to the p- and d-wave barrier and the expectation from capture theory as well as from the unitarity limit.

Figure 5 shows preliminary results for collision experiments between LiCs molecules and Cs atoms (this process is currently under closer investigation). From measured ion yields and a detection efficiency of 1% [20] one can estimate that around 5000 molecules are stored in the trap. This corresponds to roughly 10% of the number of trapped cesium atoms. As the trapping potentials for cesium and LiCs are nearly identical and the LiCs and Cs temperatures are very close, the molecular density is at least one order of magnitude smaller than the cesium density. This yields an upper limit of $5\times10^9\,\mathrm{cm}^{-3}$ for the LiCs density. It is therefore not likely that molecule-molecule collisions are observable in the current experiment. From a first evaluation, we infer that the rate coefficient is of the same order of magnitude as the rate coefficient measured in the Cs-Cs_2 case.

6. Conclusions

Currently, we are setting up a new experimental apparatus which will allow us to enter the regime of quantum degeneracy for the atomic gases. Based on the knowledge gained by the experiments presented here and by other groups in the field, we will develop a path towards the formation and trapping of LiCs molecules at high phase-space densities. The approach will rely upon magnetoassociation and STIRAP, eventually combined with photoassociation techniques. Due to the large dipole moment, we expect that the dynamics of the gas will be determined by the long-range dipolar interactions, allowing us to study new quantum phases and other many-body phenomena. As a second direction, we will move further towards "Chemistry without Entropy", using electric fields and/or shaped laser fields for quantum control. One of the big issues in this direction will be the development of appropriate detectors for state-resolved product state detection, e.g. using velocity-map imaging techniques adapted to ultracold atomic and molecular gases. It is truly amazing how much progress has been made in the quest for molecular quantum gases within the last few years, with many unforeseen discoveries on this way. In view of the many intriguing perspectives of molecular quantum gases in quantum many-body physics, quantum information and quantum chemistry, even more will be discovered in the forthcoming years.

Acknowledgments

We thank K. Mörtlbaur, C. Glück, and J. Lange for their contributions to the experiment. Inspiring discussions with R. Côté, P. Pellegrini, A. Pashov and E. Tiemann are thankfully acknowledged. The experimental work is supported by the DFG under WE2661/6-1 in the framework of the Collaborative Research Project QuDipMol within the ESF EUROCORES EuroQUAM program, and by the Heidelberg Center for Quantum Dynamics.

[1] Nikolov A N, Ensher J R, Eyler E E, Wang H, Stwalley W C and Gould P L 2000 *Phys. Rev. Lett.* **84** 246
[2] Sage J M, Sainis S, Bergeman T and DeMille D 2005 *Phys. Rev. Lett.* **94** 203001
[3] Viteau M, Chotia A, Allegrini M, Bouloufa N, Dulieu O, Comparat D and Pillet P 2008 *Science* **321** 232
[4] Deiglmayr J, Grochola A, Repp M, Mörtlbauer K, Glück C, Lange J, Dulieu O, Wester R and Weidemüller M 2008 *Phys. Rev. Lett.* **101** 133004
[5] Lang F, Winkler K, Strauss C, Grimm R and Denschlag J H 2008 *Phys. Rev. Lett.* **101** 133005
[6] Ni K K, Ospelkaus S, de Miranda M H G, Pe'er A, Neyenhuis B, Zirbel J J, Kotochigova S, Julienne P S, Jin D S and Ye J 2008 *Science* **322** 231
[7] Danzl J G, Mark M J, Haller E, Gustavsson M, Hart R, Aldegunde J, Hutson J M and Nagerl H C 2010 *Nat Phys* **6** 265–270
[8] Baranov M 2008 *Physics Reports* **464** 71
[9] Lahaye T, Menotti C, Santos L, Lewenstein M and Pfau T 2009 *Rep. Progr. Phys.* **72** 126401
[10] Micheli A, Brennen G K and Zoller P 2006 *Nat. Phys.* **2** 341
[11] Pupillo G, Griessner A, Micheli A, Ortner M, Wang D W and Zoller P 2008 *Phys. Rev. Lett.* **100** 050402
[12] Rabl P, DeMille D, Doyle J M, Lukin M D, Schoelkopf R J and Zoller P 2006 *Phys. Rev. Lett.* **97** 033003
[13] Zelevinsky T, Kotochigova S and Ye J 2008 *Phys. Rev. Lett.* **100** 043201
[14] Tscherbul T V and Krems R V 2006 *Phys. Rev. Lett.* **97** 083201
[15] Aymar M and Dulieu O 2005 *J. Chem. Phys.* **122** 204302
[16] Deiglmayr J, Grochola A, Repp M, Dulieu O, Wester R and Weidemüller M 2010 *Phys. Rev. A* **82** 032503
[17] Thorsheim H R, Weiner J and Julienne P S 1987 *Phys. Rev. Lett* **58** 2420
[18] Jones K M, Tiesinga E, Lett P D and Julienne P S 2006 *Rev. Mod. Phys.* **78** 483
[19] Ketterle W, Davis K B, Joffe M A, Martin A and Pritchard D E 1993 *Phys. Rev. Lett.* **70** 2253–2256
[20] Deiglmayr J, Repp M, Grochola A, Mörtlbauer K, Glück C, Lange J, Dulieu O, Wester R and Weidemüller M 2009 *Faraday Discussions* **142** 335
[21] Kraft S, Mikosch J, Staanum P, Deiglmayr J, Lange J, Fioretti A, Wester R and Weidemüller M 2007 *Appl. Phys. B* **89** 453–457
[22] Wang D, Kim J T, Ashbaugh C, Eyler E E, Gould P L and Stwalley W C 2007 *Phys. Rev. A* **75** 032511
[23] Staanum P, Kraft S D, Lange J, Wester R and Weidemüller M 2006 *Phys. Rev. Lett.* **96** 023201
[24] Molof R W, Schwartz H L, Miller T M and Bederson B 1974 *Phys. Rev. A* **10** 1131
[25] Deiglmayr J, Aymar M, Wester R, Weidemüller M and Dulieu O 2008 *J. Chem. Phys.* **129** 064309

[26] Bernath P F 2005 *Spectra of Atoms and Molecules; 2nd edition* (Oxford: Oxford University Press, USA)
[27] Townes C H and Schawlow A L 1955 *Microwave Spectroscopy* (McGraw-Hill Book Company)
[28] Staanum P, Pashov A, Knöckel H and Tiemann E 2007 *Phys. Rev. A* **75** 042513–9
[29] LeRoy R 2007 Level 8.0: A computer program for solving the radial schrödinger equation for bound and quasibound levelsdiegm online at http://leroy.uwaterloo.ca/programs/
[30] Deiglmayr J, Pellegrini P, Grochola A, Repp M, Côté R, Dulieu O, Wester R and Weidemüller M 2009 *New Journal of Physics* **11** 055034
[31] Hansson A and Watson J K G 2005 *J. Mol. Spec.* **233** 169
[32] Vanhaecke N and Dulieu O 2007 *Molecular Physics* **105** 1723–1731
[33] Buhmann S Y, Tarbutt M, Scheel S and Hinds E A 2008 *Phys. Rev. A* **78** 052901
[34] Mukaiyama T, Abo-Shaeer J R, Xu K, Chin J K and Ketterle W 2004 *Phys. Rev. Lett.* **92** 180402 (pages 4)
[35] Chin C, Kraemer T, Mark M, Herbig J, Waldburger P, Nägerl H C and Grimm R 2005 *Phys. Rev. Lett.* **94** 123201
[36] Quéméner G, Honvault P and Launay J M 2004 *Eur. Phys. J. D* **30** 201
[37] Quéméner G, Launay J M and Honvault P 2007 *Phys. Rev. A* **75** 050701–4
[38] Idziaszek Z and Julienne P S 2010 *Phys. Rev. Lett.* **104** 113202
[39] Zahzam N, Vogt T, Mudrich M, Comparat D and Pillet P 2006 *Phys. Rev. Lett.* **96** 023202

Journal of Physics: Conference Series **264** (2011) 012015 doi:10.1088/1742-6596/264/1/012015

Ultracold and dense samples of ground-state molecules in lattice potentials

Hanns-Christoph Nägerl, Manfred J. Mark, Elmar Haller, Mattias Gustavsson, Russell Hart, and Johann G. Danzl

Institut für Experimentalphysik und Forschungszentrum für Quantenphysik, Universität Innsbruck, 6020 Innsbruck, Austria

E-mail: christoph.naegerl@uibk.ac.at

Abstract. We produce an ultracold and dense sample of rovibronic ground state Cs_2 molecules close to the regime of quantum degeneracy, in a single hyperfine level, in the presence of an optical lattice. The molecules are individually trapped, in the motional ground state of an optical lattice well, with a lifetime of 8 s. For preparation, we start with a zero-temperature atomic Mott-insulator state with optimized double-site occupancy and efficiently associate weakly-bound dimer molecules on a Feshbach resonance. Despite extremely weak Franck-Condon wavefunction overlap, the molecules are subsequently transferred with >50% efficiency to the rovibronic ground state by a stimulated four-photon process. Our results present a crucial step towards the generation of Bose-Einstein condensates of ground-state molecules and, when suitably generalized to polar heteronuclear molecules such as RbCs, the realization of dipolar many-body quantum-gas phases in periodic potentials.

1. Introduction

Over the last 15 years there has been tremendous progress in controlling atomic gaseous quantum matter at ultralow temperatures in laboratory experiments. At the same time there have been spectacular advances in the theoretical understanding of ultracold matter. For example, quantum degenerate atomic samples have been loaded into optical lattice potentials, allowing the realization of strongly-correlated many-body systems and enabling the direct observation of quantum phase transitions with essentially full control over the parameter space [1–4]. Also, quantum gas mixtures have been at the focus of experimental and theoretical interest. For example, fermionic spin mixtures have enabled ground-breaking experiments on the BEC-BCS crossover (for an overview, see Ref. [5, 6]). Bose-Bose and Fermi-Bose quantum gas mixtures have recently attracted further widespread attention as a starting point for efficient creation of high phase-space density samples of weakly-bound molecules [7, 8] created on Feshbach resonances and, after molecular state transfer, the creation of ultracold and dense samples of deeply bound ground-state molecules [9–11]. Such molecular samples promise myriad possibilities for new directions of research in fields such as ultracold chemistry, quantum information science, precision metrology, Bose-Einstein condensation, and quantum many-body physics [12–17] (for recent overview articles on the status of the field and on possible experiments, see Refs. [18–23]). Ultracold samples of homonuclear molecules such as Cs_2 would allow novel studies of few-body collisional physics [24–28], would enable Bose-Einstein condensation of ground state molecules [29], and would open up new terrain for high resolution spectroscopy [30, 15, 16]. Heteronuclear molecules such as KRb and RbCs, with sizable permanent electric dipole moments of the order of 1 Debye in their rovibronic ground state [31, 32], feature long-range and anisotropic dipole-dipole interactions. Therefore, ultracold samples of polar molecules open up completely new avenues for research. For example, at ultralow temperatures, scattering and reaction dynamics are expected to be dominated by dipolar effects [28, 33–39]. In fact, recent experiments on KRb have reached this regime [40]. A large variety of novel quantum phases has been proposed for bosonic dipolar gases that are confined to two- or three-dimensional lattice potentials [17, 41–51]. In particular, strongly coupled ladder systems of strongly-interacting one-dimensional (1D) gases should become accessible [52–56], allowing us to greatly extend recent work in our group on strongly interacting 1D systems [57, 4] to the case of dipolar systems.

Here, we review and extend our work [11] on producing ultracold and dense samples of ground state Cs_2 molecules individually trapped at the sites of an optical lattice. First, we create ultracold samples of weakly bound Feshbach molecules out of an atomic Mott-insulator state by means of a Feshbach resonance [58, 59]. While maintaining ultralow temperatures and state fidelity, the molecules are then transferred into the ro-vibrational ground state by a stimulated four-photon transition using the STIRAP technique [60]. Recently, the STIRAP technique has been successfully implemented with high efficiencies on ultracold and dense samples of Feshbach molecules for transfer into the rovibrational ground state of the triplet $^3\Sigma$-potential for Rb_2 molecules [61], for transfer into deeply bound rovibrational levels of the singlet $^1\Sigma$-potential for Cs_2 molecules [9] and also, in a four-photon process, into the rovibronic ground state [11] of this molecule, and for transfer into the rovibronic ground state for polar KRb molecules [10].

2. Experimental procedure

Our molecular quantum-gas preparation procedure is summarized in Fig. 1. We first load a pure Bose-Einstein condensate (BEC) of about 10^5 Cs atoms [62] into a three-dimensional, far detuned optical lattice. By raising the lattice depth and adjusting the external trap confinement we drive the superfluid-to-Mott-insulator (SF-MI) phase transition [3] to localize pairs of atoms at individual lattice sites. The atomic number distribution in the Mott-insulator state is inhomogeneous as a result of the external harmonic confinement. Shells with a constant number of precisely n atoms per lattice site, where $n = 1, 2, 3 \ldots$, are separated by narrow superfluid regions [3]. By appropriately increasing the strength of the external confinement and hence compressing the sample during lattice loading, we aim to create a $(n = 2)$-Mott shell with maximum extent in the central region of the lattice in order to obtain the highest number of lattice sites at which there are precisely two atoms. With up to 45(2)% of the atoms at doubly-occupied lattice sites we come close to the theoretical limit of 53% given the parabolic density profile of the BEC [63]. The atom pairs reside in the motional ground state at each well and are then associated [64] with 94(1)% probability [11] to Cs_2 Feshbach molecules, which are subsequently transferred via a series of avoided level crossings to the weakly-bound level $|1\rangle$, the starting level for the optical transfer [65, 9, 66], see Fig. 2. We typically set the depth of the optical lattice potential to 20 E_R for atoms, corresponding to 80 $\tilde{E}_R$ for Feshbach molecules with double the atomic polarizability and double the mass, where E_R and $\tilde{E}_R$ are the atomic and molecular photon recoil energies, respectively. Atoms at singly-occupied sites are removed by a combination of microwave and optical excitation [64]. We now have a pure molecular sample with a high occupation of about 85(3)% in the central region of the lattice. Each molecule is in the motional ground state of its respective lattice well and perfectly shielded from collisional loss.

The molecular population is transferred to the rovibronic ground state by coherent optical transfer using the Stimulated Raman Adiabatic Passage (STIRAP) technique. STIRAP allows highly efficient population transfer between molecular levels with exquisite state selectivity, since there are no spontaneous processes involved [60]. The molecular population is rotated to the

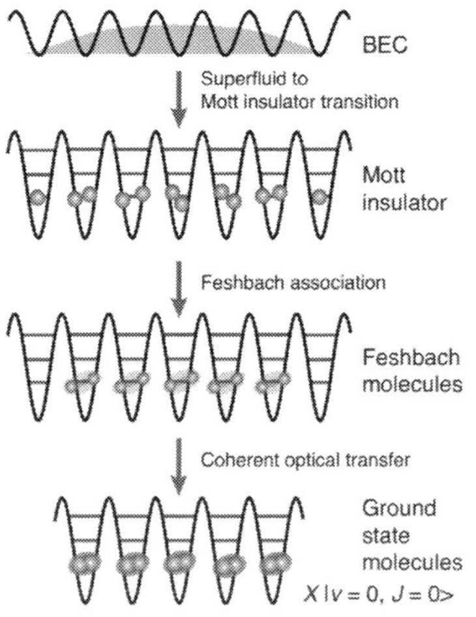

Figure 1. Molecular quantum gas preparation procedure: A BEC of Cs atoms is loaded into an optical lattice. The superfluid-to-Mott-insulator quantum phase transition is driven under conditions that maximize the number of doubly occupied lattice sites. Feshbach association creates weakly bound molecules with each molecule individually trapped in the lowest vibrational level of its respective lattice site. Atoms remaining at singly occupied sites are cleared out. The molecules are then coherently transferred by STIRAP to a specific hyperfine quantum state of the rovibronic ground state $X^1\Sigma_g^+|v=0, J=0>$. They remain trapped in the lowest lattice vibrational level.

final molecular level via a dark state which has no contribution from lossy electronically excited levels. STIRAP is effected by a sequence of overlapping laser pulses, which adiabatically changes the character of the dark state from the initially populated molecular level to the target level. To satisfy adiabaticity, the product of the STIRAP transfer time and the Rabi frequencies for the STIRAP transitions needs to be sufficiently high. For transfer from the extremely weakly bound molecular levels, populated by Feshbach association, to the rovibronic ground state, this requirement poses a considerable challenge. On the one hand, there is a large mismatch in internuclear separation that has to be bridged. On the other hand, the lasers involved in molecular state transfer need to be phase coherent with respect to each other on the timescale of the STIRAP process. Long STIRAP transfer times, while beneficial for adiabaticity, pose stringent requirements on the relative phase stability of the transfer lasers operating at widely different wavelengths.

We have identified a transfer scheme that allows us to efficiently transfer the molecular population to the rovibronic ground state. A total of four molecular transitions, driven by lasers $L_1 \ldots L_4$, couple the initial weakly bound level $|1>$ with the rovibronic ground state $|5>$ via two intermediate electronically excited state levels $|2>, |4>$ and one intermediate ground state level $|3>$, as illustrated in Fig. 2a and 2b. The four-photon (4p) scheme is preferred over a two-photon scheme to overcome otherwise prohibitively low Franck-Condon (FC) factors. When compared to heteronuclear molecules, homonuclear molecules, such as Cs_2, feature an additional symmetry with respect to exchange of the identical nuclei, the so-called *gerade/ungerade* (g/u) symmetry, which imposes additional selection rules for optical transitions and hence restricts the excited state potentials that can be employed in the optical transfer. In addition, a resonant term in the interaction leads to a $1/R^3$ scaling of the excited state potentials as opposed to the $1/R^6$ scaling in the ground state, implying that ground and excited state potentials do not line up favorably for transfer from weakly to deeply bound levels. While adding some complexity to the system, the four-photon scheme allows acceptable, albeit low, FC factors for coherent state transfer. Specifically, both excited state intermediate levels $|2>$ and $|4>$belong to the mixed singlet/triplet $(A^1\Sigma_u^+ - b^3\Pi_u)0_u^+$ coupled system, as illustrated in Fig. 2a. The intermediate ground state level $|3>$ is bound by $\sim hc \times 1060$ cm^{-1} or $\sim h \times 32$ THz, where h is Planck's constant and c is the speed of light, and has vibrational quantum number $v = 73$. We choose either $J = 0$ or $J = 2$, where J is the rotational quantum number. In the first transition, the singlet $X^1\Sigma_g^+$ component of the initial Feshbach level enables coupling to the 0_u^+ system. In contrast to heteronuclear molecules, where singlet/triplet mixing in the excited state can be favourably exploited [67] in

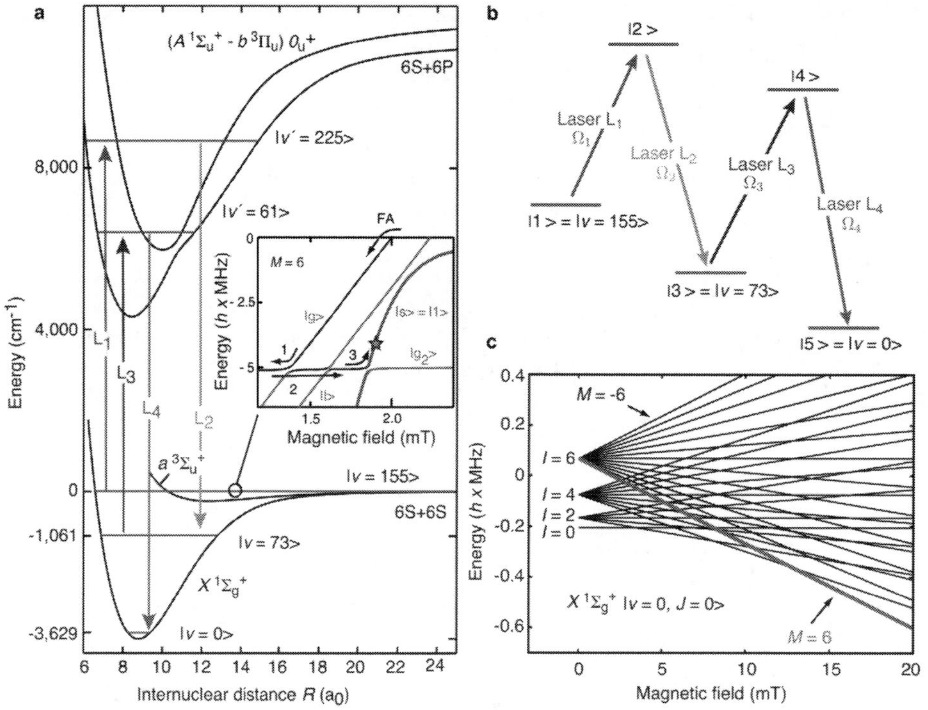

Figure 2. Molecular potentials and level schemes for transfer to the rovibronic ground state: a, Molecular potentials. The initial Feshbach level (not resolved near threshold) of mixed $(a^3\Sigma_u^+ - X^1\Sigma_g^+)$ character is linked to the rovibronic ground state by 4 laser transitions via two excited state levels belonging to the $(A^1\Sigma_u^+ - b^3\Pi_u)0_u^+$ coupled system and one ground state intermediate level $|v = 73\rangle$. The transition wavelengths for lasers L_i, where $i = 1, 2, 3, 4$, are near 1126 nm, 1006 nm, 1351 nm, and 1003 nm, respectively. Inset: Zeeman diagram for the molecular levels near threshold. Molecules are created by Feshbach association (FA) and the molecular population is transferred to the desired starting level for STIRAP $|s\rangle = |1\rangle$ ($\star$) via avoided level crossings by slow (1,3) and fast (2) magnetic field ramps. All weakly bound levels are characterized by $M = 6$. **b,** Schematic 5-level scheme linking the initial weakly bound level $|1\rangle$ to the rovibronic ground state $|5\rangle$ by lasers L_i with Rabi frequencies Ω_i. **c,** Zeeman diagram showing the calculated hyperfine structure of the rovibronic ground state $|5\rangle = |v = 0, J = 0\rangle$. Coherent optical transfer selectively populates the level $|I = 6, M_I = 6\rangle$ indicated in red. This is the lowest energy absolute ground state at magnetic fields above $\sim$ 13 mT. There are no avoided crossings between hyperfine sublevels. Calculation taken from Ref. [11].

combination with excitation from the dominant $a^3\Sigma^+$ component of Feshbach levels, a transition from the $a^3\Sigma_u^+$ state to 0_u^+ levels is forbidden here. We have performed ultra-high-resolution molecular spectroscopy to identify the previously unknown excited state intermediate molecular levels in the four-photon scheme [68, 69]. These have highly mixed singlet/triplet character. Interestingly, for the second two-photon transition, the excited state intermediate level $|4\rangle$ is primarily of triplet character, giving a more balanced distribution of FC factors for lasers L_3 and L_4 than neighboring levels of primarily singlet character.

Our experimental procedure ensures that only a single quantum state is populated in $X|v = 0, J = 0\rangle$. The rovibrational ground state of the $^1\Sigma$ electronic ground state in alkali dimers comprises a series of hyperfine sublevels separated by a very small energy at low magnetic fields [70, 71]. In Cs_2, there are 28 hyperfine sublevels in $|v=0, J=0>$ separated by $\sim h \times 270$ kHz at zero field [11], as illustrated in Fig. 2c. These correspond to the allowed values for the total molecular nuclear spin $I = 0, 2, 4, 6$ with the $(2I+1)$ projections on the magnetic field. In homonuclear dimers there are no avoided crossings between hyperfine sublevels for $J = 0$ levels [71]. For comparison, the inset in Fig. 2a shows the weakly bound molecular levels near the dissociation threshold. We prepare the atomic BEC in the lowest atomic hyperfine sublevel $|F_a = 3, m_{F_a} = 3>$, where F_a and m_{F_a} are the total atomic spin and its projection on the magnetic field. Feshbach association and transfer between weakly bound levels via avoided crossings conserve [59] the total projection quantum number $M = m_{Fa_1} + m_{Fa_2} = 6$. In the coherent optical transfer, we exploit the selection rule $\Delta M = 0$ for linear polarization parallel to the quantization axis, such that the rovibronic ground state molecules inherit the $M = 6$ total angular momentum projection. From the Zeeman diagram in Fig. 2c it is immediately clear that there is only a single $M = 6$ quantum level in the rovibronic ground state, namely the $|I = 6, M_I = 6>$ level, and hence a single, well-defined quantum state is selectively populated by our optical state transfer. Importantly, the $|I = 6, M_I = 6>$ quantum state is an energetically excited hyperfine sublevel at low magnetic fields at which the STIRAP transfer is carried out, but it constitutes the lowest hyperfine sublevel and hence the lowest energy absolute ground state of the Cs dimer at magnetic fields above ~ 13 mT.

To ensure phase coherence of the STIRAP lasers on the timescale of the optical transfer, the lasers are locked to narrow linewidth optical resonators for short term stability and referenced to an optical frequency comb for long-term stability and reproducibility.

3. Results

There are two possibilities to transfer the molecular population from $|1>$ to $|5>$. In sequential STIRAP (s-STIRAP), the molecular population is first transferred to the intermediate ground state level $|3>$ in a conventional two-photon process and then in a second two-photon step to $|v=0, J=0>$. In contrast, 4-photon STIRAP (4p-STIRAP) transfers the molecular population directly into $|v=0, J=0>$ via a 4p-dark state [72], which has no contribution from the two excited state intermediate levels $|2>$ and $|4>$. A recent analysis in the context of ultracold molecules can be found in Ref. [73].

We first investigate transfer in the near quantum degenerate regime to the intermediate ground state level $|3> = |v=73, J=2>$ in the absence of an optical lattice [9]. This presents the crucial transfer step because it links the long-range molecular levels near threshold with the tightly, i.e., chemically bound levels deep down in the molecular potential, bridging a binding energy difference of ~ 1060 cm^{-1} and most of the mismatch in internuclear distance. As shown in Fig. 3a we monitor the number of Feshbach molecules as a function of STIRAP time τ during transfer to $|v=73, J=2>$, with the pulse sequence schematically depicted in the lower panel. After the STIRAP pulse sequence, no molecules remain in $|1>$. For detection of deeply bound molecules, they are transferred back to the original weakly bound level by a second, reverse pulse sequence. Molecules in $|1>$ are dissociated to atoms, which are then detected by standard absorption imaging. A large proportion of the molecules returns to $|1>$ after round-trip STIRAP, corresponding to a single pass transfer efficiency of >80%. The coherent nature of the transfer was demonstrated in a Ramsey-type experiment and it was shown that the STIRAP process does not lead to heating of the sample on the nano-Kelvin scale [9].

Coherent state transfer in an optical lattice poses additional challenges that have to be met: (1) The lattice light might lead to excitation of molecules to electronically excited states and hence to loss out of the three- or five-level system. (2) The lattice light might create unwanted coherences competing with the STIRAP process, e.g., between an electronically excited intermediate level and an additional ground state level. (3) In order to ensure control

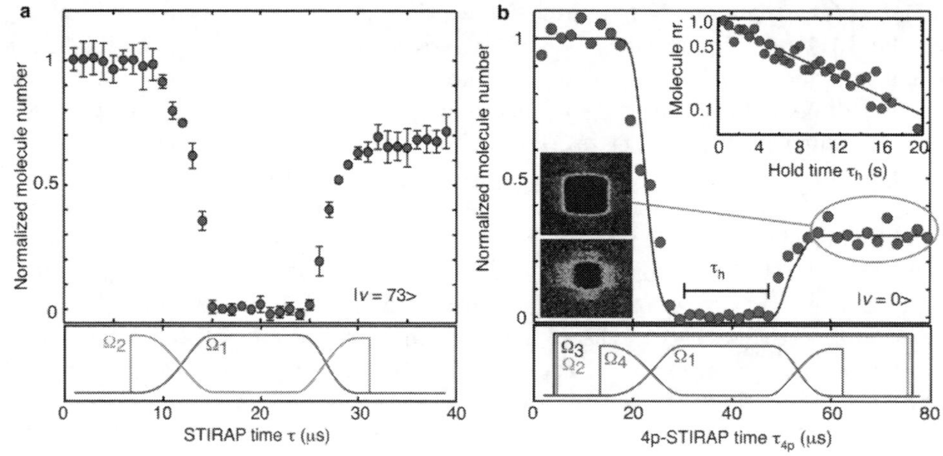

Figure 3. Transfer to the intermediate ground state level $|v = 73>$ and to the rovibronic ground state $|v = 0, J = 0>$: a, Transfer to $|v = 73, J = 2>$ and schematic pulse sequence for the Rabi frequencies Ω_i. The number of Feshbach molecules in $|1>$ is monitored as a function of STIRAP time τ after which the transfer is interrupted. As the Rabi frequencies are ramped, the molecules are transferred to $|v = 73>$ and no molecules are detected in $|1>$. For detection, the sequence is reversed and a large proportion of the molecules returns to $|1>$. Data taken from Ref. [9]. **b,** Transfer to the rovibronic ground state $X^1\Sigma_g^+|v = 0, J = 0>$ by 4-photon STIRAP with schematic pulse sequence. The molecules are transferred in a single step to $|v = 0>$ as the Rabi frequencies for lasers L_1 and L_4 are ramped while L_2 and L_3 continuously couple all intermediate levels. For detection, the pulse sequence is reversed. The solid line is a model calculation. Upper right inset: Normalized number of ground state molecules as a function of hold time τ_h in the lattice. STIRAP lasers were turned off during τ_h and the lattice depth was adiabatically reduced after transfer to $|v = 0, J = 0>$ to reduce off-resonant light scattering. Upper left inset: Absorption image after STIRAP to $|v = 0, J = 0>$ and back showing the momentum distribution after time of flight, corresponding to an average of the circled data points. Almost all the molecular population resides in the first Brillouin zone characterized by its rectangular shape. Lower left inset: Momentum distribution after transfer to $|v = 0, J = 2>$ and back with resonant modulation of lattice depth during τ_h, showing molecular population transferred to higher lattice bands. Data taken from Ref. [11].

over the motional state of the ground-state molecules in the optical lattice, the lattice has to be operated near the magic wavelength that gives equal light shifts for the initial and the final molecular levels. In order to realize lattice-based STIRAP to the rovibrational ground state, we first implemented transfer to $X|v = 73, J = 2>$ in the optical lattice [66]. Transfer efficiencies similar to the case without an optical lattice were achieved. The lifetime of the molecular population in $|v = 73>$ was on the order of 20 ms, limited by off-resonant excitation by the lattice light. This lifetime is orders of magnitude longer than the timescale of the STIRAP transfer, allowing us to use this deeply bound level as an intermediate state for lattice-based STIRAP transfer to $|v = 0>$.

Figure 3b summarizes our central results on transfer to the rovibronic ground state in an optical lattice. We perform 4p-STIRAP and monitor the number of molecules in $|1>$ as a function of 4p-STIRAP time τ_{4p}. Lasers L_2 and L_3 continuously couple all intermediate levels $|2>$, $|3>$, and $|4>$ to one effective intermediate level [74] and STIRAP is carried out by varying only the Rabi frequencies Ω_1 and Ω_4 of the first and last laser. Population in the intermediate ground state level |3> can in principle be minimized during transfer by choosing $\Omega_2, \Omega_3 \gg \Omega_1, \Omega_4$. The molecules are directly transferred to $X^1\Sigma_g^+|v=0, J=0\rangle$. No molecules are detected during the hold time τ_h. The reverse pulse sequence transfers a large proportion of the molecules back to |1>, corresonding to a single pass transfer efficiency of ~ 55 %. We have achieved similar transfer efficiencies of 55 % - 60 % with s-STIRAP.

Remarkably, motional state control is maintained during STIRAP to $|v=0\rangle$. During the STIRAP process, the wavefunction of the motional ground state in the lattice potential experienced by molecules in level |1> is projected onto the motional wavefunctions associated with the lattice potential experienced by molecules in |5 >. Only if the lattice wavelength is chosen such that the dynamical polarizabilities for |1> and |5> and hence the respective lattice potentials are closely matched will this projection lead to ground state molecules that are confined to the lowest vibrational level at the individual lattice sites. The upper left inset in Fig. 3b shows an absorption image after STIRAP to $|v=0\rangle$ and back after time of flight to reveal the momentum distribution of the molecules. The rectangular shape of the first Brillouin zone can clearly be seen. We deduce that 92(3) % of the molecular population resides in the lowest lattice vibrational level. For comparison, the lower left inset in Fig. 3b shows molecular population transferred to higher lattice vibrational levels by amplitude modulation of the lattice during hold time τ_h with a frequency that matches the energy spacing between two lattice vibrational levels. By applying lattice amplitude and phase modulation and varying the modulation frequency we can map out the energy structure of the lattice bands for the $|v=0\rangle$-molecules and deduce their dynamical polarizability. We find that the polarizability of the $|v=0\rangle$-molecules is 1.05 × the polarizability of the Feshbach molecules [11], such that the magic wavelength condition is well fulfilled. We have thus created a state where the molecules are controlled at the level of single quantum states both with respect to all internal as well as the external degrees of freedom. In the central region of the lattice, about every other lattice site is filled with a molecule in the desired quantum state.

We measure the lifetime of the ground state molecules in the lattice by varying the hold time τ_h for up to 20 s. The right inset in Fig. 3b shows the number of rovibronic ground state molecules as a function of τ_h. We observe a long $1/e$ lifetime of 8.1(6) s, limited by off-resonant excitation to excited state molecular levels by the lattice light.

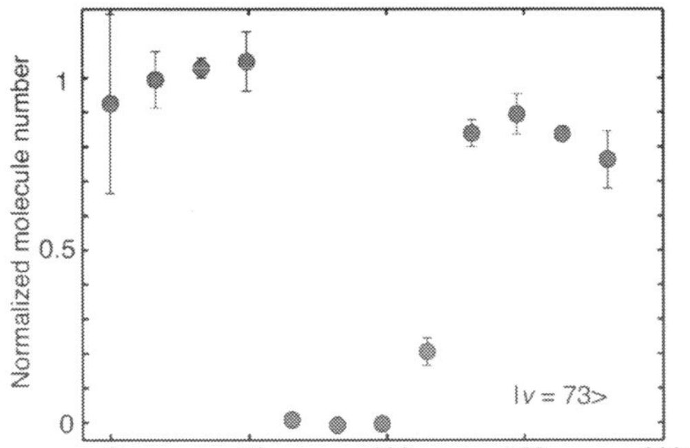

Figure 4. Improved STIRAP efficiency: Transfer to the intermediate ground state level $|3\rangle=|v{=}73, J{=}2\rangle$ with improved STIRAP laser setup. A substantial increase in transfer efficiency as compared to the data in Fig. 3a is achieved by increasing the laser phase stability.

Our simulations, based on the time dependent solution of the 5-level quantum master equation including laser phase noise and loss out of the 5-level system due to spontaneous emission, show that the STIRAP transfer efficiency is currently limited by finite available laser power to drive the extremely weak transitions in the 4p-scheme and by phase fluctuations of the STIRAP lasers. In order to optimize sample purity and hence phase-space density for quantum-state resolved collisional studies and the preparation of a BEC of ground state molecules, we have recently optimized both the short term phase coherence and the long term stability of our transfer lasers.

For short term stability on the timescale of an individual STIRAP transfer process, a Pound-Drever-Hall lock with a feedback bandwidth of several MHz suppresses fast noise. In addition, we have modified the diode laser external grating feedback setup to essentially increase the laser resonator length to reduce phase noise. These improvements are combined with a slow phase lock of all STIRAP lasers to a common reference laser via the optical frequency comb. For this, we have implemented a transfer oscillator scheme, which eliminates the correlated noise contributions from the optical frequency comb and allows the transfer of the phase information of the reference laser to distinct regions of the optical spectrum [75, 76]. Testing the improved locking scheme on the fist STIRAP step to $|v{=}73, J{=}2>$, we have been able to improve the single pass efficiency by a substantial amount from >80% to >90%, as evidenced in Fig. 4. This indicates that the limitations to transfer efficiency are mainly technical. Additional technical improvements are expected to improve this efficiency further and a similar increase is expected for the second STIRAP transition.

4. Conclusion

We have shown that it is possible to prepare with high fidelity a sample of rovibronic ground-state molecules at ultralow temperatures in the presence of an optical lattice. If the collisional properties of the ground state molecules turn out to be sufficiently favorable, the preparation of a BEC of ground state molecules should be possible. The most important obstacle to BEC formation would be an open channel for inelastic collisions in a dimer-dimer collision. However, it is known from ab-initio calculations that the channel for the formation of a trimer and an atom in a dimer-dimer collision is energetically closed for the alkali dimers [77, 78]. We believe that our results can be generalized to the case of heteronuclear diatomic molecules such as RbCs. We have dedicated a separate experimental apparatus to this molecule [79]. It has the advantage that it is chemically stable under a two-body collision, i.e., the reaction RbCs + RbCs $\rightarrow$ Rb_2 + Cs_2 is endothermic when RbCs is in its rovibronic ground state [78]. In contrast, the molecule KRb is not stable [80] under the process KRb + KRb $\rightarrow$ K_2 + Rb_2. In principle, we expect that at ultralow temperatures and in the lowest hyperfine sublevel of the rovibronic ground state of the RbCs molecule [70] only molecular three-body collisions present a possible limitation to performing experiments with dipolar many-body systems. As before, to optimize the overall molecule production efficiency by minimizing collisional loss during the preparation procedure, we will implement the Feshbach association process and the ground state transfer in the presence of the optical lattice with precisely one molecule per lattice site. Evidently, the preparation of the initial two-atom Mott-insulator states is more involved than for the case of Cs_2 [81, 82]. The central question is how to generate a many-body state in the lattice for which the number of Rb-Cs pairs as precursors to RbCs molecules is maximized. In view of considerable Cs-Cs and in particular Rb-Cs three-body loss [79] we will load the lattice with two spatially separated BECs, one BEC of ^{87}Rb atoms and one BEC of ^{133}Cs atoms. The two BECs initially reside in two separate dipole traps. We will exploit the fact that the Rb-Cs interaction can be controlled near a Rb-Cs interspecies Feshbach resonance [83]. The central idea is to first freeze out the Cs sample using the SF-MI phase transition in such a way that at first a state with precisely one Cs atom per lattice site is generated. The Rb sample, much more mobile as a result of a lower polarizability for the Rb atoms, is then brought to overlap with the Cs sample while being still superfluid. For small repulsive or even weakly attractive Rb-Cs interaction one expects that each Cs atom will then preferentially pair up with precisely one Rb atom, giving ideal conditions for molecule production and ground-state transfer.

5. Acknowledgments

We thank H. Ritsch, N. Bouloufa, O. Dulieu, J. Aldegunde, J. M. Hutson, H. Salami, T. Bergeman, S. Dürr, and K. Bergmann for valuable discussions and H. Telle, H. Schnatz, B.

Lipphardt, and J. Alnis for sharing technical expertise. We are indebted to R. Grimm for generous support. We gratefully acknowledge funding by the Austrian Ministry of Science and Research (Bundesministerium für Wissenschaft und Forschung) and the Austrian Science Fund (Fonds zur Förderung der wissenschaftlichen Forschung) in the form of a START prize grant and by the European Science Foundation within the framework of the EuroQUASAR collective research project QuDeGPM (Project I 153-N16) and within the framework of the EuroQUAM collective research project QuDipMol (Project I 124-N16). R.H. was supported by a Marie Curie International Incoming Fellowship within the 7th European Community Framework Programme.

References

[1] Jaksch D, Bruder C, Cirac J I, Gardiner C W and Zoller P 1998 *Phys. Rev. Lett.* **81** 3108–3111
[2] Greiner M, Mandel O, Esslinger T, Hänsch T W and Bloch I 2002 *Nature* **415** 39–44
[3] Bloch I, Dalibard J and Zwerger W 2008 *Rev. Mod. Phys.* **80** 885–964
[4] Haller E, Hart R, Mark M J, Danzl J G, Reichsöllner L, Gustavsson M, Dalmonte M, Pupillo G and Nägerl H C 2010 *Nature* **466** 597–600
[5] Inguscio M, Ketterle W and Salomon C (eds) 2008 *Ultracold Fermi Gases , Proceedings of the International School of Physics "Enrico Fermi", Course CLXIV, Varenna, June 2006* (IOS Press, Amsterdam)
[6] Giorgini S, Pitaevskii L P and Stringari S 2008 *Rev. Mod. Phys.* **80** 1215–1274
[7] Regal C A, Ticknor C, Bohn J L and Jin D S 2003 *Nature* **424** 47–50
[8] Herbig J, Kraemer T, Mark M, Weber T, Chin C, Nägerl H C and Grimm R 2003 *Science* **301** 1510–1513
[9] Danzl J G, Haller E, Gustavsson M, Mark M J, Hart R, Bouloufa N, Dulieu O, Ritsch H and Nägerl H C 2008 *Science* **321** 1062–1066
[10] Ni K K, Ospelkaus S, de Miranda M H G, Pe'er A, Neyenhuis B, Zirbel J J, Kotochigova S, Julienne P S, Jin D S and Ye J 2008 *Science* **322** 231–235
[11] Danzl J G, Mark M J, Haller E, Gustavsson M, Hart R, Aldegunde J, Hutson J M and Nägerl H C 2010 *Nature Phys.* **6** 265–270
[12] Krems R 2005 *Int. Rev. Phys. Chem.* **24** 99–118
[13] Krems R 2008 *Phys. Chem. Chem. Phys.* **10** 4079–4092
[14] DeMille D 2002 *Phys. Rev. Lett.* **88** 067901
[15] DeMille D, Sainis S, Sage J, Bergeman T, Kotochigova S and Tiesinga E 2008 *Phys. Rev. Lett.* **100** 043202
[16] Zelevinsky T, Kotochigova S and Ye J 2008 *Phys. Rev. Lett.* **100** 043201
[17] Góral K, Santos L and Lewenstein M 2002 *Phys. Rev. Lett.* **88** 170406
[18] Baranov M A 2008 *Physics Reports* **464** 71–111
[19] Pupillo G, Micheli A, Büchler H P and Zoller P 2009 chap. 12 in *Cold molecules: Theory, experiment, applications* ed Krems R V, Stwalley W C and Friedrich B (CRC Press, Boca Raton)
[20] Lahaye T, Menotti C, Santos L, Lewenstein M and Pfau T 2009 *Rep. Prog. Phys.* **72** 126401
[21] Carr L D, DeMille D, Krems R V and Ye J 2009 *New J. Phys.* **11** 055049
[22] Friedrich B and Doyle J M 2009 *ChemPhysChem* **10** 604–623
[23] Dulieu O and Gabbanini C 2009 *Rep. Prog. Phys.* **72** 086401
[24] Chin C, Kraemer T, Mark M, Herbig J, Waldburger P, Nägerl H C and Grimm R 2005 *Phys. Rev. Lett.* **94** 123201
[25] Kraemer T, Mark M, Waldburger P, Danzl J G, Chin C, Engeser B, Lange A D, Pilch K, Jaakkola A, Nägerl H C and Grimm R 2006 *Nature* **440** 315–318
[26] Staanum P, Kraft S D, Lange J, Wester R and Weidemüller M 2006 *Phys. Rev. Lett.* **96** 023201
[27] Zahzam N, Vogt T, Mudrich M, Comparat D and Pillet P 2006 *Phys. Rev. Lett.* **96** 023202
[28] Idziaszek Z, Quéméner G, Bohn J L and Julienne P S 2010 *Phys. Rev. A* **82** 020703(R)
[29] Jaksch D, Venturi V, Cirac J I, Williams C J and Zoller P 2002 *Phys. Rev. Lett.* **89** 040402
[30] Flambaum V V and Kozlov M G 2007 *Phys. Rev. Lett.* **99** 150801
[31] Kotochigova S and Tiesinga E 2005 *J. Chem. Phys.* **123** 174304
[32] Deiglmayr J, Aymar M, Wester R, Weidemüller M and Dulieu O 2008 *J. Chem. Phys.* **129** 064309
[33] Micheli A, Pupillo G, Büchler H P and Zoller P 2007 *Phys. Rev. A* **76** 043604
[34] Micheli A, Idziaszek Z, Pupillo G, Baranov M A, Zoller P and Julienne P S 2010 *Phys. Rev. Lett.* **105** 073202
[35] Ticknor C 2010 *Phys. Rev. A* **81** 042708
[36] Quéméner G and Bohn J L 2010 *Phys. Rev. A* **81** 022702
[37] Quéméner G and Bohn J L 2010 *Phys. Rev. A* **81** 060701
[38] Idziaszek Z and Julienne P S 2010 *Phys. Rev. Lett.* **104** 113202
[39] Li Z, Alyabyshev S V and Krems R V 2008 *Phys. Rev. Lett.* **100** 073202

[40] Ni K K, Ospelkaus S, Wang D, Quemener G, Neyenhuis B, de Miranda M H G, Bohn J L, Ye J and Jin D S 2010 *Nature* **464** 1324–1328
[41] Barnett R, Petrov D, Lukin M and Demler E 2006 *Phys. Rev. Lett.* **96** 190401
[42] Büchler H P, Demler E, Lukin M, Micheli A, Prokof'ev N, Pupillo G and Zoller P 2007 *Phys. Rev. Lett.* **98** 060404
[43] Menotti C, Trefzger C and Lewenstein M 2007 *Phys. Rev. Lett.* **98** 235301
[44] Yi S, Li T and Sun C P 2007 *Phys. Rev. Lett.* **98** 260405
[45] Danshita I and Sá de Melo C A R 2009 *Phys. Rev. Lett.* **103** 225301
[46] Cooper N R and Shlyapnikov G V 2009 *Phys. Rev. Lett.* **103** 155302
[47] Capogrosso-Sansone B, Trefzger C, Lewenstein M, Zoller P and Pupillo G 2010 *Phys. Rev. Lett.* **104** 125301
[48] Pollet L, Picon J D, Büchler H P and Troyer M 2010 *Phys. Rev. Lett.* **104** 125302
[49] Li Q, Hwang E H and Sarma S D 2010 Preprint at arXiv:1009.4204
[50] Potter A C, Berg E, Wang D W, Halperin B I and Demler E 2010 Preprint at arXiv:1007.5061
[51] Pikovski A, Klawunn M, Shlyapnikov G and Santos L 2010 Preprint at arXiv:1008.3264
[52] Dalla Torre E G, Berg E and Altman E 2006 *Phys. Rev. Lett.* **97** 260401
[53] Kollath C, Meyer J S and Giamarchi T 2008 *Phys. Rev. Lett.* **100** 130403
[54] Huang Y P and Wang D W 2009 *Phys. Rev. A* **80** 053610
[55] Klawunn M, Duhme J and Santos L 2010 *Phys. Rev. A* **81** 013604
[56] Dalmonte M, Pupillo G and Zoller P 2010 *Phys. Rev. Lett.* **105** 140401
[57] Haller E, Gustavsson M, Mark M J, Danzl J G, Hart R, Pupillo G and Nägerl H C 2009 *Science* **325** 1224–1227
[58] Köhler T, Góral K and Julienne P S 2006 *Rev. Mod. Phys.* **78** 1311–1361
[59] Chin C, Grimm R, Julienne P S and Tiesinga E 2010 *Rev. Mod. Phys.* **82** 1225–1286
[60] Bergmann K, Theuer H and Shore B W 1998 *Rev. Mod. Phys.* **70** 1003–1025
[61] Lang F, Winkler K, Strauss C, Grimm R and Hecker Denschlag J 2008 *Phys. Rev. Lett.* **101** 133005
[62] Kraemer T, Herbig J, Mark M, Weber T, Chin C, Nägerl H C and Grimm R 2004 *Appl. Phys. B* **79** 1013–1019
[63] Volz T, Syassen N, Bauer D, Hansis E, Dürr S and Rempe G 2006 *Nature Phys.* **2** 692–695
[64] Thalhammer G, Winkler K, Lang F, Schmid S, Grimm R and Hecker Denschlag J 2006 *Phys. Rev. Lett.* **96** 050402
[65] Mark M, Ferlaino F, Knoop S, Danzl J G, Kraemer T, Chin C, Nägerl H C and Grimm R 2007 *Phys. Rev. A* **76** 042514
[66] Danzl J G, Mark M J, Haller E, Gustavsson M, Hart R, Liem A, Zellmer H and Nägerl H C 2009 *New J. Phys.* **11** 055036
[67] Stwalley W C 2004 *Eur. Phys. J. D* **31** 221–225
[68] Danzl J G, Mark M J, Haller E, Gustavsson M, Bouloufa N, Dulieu O, Ritsch H, Hart R and Nägerl H C 2009 *Faraday Discuss.* **142** 283–295
[69] Mark M J, Danzl J G, Haller E, Gustavsson M, Bouloufa N, Dulieu O, Salami H, Bergeman T, Ritsch H, Hart R and Nägerl H C 2009 *Appl. Phys. B* **95** 219–225
[70] Aldegunde J, Rivington B A, Żuchowski P S and Hutson J M 2008 *Phys. Rev. A* **78** 033434
[71] Aldegunde J and Hutson J M 2009 *Phys. Rev. A* **79** 013401
[72] Malinovsky V S and Tannor D J 1997 *Phys. Rev. A* **56** 4929–4937
[73] Kuznetsova E, Pellegrini P, Côté R, Lukin M D and Yelin S F 2008 *Phys. Rev. A* **78** 021402(R)
[74] Vitanov N, Shore B and Bergmann K 1998 *Eur. Phys. J. D* **4** 15–29
[75] Stenger J, Schnatz H, Tamm C and Telle H R 2002 *Phys. Rev. Lett.* **88** 073601
[76] Telle H, Lipphardt B and Stenger J 2002 *Appl. Phys. B* **74** 1–6
[77] Guérout R, Soldán P, Aymar M, Deiglmayr J and Dulieu O 2009 *Int. J. Quant. Chem.* **109** 3387–3398
[78] Żuchowski P S and Hutson J M 2010 *Phys. Rev. A* **81** 060703(R)
[79] Lercher A D, Takekoshi T, Debatin M, Schuster B, Rameshan R, Ferlaino F, Grimm R and Nägerl H C 2010 In preparation
[80] Ospelkaus S, Ni K K, Wang D, de Miranda M H G, Neyenhuis B, Quemener G, Julienne P S, Bohn J L, Jin D S and Ye J 2010 *Science* **327** 853–857
[81] Damski B, Santos L, Tiemann E, Lewenstein M, Kotochigova S, Julienne P and Zoller P 2003 *Phys. Rev. Lett.* **90** 110401
[82] Freericks J K, Maśka M M, Hu A, Hanna T M, Williams C J, Julienne P S and Lemański R 2010 *Phys. Rev. A* **81** 011605
[83] Pilch K, Lange A D, Prantner A, Kerner G, Ferlaino F, Nägerl H C and Grimm R 2009 *Phys. Rev. A* **79** 042718

Journal of Physics: Conference Series **264** (2011) 012016 doi:10.1088/1742-6596/264/1/012016

Bose-Bose mixtures in reduced dimensions

F Minardi[1,2], G Barontini[1], J Catani[1,2], G Lamporesi[1], Y Nishida[3] and M Inguscio[1,2]

[1] LENS and Università di Firenze, Via Nello Carrara 1, 50019 Sesto Fiorentino, Italy
[2] INO-CNR, Via Giovanni Sansone 1, 50019 Sesto Fiorentino, Italy
[3] Center for Theoretical Physics, Massachusetts Institute of Technology, Cambridge, Massachusetts 02139, USA

E-mail: `minardi@lens.unifi.it`

Abstract. The two-body scattering is greatly modified in reduced dimensions. With ultracold atoms, low dimensional configurations are routinely accessible thanks to the use of optical lattices which allow confinements sufficiently strong to freeze the motion along chosen directions. With two different atomic species, we use a species-selective optical potential, in the form of a standing wave, to confine only one species in 2D disks and study the scattering between particles existing in different dimensions, i.e., we realize a 2D-3D *mix-dimensional configuration*, reminiscent of a brane world.

We review the scattering theory specific to this configuration and derive an effective scattering length $a_{\rm eff}$ in terms of the free-space scattering length a and the confinement parameters. We detect experimentally the enhancement of inelastic collisions arising at particular values of a and relate these values to the divergences of $a_{\rm eff}$. Unlike the confinement-induced resonances predicted and observed for identical particles, our mixed-dimensional resonances occur in a series of several resonances, because the relative and centre-of-mass motion are coupled.

1. Introduction

Some of the most spectacular developments witnessed in the domain of quantum degenerate gases have been made possible by the control of interatomic interactions achieved by means of scattering resonances. Altering the two-body scattering length has been instrumental both for the creation of strongly correlated many-body states [1] and for the investigation of few-body physics, most notably for the observation of Efimov resonances [2].

While Fano-Feshbach resonances represent the most widely used method to modify the scattering length between two atoms, more than a decade ago M. Olshanii realized that strong confinement also affects the scattering in a non-trivial manner [3]. The most dramatic prediction was that the effective coupling strength of two identical particles tightly confined along two directions, and therefore moving only in 1D, diverges for one particular value of the transverse-confinement length to the scattering length ratio, ℓ/a [3]. Such a scattering resonance was therein termed a *confinement-induced resonance* (CIR). The work of Olshanii also provided a much needed interpretation of the CIR [4], as a resonance arising from the degeneracy of the two colliding atoms, both in the ground state of the transverse potential, with the bound state in the closed channel formed by the set of excited levels of the transverse potential. The scattering of identical particles moving on a 2D plane also displays non-trivial features [5, 6, 7]:

the coupling strength diverges for a single negative value of the free-space scattering length, $\ell/|a| = (1/\sqrt{2\pi})\ln(1/q^2\ell^2)$, which however depends on the relative momentum q.

Following the above-mentioned works, the modifications of the scattering in reduced dimensions was later confirmed by experimental observations. First, Moritz et al. observed confinement-induced molecules, i.e., a 1D bound state existing even at negative values of the scattering length $a < 0$ for which, in the absence of the external confinement, no weakly bound state exists [8]. The same group also studied the modifications of p-wave interactions in low dimensionality [9]. Haller et al. observed the presence of a CIR and employed it to enhance the interactions of a 1D degenerate gas of Cs atoms [10]. The dramatic increase of atomic interactions afforded by the CIR led to the creation of a strongly correlated Tonks-Girardeau gas of repulsive 1D bosons and of a metastable, strongly-correlated state of attractive bosons, the so-called "super-Tonks gas" [10]. Finally, a CIR between two spin states of ^{6}Li atoms has been observed in a quasi-2D gas confined in a disk-shaped trap [11].

In this work, we describe a novel type of resonances arising between two different particles when their confinement is strongly asymmetric. Specifically, we studied a system of K and Rb atoms, such that the K atoms are confined in an array of 2D disks whereas the Rb atoms move freely in 3D. We will refer to these resonances as *mix-dimensional resonances* (MDR) [12]. Such resonances occur whenever the effective scattering length a_{eff} between the two colliding particles diverges. Although a_{eff} pertains to the two-body collisions, we detect the divergence of a_{eff} by monitoring the entity of three-body inelastic collisions that expel the collisional partners from the trap. Throughout, our working assumption is that, like in free-space, inelastic 3-body collisions are greatly enhanced as $a_{\text{eff}} \to \infty$.

A crucial difference with respect to all the above mentioned systems is that the asymmetric confinement couples the relative motion with the centre-of-mass (COM) motion. Indeed, for two particles subject to equal harmonic confinement, the COM motion is completely decoupled and, as such, can be ignored.

Closely related to our work and appearing soon afterwards, is a recent experiment where the CIR between identical Cs atoms is found to split in two separate resonances as the 2D optical lattice is transformed into a 1D lattice, i.e., the system undergoes a dimensional crossover from 1D tubes to 2D disks [13]. The interpretation of the resonance splitting proposed in [13] is under debate [14].

2. Scattering in mixed dimensions

We apply here the theory of scattering to two particles of unequal masses m_1 and m_2, the first being confined along the z-axis by a strong potential, the second being free. First, we consider the case of harmonic confinement and the interactions are modelled by a delta-like pseudo-potential, similar to the work of Massignan and Castin [15]. We then introduce the effective range correction to the pseudo-potential approximation, since this turns out to be important for our experiment. Finally, we replace the harmonic potential with the 1D lattice: while the effective scattering length is no longer available, we can still identify the values of a that make a_{eff} diverge by using a simple argument relating the energy of closed and open channels. Although an approximation, such an argument holds in the case of harmonic confinement: it yields the position of MDRs in perfect agreement with the formal theory. Thus, we use it for the case of 1D lattice confinement as well and compare the predictions with the observed data.

2.1. Harmonic confinement

The relevant Schrödinger equation is given by

$$\left[-\frac{\hbar^2}{2m_1}\nabla^2_{\boldsymbol{r}_1} - \frac{\hbar^2}{2m_2}\nabla^2_{\boldsymbol{r}_2} + \frac{1}{2}m_1\omega^2 z_1^2 + gV\right]\psi(\boldsymbol{r}_1, \boldsymbol{r}_2) = E\psi(\boldsymbol{r}_1, \boldsymbol{r}_2) \qquad (1)$$

where $g = 2\pi\hbar^2 a/\mu$, a being the s-wave scattering length, μ the reduced mass, and V the pseudo-potential

$$V\psi(\boldsymbol{r}_1, \boldsymbol{r}_2) = \delta(\boldsymbol{r}_1 - \boldsymbol{r}_2)\frac{\partial}{\partial|\boldsymbol{r}_1 - \boldsymbol{r}_2|}(|\boldsymbol{r}_1 - \boldsymbol{r}_2|\psi(\boldsymbol{r}_1 - \boldsymbol{r}_2)) \equiv \chi(\boldsymbol{r})|_{\boldsymbol{r}=\boldsymbol{r}_1=\boldsymbol{r}_2}. \tag{2}$$

Due to the translational symmetry, the COM motion in the xy plane can be eliminated and the Schrödinger equation reduces to

$$\left[-\frac{\hbar^2}{2m_1}\frac{\partial^2}{\partial z_1^2} + \frac{1}{2}m_1\omega^2 z_1^2 - \frac{\hbar^2}{2m_2}\frac{\partial^2}{\partial z_2^2} - \frac{\hbar^2}{2\mu}\nabla_\rho^2 + gV\right]\psi(z_1, z_2, \boldsymbol{\rho}) = E\psi(z_1, z_2, \boldsymbol{\rho}), \tag{3}$$

where $\boldsymbol{\rho} = (x_1 - x_2, y_1 - y_2)$. The general solution of this equation can be written as

$$\psi(z_1, z_2, \boldsymbol{\rho}) = \psi_0(z_1, z_2, \boldsymbol{\rho}) + g\int_{-\infty}^{\infty} dz' G_E(z_1, z_2, \boldsymbol{\rho}; z', z', \boldsymbol{0})\chi(z', \boldsymbol{0}) \tag{4}$$

with ψ_0 the solution of the non-interacting Schrödinger equation ($V = 0$) and G_E the two-particle retarded Green's function:

$$\begin{aligned} G_E(z_1, z_2, \boldsymbol{\rho}; z_1', z_2', \boldsymbol{\rho}') &= \sum_{n=0}^{\infty}\int_{-\infty}^{\infty}\frac{d^2\boldsymbol{k}dk_z}{(2\pi)^3}\frac{\langle z_1, z_2, \boldsymbol{\rho}|\boldsymbol{k}, n, k_z\rangle\langle\boldsymbol{k}, n, k_z|z_1', z_2', \boldsymbol{\rho}'\rangle}{E - \left[\frac{(\hbar\boldsymbol{k})^2}{2\mu} + \left(n + \frac{1}{2}\right)\hbar\omega + \frac{(\hbar k_z)^2}{2m_2}\right] + i0^+} \\ &= -\frac{2\mu}{\hbar^2}\sum_n \frac{e^{-\sqrt{2m_2(n+1/2)\hbar\omega - 2m_2E - i0^+}|\tilde{\boldsymbol{r}} - \tilde{\boldsymbol{r}}'|/\hbar}}{4\pi|\tilde{\boldsymbol{r}} - \tilde{\boldsymbol{r}}'|}\phi_n(z_1)\phi_n^*(z_1'), \end{aligned} \tag{5}$$

where $\phi_n(z)$ are the wave functions of the 1D harmonic oscillator with mass m_1 defined as

$$\phi_n(z) = \left(\frac{1}{2^n n!\sqrt{\pi}\ell}\right)^{1/2} e^{-\frac{z^2}{2\ell^2}} H_n\left(\frac{z}{\ell}\right), \quad \ell = \sqrt{\frac{\hbar}{m_1\omega}} \tag{6}$$

H_n being the Hermite polynomials. We have also defined $\tilde{\boldsymbol{r}} \equiv \left(-\sqrt{\frac{\mu}{m_2}}\boldsymbol{\rho}, z_2\right)$, rescaled and anisotropic coordinates of the free atom relative to the confined atom.

We now consider the low-energy scattering for which $E - \hbar\omega/2 = (\hbar^2 k_z^2)/(2m_2) + (\hbar^2 k_\rho^2)/(2\mu) \ll \hbar\omega$. Then ψ_0 becomes

$$\psi_0(z_1, \tilde{\boldsymbol{r}}) = Ce^{i\boldsymbol{k}\cdot\tilde{\boldsymbol{r}}}\phi_0(z_1), \quad \boldsymbol{k} \equiv (\boldsymbol{k}_\rho, k_z) \tag{7}$$

We can also write the asymptotic form of ψ at a large separation $|\tilde{\boldsymbol{r}}| \gg \ell$ as

$$\psi(z_1, \tilde{\boldsymbol{r}}) \to C\left[e^{i\boldsymbol{k}\cdot\tilde{\boldsymbol{r}}} + \frac{e^{ik\tilde{r}}}{\tilde{r}}f(k_z')\right]\phi_0(z_1), \tag{8}$$

where $f(k_z')$ with $\boldsymbol{k}' \equiv k\hat{\tilde{\boldsymbol{r}}}$ defines the two-body scattering amplitude in the 2D-3D mixed dimensions:

$$f(k_z') \equiv -\frac{1}{C}\int dz' e^{-ik_z'z'}\phi_0^*(z')\chi(z'). \tag{9}$$

We note that χ has an implicit k_z dependence and both χ and f depend on k through the Green's function (5). The unknown function χ can be determined by substituting the solution

Journal of Physics: Conference Series **264** (2011) 012016 doi:10.1088/1742-6596/264/1/012016

(4) into the Bethe-Peierls boundary condition. We isolate the regular part of the Green's G_E and denote it with $\mathcal{G}$:

$$G_E(z_1, z_2, \boldsymbol{\rho}; z', z', \mathbf{0})\big|_{z_1,z_2\to z;\boldsymbol{\rho}\to\mathbf{0}} \equiv -\frac{\mu}{2\pi\hbar^2\sqrt{(z_1-z_2)^2+\rho^2}}\delta(z-z') + \mathcal{G}(z;z'), \tag{10}$$

and obtain

$$\frac{1}{a}\chi(z) = Ce^{ik_z z}\phi_0(z) + \frac{2\pi\hbar^2}{\mu}\int dz'\mathcal{G}(z;z')\chi(z'). \tag{11}$$

This integral equation determines χ/C, which in turn provides f from Eq. (9).

Because the system has reflection symmetry about the z-axis, f, χ, and $\mathcal{G}$ can be decomposed into their even and odd-parity components, i.e., $f_\pm(k'_z) = [f(k'_z) \pm f(-k'_z)]/2$, $\chi_\pm(z) = [\chi(z) \pm \chi(-z)]/2$ and $\mathcal{G}_\pm(z;z') = [\mathcal{G}(z;z') \pm \mathcal{G}(z;-z')]/2$. At the lowest order in k, we only need to consider even components. At finite temperature, odd components might, however, play a role as they feature an independent set of resonances. From the explicit calculation using the Green's function in Eq. (5), we can show that f_+ has the following low-energy expansion:

$$\lim_{k\to 0} f_+ = -\frac{1}{\frac{1}{a_{\text{eff}}} - \frac{1}{2}r_{\text{eff}}k^2 + O(k^4) + ik\left[1+O(k^2)\right]} + O(k_z^2, k_z'^2) \tag{12}$$

where a_{eff} and r_{eff} are effective scattering length and range parameters in the even parity channel.

At zero-energy $k \to 0$, $f_+(0) = -a_{\text{eff}}$. Then from (9) and (11), we obtain the following integral equation

$$\frac{1}{a}\chi(z) = \frac{1}{a_{\text{eff}}}\phi_0(z)\int dz'\phi_0^*(z')\chi(z') + \frac{2\pi\hbar^2}{\mu}\int dz'\mathcal{G}_\pm(z;z')\big|_{k\to 0}\chi_\pm(z') \tag{13}$$

.

3. Secular equation

Now we briefly show how to turn the above integral equation (13) into a linear secular equation. We first evaluate the regular part of the Green's function $\mathcal{G}$ and then expand both $\mathcal{G}$ and χ into an orthonormal set of functions.

At low energy $E - \frac{\hbar\omega}{2} \equiv \frac{\hbar q^2}{2m_2} \le 0$, it is useful to represent G_E as

$$G_E(z_1, z_2, \boldsymbol{\rho}; z_1', z_2', \boldsymbol{\rho}') = -\langle z_1, z_2, \boldsymbol{\rho}|\int_0^\infty d\tau\hbar^{-1}e^{(E-H_0)\tau/\hbar}|z_1', z_1', \boldsymbol{\rho}'\rangle =$$

$$= -\int_0^\infty d\tau\hbar^{-1}e^{\frac{\hbar^2 q^2}{2m_2}\tau}\sqrt{\frac{m_1\omega\, e^{\omega\tau}}{2\pi\hbar\sinh\omega\tau}}\sqrt{\frac{m_2}{2\pi\hbar\tau}}\frac{\mu}{2\pi\hbar\tau}e^{-\frac{m_1\omega}{2\hbar}\frac{(z_1+z_1')^2\cosh\omega\tau - 2z_1z_1'}{\sinh\omega\tau} - \frac{m_2}{2\hbar\tau}(z_2-z_2')^2 - \frac{\mu}{2\hbar\tau}(\boldsymbol{\rho}-\boldsymbol{\rho}')^2}. \tag{14}$$

Therefore

$$\mathcal{G}(z;z') = -\int_0^\infty d\tau\ \hbar^{-1}e^{\frac{\hbar q^2}{2m_2}\tau}\sqrt{\frac{m_1\omega\, e^{\omega\tau}}{2\pi\hbar\sinh\omega\tau}}\sqrt{\frac{m_2}{2\pi\hbar\tau}}\frac{\mu}{2\pi\hbar\tau}e^{-\frac{m_1\omega}{2\hbar}\frac{(z+z')^2\cosh\omega\tau-2zz'}{\sinh\omega\tau} - \frac{m_2}{2\hbar\tau}(z-z')^2}$$
$$+\int_0^\infty d\tau\hbar^{-1}\left(\frac{\mu}{2\pi\hbar\tau}\right)^{3/2}\delta(z-z'). \tag{15}$$

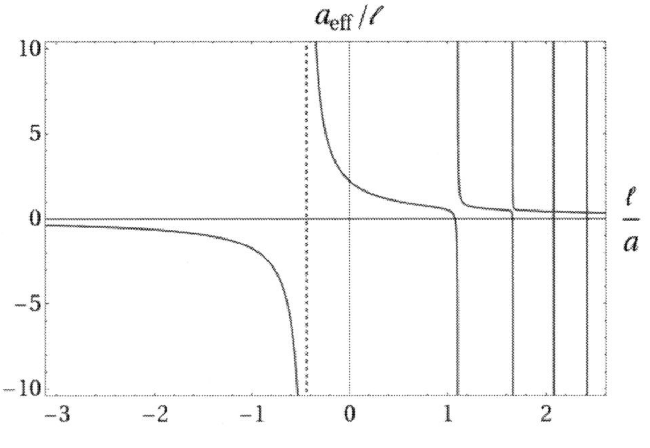

Figure 1. Effective scattering length $a_{\rm eff}/\ell$ versus the inverse free-space scattering length $1/a$, calculated for the mass ratio specific to our experiment $m_1/m_2 = 41/87$. Both $a_{\rm eff}$ and a are given in units of the harmonic oscillator length ℓ (see text).

and

$$\mathcal{G}_+(z;z') = -\int_0^\infty d\tau \hbar^{-1} e^{\frac{\hbar q^2}{2m_2}\tau} \sqrt{\frac{m_1\omega\, e^{\omega\tau}}{2\pi\hbar \sinh\omega\tau}} \sqrt{\frac{m_2}{2\pi\hbar\tau}} \frac{\mu}{2\pi\hbar\tau} \cosh\left[\left(\frac{m_1\omega}{\hbar\sinh\omega\tau} + \frac{m_2}{\hbar\tau}\right) zz'\right] \times$$
$$e^{-\left(\frac{m_1\omega\cosh\omega\tau}{\hbar\sinh\omega\tau} + \frac{m_2}{\hbar\tau}\right)\frac{z^2+z'^2}{2}} \tag{16}$$
$$+\int_0^\infty d\tau\hbar^{-1}\left(\frac{\mu}{2\pi\hbar\tau}\right)^{3/2} \frac{\delta(z-z')+\delta(z+z')}{2}.$$

We now evaluate the matrix elements of $\mathcal{G}_+$ with respect to the orthonormal basis of 1D harmonic oscillator eigenfunctions $\phi_n(z)$. A lengthy but straightforward calculation leads to

$$M_{ij} \equiv \frac{2\pi\hbar^2}{\mu}\int dz dz' \phi_i(z)\mathcal{G}_+(z;z')\big|_{q=0}\phi_j(z')$$
$$= \frac{1}{\ell}\frac{1}{2\sqrt{\pi}}\sqrt{\frac{m_2}{m_1}}\frac{(-1)^{\frac{i}{2}+\frac{j}{2}+1}}{\left(\frac{i}{2}\right)!\left(\frac{j}{2}\right)!}\sqrt{\frac{i!\,j!}{2^{i+j}}}\int_0^\infty \frac{\omega d\tau}{(\omega\tau)^{3/2}}\sqrt{\frac{e^{\omega\tau}}{\sinh\omega\tau}}\left(\frac{1}{X^2-Y^2}\right)^{1/2}$$
$$\times\left(1-\frac{X}{X^2-Y^2}\right)^{\frac{i}{2}+\frac{j}{2}} {}_2F_1\left(-\frac{i}{2},-\frac{j}{2},\frac{1}{2},\left(\frac{Y}{X^2-X-Y^2}\right)^2\right) + \delta_{ij}\frac{1}{\ell}\sqrt{\frac{\mu}{2\pi m_1}}\int_0^\infty \frac{\omega d\tau}{(\omega\tau)^{3/2}} \tag{17}$$

for i, j = even integers, where ${}_2F_1$ is the hypergeometric function, $2X \equiv 1 + m_2/(m_1\omega\tau) + 1/\tanh\omega\tau$ and $2Y \equiv m_2/(m_1\omega\tau) + 1/\sinh\omega\tau$. Expanding also χ_+ in terms of ϕ_n, the integral equation (13) reduces to the following secular equation

$$\frac{1}{a}b_i = \left[\frac{1}{a_{\rm eff}}\delta_{ij}\delta_{i0} + M_{ij}\right] b_j \tag{18}$$

In figure 1 we show the numerical solution of the above secular equation. Inspection of (17) reveals that, once the scattering lengths are expressed in units of the harmonic oscillator length $\ell = \sqrt{\hbar/m_1\omega}$, the solution depends only on the m_1/m_2 mass ratio.

We notice that in addition to one resonance, i.e., divergence of $a_{\rm eff}$ at $a < 0$, there is an infinite series of resonances for positive a values, in stark contrast to the case of CIR, where only one resonance is expected. In the limit of large mass imbalance $m_1 \gg m_2$, it can be shown that the positive solution of the secular equations are given by

$$\hbar\omega\sqrt{\frac{m_1}{m_1+m_2}}(n'+\frac{1}{2}) - \frac{\hbar^2}{2\mu a^2} = \frac{1}{2}\hbar\omega, \tag{19}$$

for $n' = 2, 4, \ldots$. Actually, even for the small mass ratio of our experiment $m_1/m_2 = 41/87$, the above equation (19) is very accurate. Based on this equation, we propose the interpretation of MDR given below.

3.1. *Effective range corrections*

The pseudo-potential approximation is insufficient to accurately describe the binding energy of the closed-channel state in the range experimentally explored. Therefore, following [16] we introduce the *effective range* correction by replacing the inverse scattering length $1/a$ with the energy-dependent quantity

$$\frac{1}{\tilde{a}(E_c)} = \frac{1}{a} - \frac{\mu r_0}{\hbar^2}\left[E + \frac{\hbar^2}{2(m_1+m_2)}\frac{\partial^2}{\partial z^2} - \frac{1}{2}m_1\omega^2 z_1^2\right] \tag{20}$$

where r_0 denotes the effective range of the interaction potential: in our case, $r_0 = 168.37a_0$ (a_0 Bohr radius) is evaluated by fitting the measured binding energy [17] with the formula [16]

$$E_b = \frac{\hbar^2}{\mu r_0^2}\left(1 - \frac{r_0}{a} - \sqrt{1 - 2\frac{r_0}{a}}\right). \tag{21}$$

As a consequence, in the secular equation (18) we must replace $M_{i,j}$ with the following matrix:

$$\tilde{M}_{ij} = M_{i,j} + \frac{\mu r_0}{\hbar^2}\int_{-\infty}^{\infty} dz\phi_i^*(z)\left[\frac{\hbar\omega}{2} + \frac{\hbar^2}{2(m_1+m_2)}\frac{\partial^2}{\partial z^2} - \frac{1}{2}m_1\omega^2 z^2\right]\phi_j(z). \tag{22}$$

3.2. *1D lattice confinement*

The above analysis refers to harmonic confinement; experimentally the K atoms are confined by a 1D lattice, i.e., in an array of 2D disks. In the direction orthogonal to the disks, the confinement can be assumed to be harmonic if the individual sites are isolated from their neighbours. It turns out that this is not the case, for the lattice strengths considered in the experiment: on one hand, the tunneling time of K atoms is of the same order (or shorter) than the experiment duration; on the other hand, the K atoms start from delocalized states before the 1D lattice turns on. For this reason, we modify the above results to include lattice physics, i.e., we replace the harmonic oscillator levels with energy bands. In doing so, we lack a derivation of a_{eff} but we can still invoke the energy degeneracy of open and closed channels to find the values of a that make $a_{\text{eff}} \to \infty$.

In the open channel, the incoming state can be written in terms of mixed coordinates: the coordinates of individual atoms (z_1, z_2) along the direction of the 1D lattice plus the COM and relative coordinates along the orthogonal directions (X, Y, r_x, r_y). The incoming state is a product of plane waves in (X, Y, r_x, r_y), a plane wave of momentum p in z_2 and a 0−th band Bloch wave $\psi_{q,n=0}(z_1)$. In the closed channel, instead, it is more convenient to use COM and relative coordinates along all directions: the states that couple to the open channel are products of the same plane waves of incoming channel in (X, Y), times a n'−th band Bloch wave $\psi_{q+p,n'}(Z)$, times the wavefunction of the s-wave weakly bound molecule $\phi_b(|\boldsymbol{\rho}|)$. For simplicity, we drop the energy of all plane waves and write the energy degeneracy of open and closed channel states as:

$$\epsilon(q, n = 0; V_K) = \epsilon(q + p, n'; V_K) - E_b \tag{23}$$

E_b denotes the binding energy as in (21), while V_K denotes the lattice strength. Notice that the lattice strength for the weakly bound molecules is equal to that of K atoms. However, due to the mass difference, the band structures of K and KRb are completely different.

4. Species-selective confinement

A mix-dimensional configuration might seem unnatural or unrealistic, yet in atomic physics it can be easily realized, as it is extremely simple to confine one atomic species without affecting the other [18]. Such a possibility is granted by optical potentials, whose strength depends on the detuning of the trapping laser from the atomic transitions. In our specific experiment, we use a mixture of two alkali atoms, whose D_1 and D_2 transitions are actually quite close, lying in a wavelength interval ranging from 766.49 nm to 794.76 nm. Therefore the most obvious choice of tuning the trapping dipole laser next to a specific atomic transition is unsuitable, due to the strong spontaneous emission. Instead, we have chosen a wavelength corresponding to the weighted-average of the D_1 and D_2 lines of ^{87}Rb, such that the AC-Stark (light) shifts induced by the D_1 and D_2 transitions cancel out [19].

For a laser detuning much larger that the atomic hyperfine structure splittings, the AC-Stark shift induced by a linearly polarized laser beam of intensity I is given by [20]

$$U = \frac{\pi c^2}{2\omega_0^3}\Gamma I(\frac{2}{\Delta_2} + \frac{1}{\Delta_1}),$$

with $\omega_0 = (\omega_2 + \omega_1)/2$, $\omega_{1,2}$ are the D-lines transition wavelengths with the respective detunings $\Delta_{1,2} = \omega_L - \omega_{1,2}$, while Γ is the transition linewidth. Therefore for $\Delta_2 = -2\Delta_1$ the cancellation occurs. Experimentally, we optimize the cancellation of Rb light shift by minimizing the Raman-Nath diffraction induced by a pulsed 1D optical lattice, $\lambda = 790.02$nm. By this means we find an upper limit to the ratio of K to Rb potentials: $V_{Rb}/V_K < 10^{-2}$.

5. Experiment and results

We start the experiment by loading a crossed dipole trap with a cold mixture of ^{87}Rb and ^{41}K atoms in their lowest hyperfine states, $(F = 1, m_F = 1)$. Evaporation is continued by decreasing the laser beams' intensity to a temperature of the order of 300 nK. At this stage, we set the external, uniform, magnetic field to a constant value of 1.4 mT where the free-space scattering length is positive. We ramp the species-selective standing wave in 50 ms to values ranging from $s = 10$ to 25, in units of K recoil energy, and then ramp linearly the magnetic field to its final value. Thanks to two Feshbach resonances located at 3.8 mT and 7.9 mT, the magnetic field controls the interspecies scattering length [21].

The mixture is held for 65 ms to 100 ms and then we measure the number of atoms after time-of-flight absorption imaging. We scan the final magnetic field to adjust the free-space scattering length, or equivalently the binding energy. During the hold time, the number of trapped atoms decays, by inelastic losses; therefore minima in the final atom number correspond to maxima of the inelastic collision rate, that we identify as divergences of a_{eff}.

In the left panel of figure 2, we show a typical experimental scan where several peaks are visible. The magnetic field is known within 10 μT as it is routinely calibrated by Rb hyperfine spectroscopy. The free-space scattering length is obtained from the magnetic field by an empirical formula previously derived [12]

$$a(B) = 208a_0\left(1 + \frac{3.09}{B + 3.852} - \frac{4.992}{B - 3.837} - \frac{0.164}{B - 7.867}\right). \tag{24}$$

where B is the magnetic field expressed in mT.

From the magnetic field scans, we extract the position of peaks, i.e., atom number minima, at different lattice strengths. We then compare these values with the predictions of the above theoretical analysis, see right panel in figure 2. It is evident that, if we consider the harmonic confinement of isolated lattice sites, the agreement is rather poor, whereas it is satisfactory when we use the lattice energy bands.

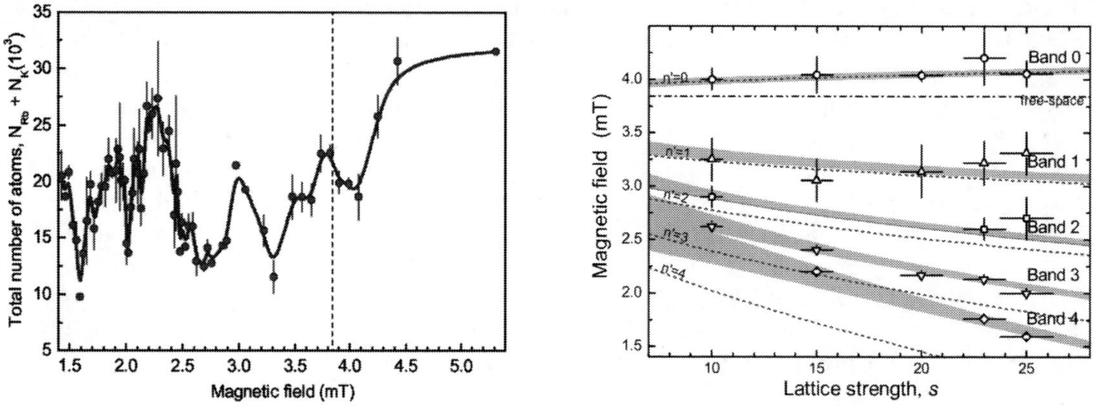

Figure 2. Left: total number of atoms, $N_{\mathrm{Rb}} + N_{\mathrm{K}}$, remaining in the trap after fixed hold time as we scan the external magnetic field, in mix-dimensional configuration with a 1D lattice at $s = 25$. The minima are interpreted as MDR. In 3D we observe a broad asymmetric minimum (not shown) around the resonant field of 3.84 mT (dashed line). Right: Magnetic field values of MDR for different 1D lattice strengths, compared to the theory predictions based on the harmonic oscillator analysis (dashed lines) and the energy band structure (color shaded areas).

6. Intepretation

To understand the origin of MDR, it is useful to first focus on the scattering of 2D identical particles: the confinement induces a single CIR, although the resonance condition depends on the relative momentum q [6]. The COM motion can be ignored since, for harmonic confinement perpendicular to the 2D plane, it decouples from the relative motion.

In the mix-dimensional case, the perpendicular confining potential affects only one particle: $2U = m_1\omega^2 z_1^2 = m_1\omega^2 Z^2 + \mu\omega^2(m_2/M)z^2 - 2\mu\omega^2 Zz$. If we neglect the last term, the relative and COM motion are decoupled and we expect a single CIR at $a = \ell'\sqrt{2\pi}/\log(\pi q^2\ell'^2)$ [6], where $\ell' = \sqrt{\hbar/(\sqrt{m_1\mu}\omega)}$ and $q^2 \simeq 2\mu k_B T/\hbar^2$. For the harmonic frequencies corresponding to $s = 10(25)$ and a relative momentum q corresponding to $T = 0.25\,\mu$K, the CIR would occur at 3.89(4.02) mT, in reasonable agreement with the peak observed at negative scattering lengths for $B > B_0 = 3.84\,$mT.

The $-\mu\omega^2 Zz$ term can be seen as a perturbation coupling the relative and COM motion which, to first order, does not alter the energy levels [15]. Thanks to this term, many COM states of the closed channel are available for the colliding atoms, hence the series of resonances. This interpretation is illustrated in figure 3. The coupling term preserves parity: thus, an initial state with vanishing Rb momentum, $e^{ipz_{Rb}}\phi_0(z_K)|_{p\to 0}$, being even, is coupled only to even $n' = 0, 2, 4\ldots$ COM states (because the molecular wavefunction $\phi_b(|\boldsymbol{r}|)$ is itself even under parity transformations).

However, as explained in Section 3.2, the harmonic oscillator description is not sufficiently accurate when compared to the data: introducing the energy bands, the qualitative interpretation remains unchanged and the quantitative agreement with data is striking. Introducing the Bloch states also allows coupling to odd n' states even for $p_{\mathrm{Rb}} = 0$, since Bloch states are not parity eigenstates. Finally, we remark that the coupling of relative and COM motion and, consequently, multiple resonances arises also for identical particles trapped by anharmonic potentials [22].

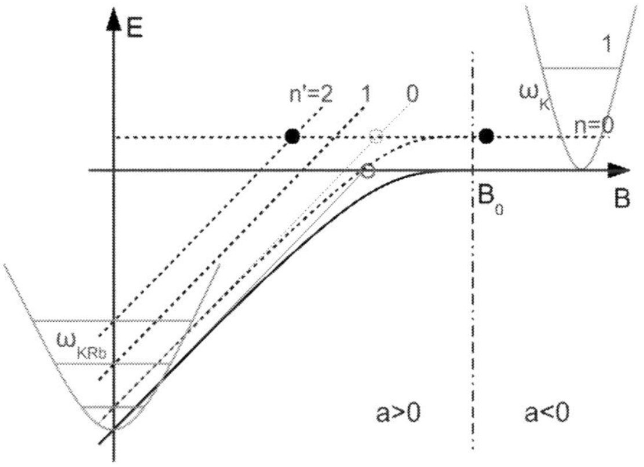

Figure 3. Pictorial representation of the states involved in MDR, in the case of harmonic confinement, see equation (19). The open channel state of colliding atoms (horizontal dashed line) is shifted by the ground state energy of the K atom $\hbar\omega_K/2$ with respect to free-space; the energies of the closed-channel bound states (sloping dashed lines) are separated by quanta of the COM motion $\hbar\omega_{\mathrm{KRb}}(n'+1/2)$, $\omega_{\mathrm{KRb}} = \omega_K\sqrt{m_K/m_{\mathrm{KRb}}}$.

7. Conclusions

We have shown that in a mix-dimensional configuration the two-body scattering is profoundly modified. The coupling of relative and COM motion multiplies the number of possible resonances, differently from the (1D or 2D) CIR of identical particles. We observed a series of resonances as we tuned the free-space scattering length and we explored different confinement strengths. We have also shown that a simple argument, based on the degeneracy between open and closed channel states, allows us to locate the mix-dimensional resonances in good agreement with the formal theory for harmonic confinement. This argument has therefore been extended to consider the energy bands of the 1D optical lattice that generates the mix-dimensional configuration: this step was crucial in order to achieve a good quantitative agreement with the data.

Besides their specific interest, mix-dimensional configurations are predicted to facilitate the observation of the Efimov effect in 3-body collisions [23]. For bosons the energy of Efimov states are closer in mixed dimensions. In the case of our heteronuclear mixture, the energy ratio of successive Efimov states is given by $E_{n+1}/E_n = \lambda^2$ with $\lambda = 131\,(3.48 \times 10^5)$ for the KRbRb (KKRb) trimers in 3D [24]. This ratio decreases to $\lambda = 27.7(59.3)$ in a 2D/3D dimensional configuration, if the heavier Rb atoms are confined. For the fermionic atoms, Efimov effect can occur at lower mass ratios: in 3D, a binary mixture of fermions features Efimov states only for $m_1/m_2 > 13.6$ in 3D, but the critical mass ratio is 6.35 in 2D/3D and 2.07 in 1D/3D, provided the heavier atom is confined [23].

Appendix A. Analogy with Kaluza-Klein modes

A mix-dimensional configuration immediately prompts us to draw an analogy with the brane world, where extra dimensions are postulated to account for the weakness of gravitational interactions. This analogy can be taken a step further as we consider Kaluza-Klein modes.

Suppose we have a massless particle in four spatial dimensions, with the extra fourth direction compactified with the radius R, meaning that the extra direction has length R and is periodic. Because of the periodic boundary conditions, the momentum in the fourth direction is quantized as $p_4 = nh/R$ where n is an integer. The dispersion relation of such a particle is $E = c(p_x^2 + p_y^2 + p_z^2 + p_4^2)^{1/2}$, which makes the discrete p_4 equivalent to a mass: the massless 4D particle appears in the 3D world as an infinite series of particles with masses that are integer multiples of $h/(Rc)$. These are Kaluza-Klein modes.

In our case, because of the confinement, the COM motion of the weakly bound molecule along the z-direction is quantized. We can view such quantized energies as a series of new molecular states whose energy, i.e., mass, is quantized as $E_n = n/(m\ell^2)$ where ℓ is the confinement length.

In the experiment, we adjust the binding energy by controlling the magnetic field and move successive molecular states to threshold, causing the observed resonances. In this sense, our molecular states are analogous to the Kaluza-Klein modes.

References

[1] Bloch I, Dalibard J and Zwerger W 2008 *Rev. Mod. Phys.* **80** 885–964
[2] Kraemer T, Mark M, Waldburger P, Danzl J G, Chin C, Engeser B, Lange A D, Pilch K, Jaakkola A, Nägerl H C and Grimm R 2006 *Nature* **440** 315
[3] Olshanii M 1998 *Phys. Rev. Lett.* **81** 938
[4] Bergeman T, Moore M G and Olshanii M 2003 *Phys. Rev. Lett.* **91** 163201
[5] Petrov D S, Holzmann M and Shlyapnikov G V 2000 *Phys. Rev. Lett.* **84** 2551–2555
[6] Petrov D S and Shlyapnikov G V 2001 *Phys. Rev. A* **64** 012706
[7] Wouters M, Tempere J and Devreese J T 2003 *Phys. Rev. A* **68** 053603
[8] Moritz H, Stöferle T, Guenter K, Köhl M and Esslinger T 2005 *Phys. Rev. Lett.* **94** 210401
[9] Günter K, Stöferle T, Moritz H, Köhl M and Esslinger T 2005 *Phys. Rev. Lett.* **95** 230401
[10] Haller E, Gustavsson M, Mark M J, Danzl J G, Hart R, Pupillo G and Nägerl H C 2009 *Science* **325** 1224
[11] Dyke P, Kuhnle, Hoinka S, Mark M, Lingham M, Rowlands W, Hannaford P and Vale C 2010 Confinement induced resonance in a quasi 2D Fermi gas international Conference of Atomic Physics, ICAP 2010, poster
[12] Lamporesi G, Catani J, Barontini G, Nishida Y, Inguscio M and Minardi F 2010 *Phys. Rev. Lett.* **104** 153202
[13] Haller E, Mark M J, Hart R, Danzl J G, Reichsollner L, Melezhik V, Schmelcher P and Nägerl H C 2010 *Phys. Rev. Lett.* **104** 153203
[14] Peng S G, Bohloul S S, Liu X J, Hu H and Drummond P D 2010 Confinement-induced resonance in quasi-one-dimensional systems under transversely anisotropic confinement (*Preprint* `cond-mat.quant-gas/1005.2794v1`)
[15] Massignan P and Castin Y 2006 *Phys. Rev. A* **74** 013616
[16] Petrov D S 2004 *Phys. Rev. Lett.* **93** 143201
[17] Weber C, Barontini G, Catani J, Thalhammer G, Inguscio M and Minardi F 2008 *Phys. Rev. A* **78** 061601(R)
[18] LeBlanc L J and Thywissen J H 2007 *Phys. Rev. A* **75** 053612
[19] Catani J, Barontini G, Lamporesi G, Rabatti F, Thalhammer G, Minardi F, Stringari S and Inguscio M 2009 *Phys. Rev. Lett.* **103** 140401
[20] Grimm R, Weidemüller M and Ovchinnikov Y 2000 *Adv. At. Mol. Opt. Phys.* **42** 95
[21] Thalhammer G, Barontini G, De Sarlo L, Catani J, Minardi F and Inguscio M 2008 *Phys. Rev. Lett.* **100** 210402
[22] Kestner J P and Duan L M 2010 *New J. Phys.* **12** 053016
[23] Nishida Y and Tan S 2008 *Phys. Rev. Lett.* **101** 170401
[24] Barontini G, Weber C, Rabatti F, Catani J, Thalhammer G, Inguscio M and Minardi F 2009 *Phys. Rev. Lett.* **103** 043201

Journal of Physics: Conference Series **264** (2011) 012017 doi:10.1088/1742-6596/264/1/012017

Towards quantum magnetism with ultracold atoms

David M. Weld and Wolfgang Ketterle
MIT-Harvard Center for Ultracold Atoms, Research Laboratory of Electronics, and Department of Physics, Massachusetts Institute of Technology, Cambridge MA 02139, USA
E-mail: `dweld@mit.edu`

Abstract. At ICAP we presented the efforts and progress at MIT towards using ultracold atoms for the realization of various forms of quantum magnetism. These efforts include a study of fermions with strong repulsive interactions in which we obtained evidence for a phase transition to itinerant ferromagnetism, the characterization of cold atom systems by noise measurements, and a new adiabatic gradient demagnetization cooling scheme which has enabled us to realize temperatures of less than 350 picokelvin and spin temperatures of less than 50 picokelvin in optical lattices. These are the lowest temperatures ever measured in any physical system.

1. Introduction

Over the last 25 years, the field of ultracold atoms has focused on motion: beam slowing, cooling atoms to low temperatures, population of a single motional state (Bose-Einstein condensation, atom lasers), superfluid motion of bosons and fermion pairs, and the suppression of motion in the superfluid-to-Mott-insulator transition. The next challenge, when all motion is frozen out, is ordering of an internal degree of freedom which can be mapped onto spin ordering. A bosonic or fermionic mixture of two or more quantum states can form magnetic phases, such as ferromagnetically or antiferromagnetically ordered states or spin liquids.

A major experimental challenge is to reach the low temperature and entropy scales required to observe these phenomena, and also to develop techniques of preparation and detection of spin-ordered phases. In this paper, we discuss the efforts of our group at MIT towards this goal. This includes a study of a two-component system of fermions with repulsive interactions, for which a phase transition to itinerant ferromagnetism has been predicted, the development of noise measurements to characterize new quantum phases, and a new refrigeration technique which works "natively" in the two-component Mott insulator and is capable of cooling below the critical temperature for spin ordering in an optical lattice.

2. Itinerant ferromagnetism in a gas of ultracold atoms

Ferromagnetism of delocalized (itinerant) fermions occurs due to repulsive interactions and the exchange energy which reduces the interaction energy for spin polarized domains due to the Pauli exclusion principle. At a critical interaction, given by the so-called Stoner criterion [1], the system spontaneously develops domains and becomes ferromagnetic. This, together with a suitable band structure in a periodic lattice, explains why certain metals, like iron and nickel, are ferromagnetic. The simplest models for ferromagnetism assume a gas of fermions with repulsive interactions, and predict, in mean-field and second-order approximation, the onset

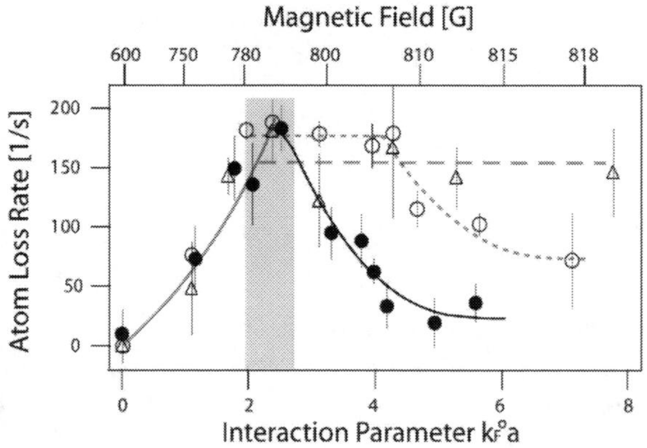

Figure 1. Atom loss rate as a probe for local spin polarization, for different temperatures. (a) $T/T_F = 0.55$ (dashed curve), (b) $T/T_F = 0.22$ (dotted curve), and $T/T_F = 0.12$ (solid curve). The atom loss rate (due to molecule formation) increases for increasing strength of interactions, until the two components of the Fermi gas separate in domains, suppressing the loss. The maximum of the loss rate occurs close to the onset of ferromagnetism. Higher temperatures appear to suppress ferromagnetism.

of ferromagnetism [2–4]. However, there has been no proof or experimental observation for ferromagnetism in a Fermi gas.

We have studied a gas of ultracold fermionic lithium atoms and increased the strength of repulsive interactions by tuning an external magnetic field close to a Feshbach resonance. We observe non-monotonic behaviour of lifetime, kinetic energy and size (see Fig. 1). This provides strong evidence for the Stoner instability, i.e., a phase-transition to a ferromagnetic state [5]. This experiment can be regarded as a quantum simulation of a simple Hamiltonian (the hard core Fermi gas), for which even the existence of a phase transition has not been proven.

In principle, the gas has several options when the strength of interactions is increased. It can form a ferromagnetic state which would have macroscopic domains or it can form a non-magnetic state where the two spin states are anti-correlated (so to speak they form microscopic fluctuating domains). Both states may show similar signature in loss rate and kinetic energy [6]. Recent Monte-Carlo simulations compared the energies of these two scenarios and concluded that for sufficiently strong interactions the ferromagnetic state is favoured [7, 8] The question remains whether this state forms faster than the paired state (which is a strongly interacting molecular condensate) which occupies the lower branch of the Feshbach resonance and has therefore lower energy. The prediction is that both instabilities (the ferromagnetic one and the pairing instability) have similar rates of formation with the pairing instability somewhat faster [9]. More experimental and theoretical work is needed to fully answer the question of whether a gas of fermions with short range interactions can undergo a phase transition to a ferromagnetic state.

In the case of itinerant ferromagnetism, the fermions still occupy different motional states which contribute to the total entropy. Therefore, the total entropy per particle can be higher than $k_B \ln 2$, which is an upper bound for the spin entropy of a magnetically ordered two-component system.

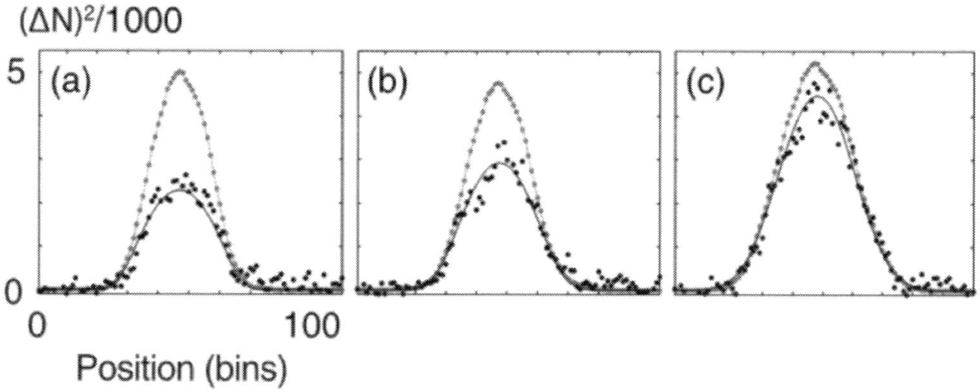

Figure 2. Comparison of observed variances (black dots) with a theoretical model (solid line) and the observed atom number (connected gray dots), at three different temperatures (a, b, and c), showing 50, 40, and 15% suppression of density fluctuations, respectively. Noise thermometry is implemented by fitting the observed fluctuations, resulting in temperatures T/T_F of 0.23, 0.33, and 0.60, in good agreement with temperatures obtained by fitting the shape of the expanded cloud.

3. Study of Density Fluctuations in Fermi gases

One frontier in the field of ultracold atoms is the realization of quantum systems with strong interactions and strong correlations. Many properties of strongly correlated systems cannot be deduced from mean density distributions. This has drawn interest toward novel ways of probing cold atoms, e.g., via RF spectroscopy, Bragg and Raman scattering, interferometric methods and by recording density correlations. Further insight into quantum systems is obtained by looking not only at expectation values, but also at fluctuations.

We have studied density fluctuations and spin fluctuations of Fermi gases. This work is motivated by the connection between density (spin) fluctuations and compressibility (magnetic susceptibility) through the fluctuation-dissipation theorem. Many interesting phases of matter have a clear signature in the compressibility or susceptibility. The crossover to a Mott or band insulator is characterized by the onset of an incompressible phase. At a ferromagnetic phase transition the susceptibility diverges, whereas in a transition to a paired or antiferromagnetic phase the susceptibility becomes exponentially small in the ratio of the pair binding energy (or antiferromagnetic gap) to the temperature.

First, we have validated our technique for determining the compressibility by applying it to the ideal Fermi gas. In recent years, several classic experiments have observed different manifestations of Pauli suppression in Fermi gases. Here we study density profiles of an ideal Fermi gas and observe Pauli suppression of density fluctuations (atom shot noise) for cold clouds deep in the quantum degenerate regime [10, 11].

For temperatures well below degeneracy, we observed strong suppression for probe volumes containing more than 10,000 atoms. Measuring the level of suppression provides noise thermometry which remains sensitive down to low temperatures since the fluctuations scale linearly in temperature (see Fig. 2).

This method was applied to strongly interacting fermions along the BEC-BCS crossover [12]. Spin fluctuations are observed by directly measuring the difference in densities for the two spin states. This was done by using a probe laser which had equal detuning from both states, but with opposite signs.

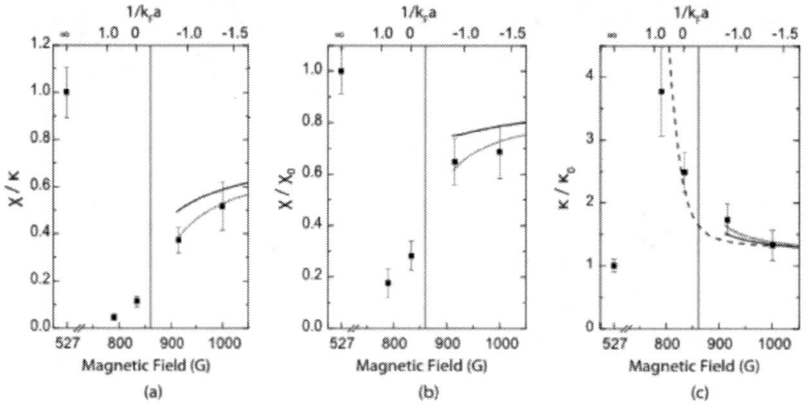

Figure 3. **(a)** The ratio χ/κ, **(b)** the normalized susceptibility χ, and **(c)** the normalized compressibility κ in the BEC-BCS crossover. The variances derived from sequences of images are converted into thermodynamic variables using the measured temperatures and a calibration factor determined from the noninteracting gas. The vertical line indicates the onset region of superfluidity, as determined via condensate fraction measurements. The curves show theoretical zero temperature estimates based on 1st (dotted) and 2nd order (solid) perturbative formulas obtained from Landau's Fermi liquid theory integrated along the line of sight, and results from a Monte Carlo calculation (dashed) for the compressibility in a homogeneous system [15].

The fluctuations in the atomic cloud led to an observed speckle pattern. The atoms imprint a phase shift into the transmitted beam. Upon propagation, this dispersive signal is transformed into an intensity speckle signal. Since the depth of the cloud is larger than the Rayleigh range associated with the resolution of the imaging systems, the transformation from phase signal to intensity signal takes place already within the atomic cloud. The random density fluctuations lead to a speckle pattern corresponding to the imaging resolution.

This new sensitive method easily resolves a tenfold suppression of spin fluctuations below shot noise due to pairing. Compressibility and magnetic susceptibility are determined from the measured fluctuations (see Fig. 3). They reproduce the expected qualitative behaviour: for the sample at unitarity and on the BEC side the spin susceptibility is strongly suppressed relative to the compressibility. This reflects the fact that the atoms form bound molecules or generalized Cooper pairs. The spin susceptibility should be exponentially small in the binding energy, while the enhanced compressibility reflects the bosonic character of the molecular condensate. This new technique is directly applicable to studying pairing and magnetic ordering of two-component gases in optical lattices [13, 14].

4. Spin gradient demagnetization cooling of lattice-trapped ultracold atoms

Magnetic ordering can occur for delocalized (itinerant) particles, as described above, and for localized particles. Many interesting new quantum phases are predicted for ultracold fermions and bosons localized in optical lattices. One reason for the substantial excitement surrounding this topic is the possibility that the physics of high-temperature d-wave superconductivity in the cuprates [16] is nothing else but the physics of a doped antiferromagnetic Hubbard model [17]. A doped antiferromagnetic Hubbard model is realized in a two-component fermionic Mott insulator at sufficiently low entropy and temperature. It looks promising to use the tools and precision of atomic physics to attack this important, long-standing, and open question of the origin of pairing in high-T_c superconductors.

However, the transition temperature to superfluidity for a doped Hubbard model is much lower than the transition for magnetic ordering. Therefore, for the time being, the experimental focus is on the achievement of spin ordering in optical lattices. This goal can be pursued in both fermionic and bosonic systems [18, 19]. In fact, both systems can in principle realize a large class of spin Hamiltonians, and each offers advantages and disadvantages for certain Hamiltonians. For example, bosonic atoms usually favour ferromagnetic phases (unless strongly spin dependent lattices and/or interactions are added), whereas fermionic atoms tend to order antiferromagnetically. All magnetically ordered phases require very low temperatures and entropy. The spin entropy per particle has to be lower than $k_B \ln 2$ for a two-component system. One advantage of bosonic systems is that lower entropies are experimentally accessible. In our group, we pursue quantum simulations of both fermionic and bosonic Hubbard models. In this section we discuss our efforts towards spin ordering in bosonic Mott insulators.

It has been shown that two-component bosonic Mott insulators support nontrivial magnetic phases [18, 19]. Exchange-stabilized correlated phases such as the XY ferromagnetic state or the antiferromagnetic state typically have a Curie or Néel temperature on the order of $1/k_B$ times the exchange energy. In an atomic Mott insulator, this energy is proportional to J^2/U, where J is the tunneling energy and U is the interaction energy. In a typical system (for example, ^{87}Rb in a lattice with a period of 532 nm and a depth sufficient to produce a Mott insulator), the resulting critical temperature for spin ordering is on the order of 200 picokelvin [20, 21]. This temperature is very cold even for the world of ultracold atoms, and in particular is lower than any temperature ever measured in any system. Clearly, new techniques of thermometry and cooling are required in order to make progress in the quest to observe spin-ordering in optical lattices.

Our approach to this problem has been to make use of the unique properties of the two-component Mott insulator to enable new methods of thermometry and cooling. Here we will discuss these unique properties before going on to discuss a recently demonstrated refrigeration technique which depends on them. The two-component Mott insulator supports the same spectrum of particle-hole excitations as the one-component Mott insulator. At sufficiently low temperatures, these excitations are localized at concentric ellipsoidal shells at the boundary between two Mott domains of definite occupation number [21, 22]. This entropy segregation, which has often been referred to as the "icing" on the "wedding cake" of the Mott insulator, has given rise to a number of proposed cooling schemes [23–30], many of which work by differentially addressing and removing the entropy-rich regions of the Mott insulator. Despite the great excitement surrounding the quest for lower temperatures in the Mott insulator, none of these techniques have been realized experimentally. One reason for this may be the difficulty of controlling and addressing the entropy-rich layers. For typical trap frequencies, atom numbers, and temperatures, the width of these layers is often as little as one or two lattice sites, and the energy spectrum of the particle-hole excitations cannot be easily controlled without greatly deforming the trapping potential.

However, the two-component Mott insulator supports an additional spectrum of excitations, more amenable to experimental control, which are not present in its one-component incarnation. These can be thought of as spin excitations. In a finite magnetic field gradient, the two spins will be pulled to opposite sides of the trap, assuming they have different magnetic moments. At zero temperature there will be a zero-width boundary between these two spin-purified regions. The spin excitations consist of atoms with spin a crossing into the spin-b region, and vice versa. The crucial point is that the energy of these excitations is linearly proportional to the applied magnetic field gradient. Since the applied gradient can be easily changed, this gives the experimenter a "handle" on the energy scale of the spin excitations which is not available in the case of particle-hole excitations. It is this "handle" on the energy spectrum of spin excitations which enables the new methods of thermometry and refrigeration that our group has

Journal of Physics: Conference Series **264** (2011) 012017 doi:10.1088/1742-6596/264/1/012017

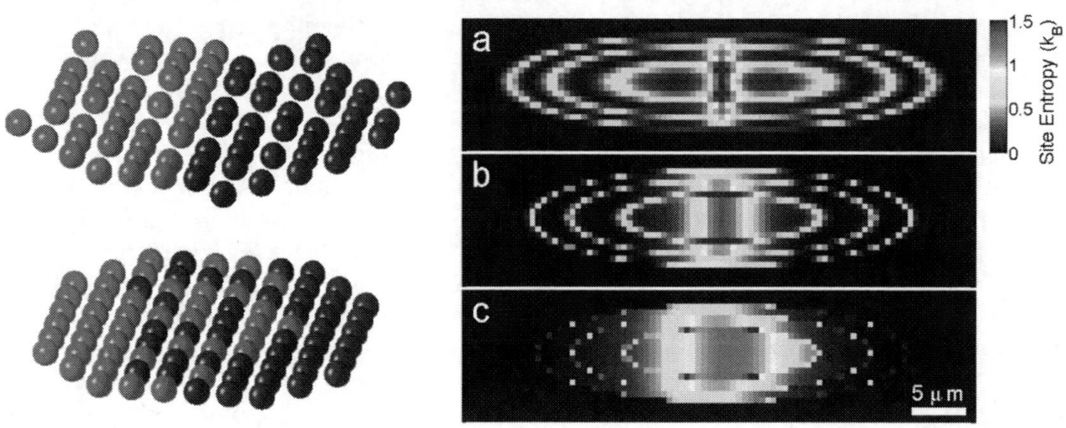

Figure 4. Illustrations of the new cooling technique. **Left:** A cartoon of spin gradient demagnetization cooling. The top image is of a sample with many particle-hole excitations but no spin excitations, and the bottom image is of a sample with no particle-hole excitations but many spin excitations. These are intended to represent the sample before and after reduction of the gradient. **Right:** Calculated effect of spin gradient demagnetization cooling on the entropy distribution. These are the results of the theoretical calculations discussed in Ref. [32]. All images are cuts through the centre of the atomic cloud, and all images are at approximately the same total entropy. The applied gradient is high in **a**, moderate in **b**, and low in **c**. The pumping of entropy from the particle-hole excitations (the rings) to the spin excitations (the stripe in the middle) is clearly visible.

recently developed. For example, at finite temperatures, there will exist a region of mixed spin between the two pure-spin regions. The width of this region will depend on the ratio between the temperature and the applied field gradient. This allows the experimenter to adjust the gradient so that the temperature one wishes to measure will correspond to an easily measurable width of the mixed-spin region. We have used this effect to demonstrate a new kind of thermometry called spin gradient thermometry [31]. The experimental simplicity of this technique is a direct result of the ability to tune the energy of spin excitations using the gradient.

Our group has also recently used the two-component Mott insulator to realize a new method of refrigeration. This technique, called spin gradient demagnetization cooling, works in a lattice and uses the spin excitations discussed above as a refrigerant. It is in some ways locally analogous to adiabatic demagnetization refrigeration in condensed matter systems [33–35]. We believe that it is capable of cooling a two-component Mott insulator of rubidium atoms below the critical temperature for spin ordering.

The new cooling method is applied to an optically trapped cloud of cold atoms in a mixture of two internal states with different magnetic moments. As discussed above, application of a suitable magnetic field gradient will result in partial spatial segregation of the two spin components, with a mixed region in the centre. This region, whose width is proportional to the temperature, comprises a spectrum of soft and easily measurable spin excitations, the energy of which can be tuned by adjusting the strength of the magnetic field gradient. The spin system can be regarded as isolated from all other degrees of freedom when the gradient is changed on a time scale faster than the spin relaxation rate. This regime enables the realization of spin distributions with a very low positive (or, if the sign of the gradient is changed, negative) temperature. Contrastingly, when the gradient is changed adiabatically, the spin system is fully equilibrated. In this regime, reduction of the gradient reduces the energy of the spin degrees of freedom, causing entropy to flow into the mixed-spin region from other degrees of freedom. This entropy redistribution (visible in Fig. 4) lowers the temperature of the whole system [32].

This technique is locally analogous to adiabatic demagnetization refrigeration, but there are important differences between the two techniques. For example, all prior realizations of demagnetization cooling (including one in an ultracold gas of chromium [35]) required spin-flips. Spin flip collisions in atomic gases are usually very slow, except for atoms with high magnetic moments such as chromium. However, spin gradient demagnetization cooling, because it uses a magnetic field gradient instead of a spatially homogeneous field, proceeds via spin transport in a system with fixed magnetization. This makes it well suited to ultracold atomic samples in optical lattices. On a practical level, it is also easier to achieve very small magnetic field gradients than very small magnetic fields. In our system, an energy resolution of $k_B \times 25$ picokelvin can be relatively easily achieved with a gradient of 1 mG/cm and an optical resolution of 5 μm, while similar energy resolution in a homogeneous system would require control of the magnetic field at the microgauss level (assuming a magnetic moment of one Bohr magneton).

We have used this method to prepare isolated spin distributions at positive and negative spin temperatures of ± 50 picokelvin, and to reduce the temperature of an apparently equilibrated sample of rubidium atoms in a Mott insulating state to 350 picokelvin [36]. These are the lowest temperatures ever measured in any system. This new cooling method not only opens up a previously inaccessible temperature regime, but also provides a realistic path to the observation of magnetic quantum phase transitions in optical lattices.

5. Conclusion

The physics of phase transitions in spin systems is extremely rich and has led to many important insights into the nature of phase diagrams. It is safe to predict that this will be an important application for ultracold atoms, probably as important as the previous studies of superfluidity in bosonic and fermionic gases, if it is possible to prepare such systems using the tools of atomic physics. The major obstacle is the realization of practical and efficient cooling schemes which work at picokelvin temperatures in optical lattices. What is needed is a technique which works as well as the magneto-optical trap at microkelvin temperatures or evaporative cooling at nanokelvin temperatures. If the solution is as simple as these techniques, then there is a lot of excitement ahead of us!

References

[1] Stoner E C 1933 *Phil. Mag.* **15** 1018

[2] Duine R A and MacDonald A H 2005 *Phys. Rev. Lett.* **95** 230403

[3] LeBlanc L J, Thywissen J H, Burkov A A and Paramekanti A 2009 *Phys. Rev. A* **80** 013607

[4] Conduit G J, Green A G and Simons B D 2009 *Phys. Rev. Lett.* **103** 207201

[5] Jo G B, Lee Y R, Choi J H, Christensen C A, Kim T H, Thywissen J H, Pritchard D E and Ketterle W 2009 *Science* **325** 1521–1524

[6] Zhai H 2009 *Phys. Rev. A* **80** 051605

[7] Chang S, Randeria M and Trivedi N e-print arXiv:1004.2680

[8] Pilati S, Bertaina G, Giorgini S and Troyer M 2010 *Phys. Rev. Lett.* **105** 030405

[9] Pekker D, Babadi M, Sensarma R, Zinner N, Pollet L, Zwierlein M W and Demler E e-print arXiv:1005.2366

[10] Sanner C, Su E J, Keshet A, Gommers R, Shin Y I, Huang W and Ketterle W 2010 *Phys. Rev. Lett.* **105** 040402

[11] Müller T, Zimmermann B, Meineke J, Brantut J P, Esslinger T and Moritz H 2010 *Phys. Rev. Lett.* **105** 040401

[12] Sanner C, Su E J, Keshet A, Huang W, Gillen J, Gommers R and Ketterle W e-print arXiv:1010.1874

[13] Recati A and Stringari S e-print arXiv:1007.4504

[14] Bruun G M, Andersen B M, Demler E and Sørensen A S 2009 *Phys. Rev. Lett.* **102** 030401

[15] Astrakharchik G E, Boronat J, Casulleras J and Giorgini S 2004 *Phys. Rev. Lett.* **93** 200404

[16] Bednorz J G and Müller K A 1986 *Z. Physik B Condensed Matter* **64** 189–193

[17] Lee P A, Nagaosa N and Wen X G 2006 *Rev. Mod. Phys.* **78** 17–85

[18] Duan L M, Demler E and Lukin M D 2003 *Phys. Rev. Lett.* **91** 090402

[19] Altman E, Hofstetter W, Demler E and Lukin M 2003 *New J. Phys.* **5** 113

[20] Capogrosso-Sansone B, Söyler Ş G, Prokof'ev N V and Svistunov B V 2010 *Phys. Rev. A* **81** 053622

[21] Ho T L and Zhou Q 2007 *Phys. Rev. Lett.* **99** 120404

[22] Gerbier F 2007 *Phys. Rev. Lett.* **99** 120405

[23] Werner F, Parcollet O, Georges A and Hassan S R 2005 *Phys. Rev. Lett.* **95** 056401

[24] Popp M, Garcia-Ripoll J J, Vollbrecht K G and Cirac J I 2006 *Phys. Rev. A* **74** 013622

[25] Capogrosso-Sansone B, Söyler Ş G, Prokof'ev N and Svistunov B 2008 *Phys. Rev. A* **77** 015602

[26] Rabl P, Daley A J, Fedichev P O, Cirac J I and Zoller P 2003 *Phys. Rev. Lett.* **91** 110403

[27] Bernier J S, Kollath C, Georges A, Leo L D, Gerbier F, Salomon C and Köhl M 2009 *Phys. Rev. A* **79** 061601

[28] Ho T L and Zhou Q 2009 *Proc. Natl. Acad. Sci. USA* **106** 6916–6920

[29] Ho T L and Zhou Q e-print arXiv:0911.5506

[30] Catani J, Barontini G, Lamporesi G, Rabatti F, Thalhammer G, Minardi F, Stringari S and Inguscio M 2009 *Phys. Rev. Lett.* **103** 140401

[31] Weld D M, Medley P, Miyake H, Hucul D, Pritchard D E and Ketterle W 2009 *Phys. Rev. Lett.* **103** 245301

[32] Weld D M, Miyake H, Medley P, Pritchard D E and Ketterle W 2010 *Phys. Rev. A* **82** 051603(R)

[33] Giauque W F 1927 *J. Amer. Chem. Soc.* **49** 1864–1870

[34] Debye P 1926 *Ann. Phys.* **386** 1154–1160

[35] Fattori M, Koch T, Goetz S, Griesmaier A, Hensler S, Stuhler J and Pfau T 2006 *Nat. Phys.* **2** 765–768

[36] Medley P, Weld D M, Miyake H, Pritchard D E and Ketterle W e-print arXiv:1006.4674

Journal of Physics: Conference Series **264** (2011) 012018 doi:10.1088/1742-6596/264/1/012018

Inelastic light scattering to probe strongly correlated bosons in optical lattices

Chiara Fort[1,2,3,4], **Nicole Fabbri**[1,4], **Leonardo Fallani**[1,2], **David Clément**[1], **Massimo Inguscio**[1,2,4]

[1]European Laboratory for Nonlinear Spectroscopy (LENS), via Nello Carrara 1, 50019 Sesto Fiorentino (FI) - Italy

[2] Dipartimento di Fisica e Astronomia Università di Firenze, via Sansone 1, 50019 Sesto Fiorentino (FI) - Italy

[3] C.N.I.S.M. UdR di Firenze

[4] INO-CNR

E-mail: fort@lens.unifi.it

Abstract. We have used inelastic light scattering to study correlated phases of an array of one-dimensional interacting Bose gases. In the linear response regime, the observed spectra are proportional to the dynamic structure factor. In particular we have investigated the superfluid to Mott insulator crossover loading the one-dimensional gases in an optical lattice and monitoring the appearance of an energy gap due to finite particle-hole excitation energy. We attribute the low frequency side of the spectra to the presence of some superfluid and normal phase fraction between the Mott insulator regions with different fillings produced in the inhomogeneous systems. In the Mott phase we also investigated excitations to higher excited bands of the optical lattice, the spectra obtained in this case being connected to the single particle spectral function. In one-dimensional systems the effect of thermal fluctuations and interactions is enhanced by the reduced dimensionality showing up in the dynamic structure factor. We measured the dynamic structure factor of an array of one-dimensional bosonic gases pointing out the effect of temperature-induced phase fluctuations in reducing the coherence length of the system.

1. Introduction

In the last few years the connection between atomic physics and condensed matter physics has become stronger and stronger thanks to the possibility of using ultracold quantum gases to study, in a relatively clean and controllable environment, phenomena that were introduced in the field of condensed matter. Seminal works in this context were the first observation of the superfluid to Mott Insulator transition with ultracold bosons [1] and fermions [2, 3] in optical lattices and the realization of Tonks-Girardeau gases [4, 5]. Soon many suggestions arose to use inelastic light scattering as a probe to investigate the dynamical properties of these strongly correlated systems [6, 7]. Inelastic particle scattering is widely used also in the context of condensed matter. For example, the first experimental observation of a condensate in superfluid ^{4}He was obtained by measuring the dynamic structure factor with neutron scattering [8]. In the context of ultracold quantum gases, inelastic scattering experiments are performed illuminating the sample with a pair of far-detuned laser beams with different frequencies (ω and ω') and different wavevectors ($\mathbf{k}$ and $\mathbf{k}'$). The atoms can then perform a two-photon transition absorbing a photon from the first laser beam and reemitting a photon with different frequency and wavevector by stimulated emission induced by the second laser beam. This process allows the transfer of an excitation to the system, with energy $\hbar(\omega - \omega')$ and momentum $\hbar(\mathbf{k} - \mathbf{k}')$ that can be changed by tuning the laser beam parameters. In the linear response regime the energy transferred to the system is proportional to the dynamic structure factor $S(\omega, \mathbf{k})$ [9].

The first experiments on inelastic light scattering (also called Bragg spectroscopy) from quantum gases were reported in [10, 11]. Bragg spectroscopy allowed measurements of the Bogoliubov excitation spectrum of a weakly-interacting Bose-Einstein condensate (BEC) [12] and phase fluctuations in an elongated condensate [13]. More recently it was extended to study correlated systems as a strongly interacting ^{85}Rb BEC across a Feshbach resonance [14] and to monitor the BEC-BCS crossover of a sample of interacting ^{6}Li fermions [15]. Bragg spectroscopy allowed also the investigation of the band structure of superfluid ultracold samples in optical lattices [16, 17, 18]. Our work extends these measurements to intra-band and inter-band transitions in the superfluid to Mott insulator crossover in a two-dimensional (2D) array of one-dimensional (1D) ultracold bosons loaded into an optical lattice. In the Mott insulator regime intra-band excitations provide access to the two-particle spectral function, the Fourier Transform of which is the dynamic structure factor, while the inter-band excitation should give information on the one-particle spectral function. We will present a detailed investigation of the Bragg spectra obtained across the transition [19] discussing open questions. We also present some preliminary results obtained by performing Bragg spectroscopy on an array of strongly correlated 1D gases pointing out the possibility to use this technique as a monitor to evaluate both phase fluctuations and correlations in 1D systems.

2. Bragg spectroscopy of 1D interacting bosons in a lattice

As suggested in [6, 7] Bragg spectroscopy can be used to measure the energy gap in the Mott insulator phase obtained by loading repulsively interacting ultracold bosons in a periodic potential realized by means of optical lattices. By varying the momentum and the energy given by the Bragg beams, one can address also the superfluid components present in an inhomogeneous wedding-cake structured system as realized in the experiments in [20].

In our experiment, we load with an exponential ramp in 140 ms a ^{87}Rb BEC in a 2D optical lattice created by two orthogonal standing waves (λ=830 nm) aligned in the $x-z$ plane. The height of the periodic potentials barriers created by the two optical lattices is 35 E_R (where $E_R = h^2/(2m\lambda^2)$ is the recoil energy acquired by the atom with mass m absorbing one photon from the optical lattice laser with wavelength λ). This creates a 2D array of independent 1D atomic tubes (as shown schematically in Fig. 1a)) longitudinally trapped by a harmonic potential. Along the y direction a third optical lattice is ramped on (Fig. 1b)) with a variable height $V_y = s_y E_R$ (with s_y ranging from 0 to 15) in order to drive the superfluid to Mott insulator transition [21]. After a few milliseconds a pair of Bragg beams (shown in Fig. 1c)) detuned by $\sim$ 200 GHz from the D2 transition of Rb are pulsed on. The direction of propagation of the Bragg beams is fixed and chosen in order to transfer a momentum $\hbar\delta k$ along the atomic tubes of $0.96\hbar k_L$ where $k_L = 2\pi/\lambda$ is the wavenumber of the optical lattice beams. This choice has been made to maximize the excitation strength of the Mott insulator phase which is proportional to $\sin^2(\pi\delta k/(2k_L))$ [7] allowing us to perform the measurements in the linear response regime. Bragg spectroscopy previously used to detect the Mott insulator with ultracold atoms, has been performed at zero momentum transfer in the non-linear regime [21]. After the Bragg pulse, we ramp the height of the three lattices down to 5 E_R where the ground state is superfluid and let the system thermalize for 5 ms [21]. We then suddenly switch off all the confining potentials and take an absorption image of the atomic density distribution after time of flight. The measured density distribution after free expansion shows an interferogram reflecting the coherence of the superfluid phase [1]. The interference pattern consists of a series of peaks separated by a distance proportional to the lattice wavevector k_L and the width of the central peak is proportional to the

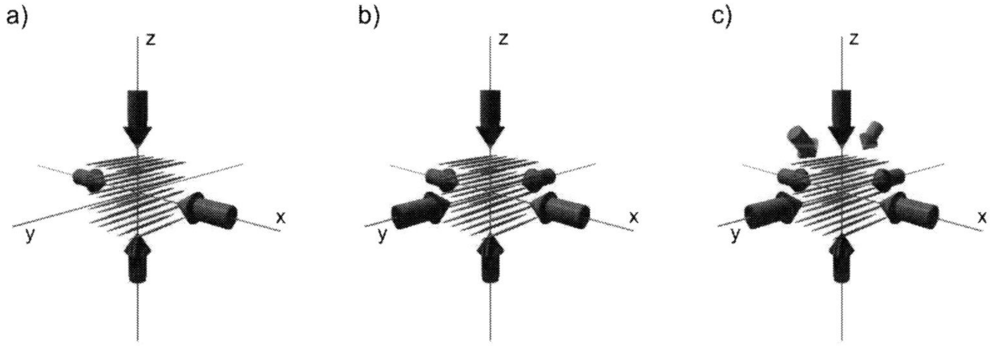

Figure 1. Sketch of the experimental system. a) An array of 1D tubes is produced by loading a BEC into a two-dimensional optical lattice; b) along the atomic tube a third optical lattice drives the superfluid to Mott insulator transition. c) Two extra laser beams allow the creation of excitations through a two-photon Bragg transition.

energy of the atomic sample. For a more detailed description of the experimental procedure the reader is referred to [22]. We measure the width of the interferogram central peak changing the frequency difference $\delta\omega = (\omega - \omega')$ of the Bragg beams and this provides us with an evaluation of the energy ΔE transferred to the system as a function of the frequency of the excitation. As $\Delta E = N\hbar\delta\omega$ (N being the number of excitations) and $N \propto S(\omega,\mathbf{k})$, the energy acquired by the system after the Bragg pulse is $\Delta E \propto \omega S(\omega,\mathbf{k})$ [9].

As the array of 1D samples is loaded into an optical lattice, a band structure arises and at fixed $\mathbf{k}$, by varying the energy difference of the Bragg beams, excitations belonging to different energy bands can be induced [17]. In particular, at lower frequencies, the excitations still belong to the lattice lower energy band (intra-band spectroscopy). In the following we will report first the measurements we have performed in this regime of frequencies, and then extend our Bragg spectroscopy measurements to the higher energy bands (inter-band spectroscopy).

2.1. Spectra across the superfluid to Mott insulator transition (intra-band spectroscopy – two-particle spectral function)

In Fig. 2 we show the energy transferred to the system of an array of 1D atomic tubes as a function of the frequency of the Bragg excitation for different values of the longitudinal lattice height s_y. With increasing s_y the ground-state of the system is expected to go from a superfluid (Fig. 2a) to the inhomogeneous Mott insulator phase (Fig. 2b,c,d) where, due to the variation of the density across our sample, we expect to have alternating regions of Mott insulator with different fillings separated by superfluid or normal regions [20]. In particular, the appearance of the first Mott insulator region (in the outer part of the trapped atomic cloud) is expected to occur at $s_y \simeq 5.5$ [23]. As shown in Fig. 2 both the amplitude of the response and the lineshape of the spectra substantially change when entering the strongly correlated Mott insulator phase. As expected, in the insulator phase the amplitude drops (Fig. 2b and c) and a multiple resonance structure arises with two clear resonances (respectively around 2 kHz and 4 kHz). The first resonance corresponds to the energy gap Δ of the insulator phase due to the finite energy necessary to promote a particle from one lattice site with n bosons to the next one with the same filling n [1]. The second resonance, at twice the frequency, can be interpreted in terms of particle-hole excitations between sites with different fillings (n and $n + 1$) due to inhomogeneity in the sample and the presence of defects across the lattice [24]. Finally, the spectra exhibit a region at low frequencies (<1 kHz) where the system can be excited. These

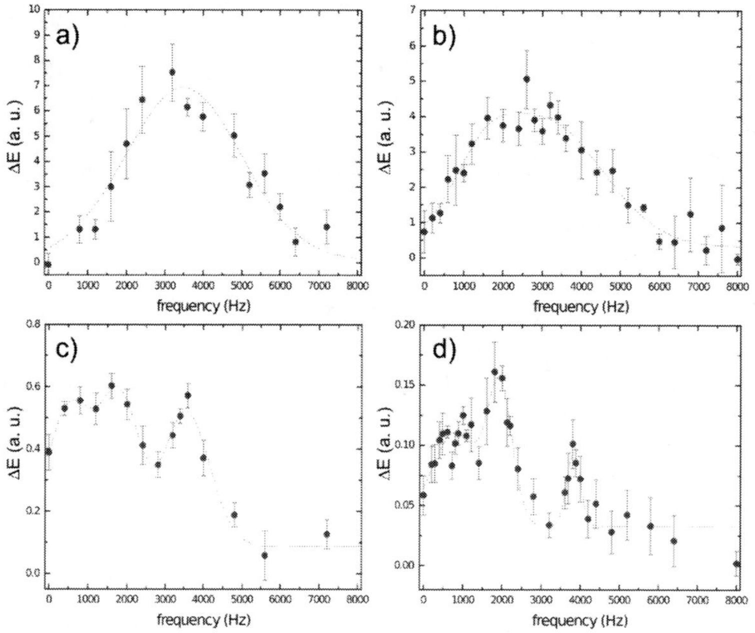

Figure 2. Bragg spectra obtained with four different values of s_y (respectively s_y=4, 7, 10, 13) across the superfluid to Mott insulator transition.

excitations can partly be explained as due to the presence of superfluid layers between Mott insulator regions with different fillings [25, 26]. As detailed in [19], we think that when strongly in the Mott regime (s=13) finite-temperature effects have to be considered in order to interpret the measured excitations in the region 500 Hz $< \omega/(2\pi) <$ 1000 Hz suggesting the possibility to use Bragg spectroscopy to estimate temperature for ultracold atoms loaded in optical lattices. The complex multi-resonance structure observed in Mott insulator phases can thus be used to monitor different components present in an inhomogeneous trapped gas system.

At $s_y < 5$ the array of 1D bosons in the optical lattice is still in the superfluid phase but close enough to the Mott phase that correlations are expected to build up. At $s_y = 4$ (spectrum in Fig.2a) the first band in the Bogoliubov approximation extends up to $\omega/(2\pi) = 3700$ Hz. This value has been calculated by considering a single atomic tube with 300 atoms which we expect to be an upper limit for the central tube in our sample [27]. The measured spectrum clearly shows excitations well above that limit in the gap between the lower band and the first excited band starting at $\omega/(2\pi) = 8100$ Hz. We suggest that the observed excitations in the band gap between the lower energy band and the first excited band in the strongly correlated superfluid phase can be a first experimental evidence of the extra gapped mode predicted in [28, 29].

2.1.1. Inter-band spectroscopy – single-particle spectral function Increasing the frequency of the Bragg excitation, we investigated a region of the spectra where $\hbar\omega > \Delta$ (Δ being the energy gap of the Mott insulator phase). In this case the excitation creates a hole in the correlated Mott insulator phase and populates single particle states in excited bands. The spectra obtained can be related to the single-particle spectral function of the system [30] if the excited single particle state is decoupled from the ground state. In Fig. 3 we show two spectra obtained populating the first and the second excited band for $s_y = 8$ (left panel) and $s_y = 10$ (right panel). The shaded areas in Fig.3 correspond to the extension of the first and second excited bands calculated using the Bogoliubov mean-field approximation for a 1D single atomic sample with 300 atoms. In particular the second excited band spectra (22 kHz< $\omega/(2\pi)$ <40 kHz) show a structured profile and we think that a detailed theoretical analysis would provide information on the dispersion relation of the hole and the density of states in the Mott phase [31].

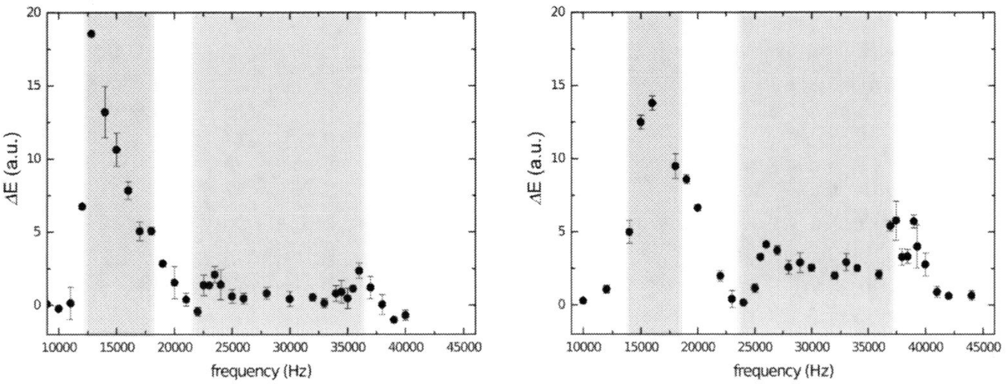

Figure 3. Bragg spectra in the MI phase (s_y = 8 and s_y = 10 for the left and right panel respectively) showing excitations to the first and second excited bands of the periodic system. Dashed areas correspond to the frequency intervals of the first and second excited bands calculated in the Bogoliubov approximation for a single 1D atomic tube with 300 atoms.

3. Bragg spectroscopy of 1D interacting bosons

In this section we present Bragg spectroscopy measurements performed on a BEC loaded into a two dimensional optical lattice (Fig. 1 a)) as a function of the height of the two lattices $s_\perp$. With increasing $s_\perp$ the system crosses from a 3D BEC, reaches an intermediate regime where an array of coupled 1D atomic tubes forms. Finally for $s_\perp > 10$ the tunneling time between the lattice sites becomes longer than the experimental time scale and we obtain an array of decoupled 1D atomic tubes.

The physics of 1D systems is very peculiar due to the fact that the interplay between strong interactions and confinement to a low-dimensional geometry amplifies the effects of quantum fluctuations, thermal fluctuations and correlations giving rise to interesting effects [32]. In a 1D homogeneous system Bose-Einstein condensation is prevented by thermal fluctuations hampering long range order. For trapped gases, the situation changes and three regimes become possible [33], namely a true condensate, a quasicondensate, and a strongly interacting regime of a trapped Tonks-Girardeau gas of impenetrable bosons [34, 35]. The study of the dynamic structure factor through Bragg spectroscopy would allow an investigation of the properties of these systems. The increase in correlations towards the fermionized Tonks-Girardeau regime is expected to strongly modify $S(\omega,\mathbf{k})$. The Lieb-Liniger model of interacting bosons in 1D predicts the presence of two types of elementary excitations [35] instead of one, as predicted by the Bogoliubov theory, and the dynamic structure factor broadens as these two modes arise [36] [37]. This effect becomes important in the strongly interacting limit which is approached when the interaction energy becomes much larger than the kinetic energy. In a 1D system the ratio of the interaction energy and kinetic energy ($\gamma = E_{int}/E_{kin}$) is given by [38]

$$\gamma = \frac{mg}{\hbar^2 n_{1D}} \tag{1}$$

where g is the interaction strength and n_{1D} is the one-dimensional atomic density.

The broadening of S(ω,**k**) for a 1D system can also be related to finite-temperature effects causing phase fluctuations on a typical length scale called "phase coherence length" L_ϕ [39, 13]

$$L_\phi = \frac{\hbar^2 n_{1D}}{m k_B T} \qquad (2)$$

where k_B is the Boltzmann constant and T is the temperature. Phase fluctuations result in a momentum spread

$$\Delta p = \frac{\hbar}{L_\phi} \qquad (3)$$

which contribute to the broadening of the Bragg spectrum.

In Fig. 4 we present three Bragg spectra obtained for an inhomogeneous array of 1D atomic tubes for $s_\perp = 10$, 20 and 45, respectively. As $s_\perp$ increases we observe a shift of the centre of the resonance towards higher frequency and a broadening. An estimation of γ [40] for our

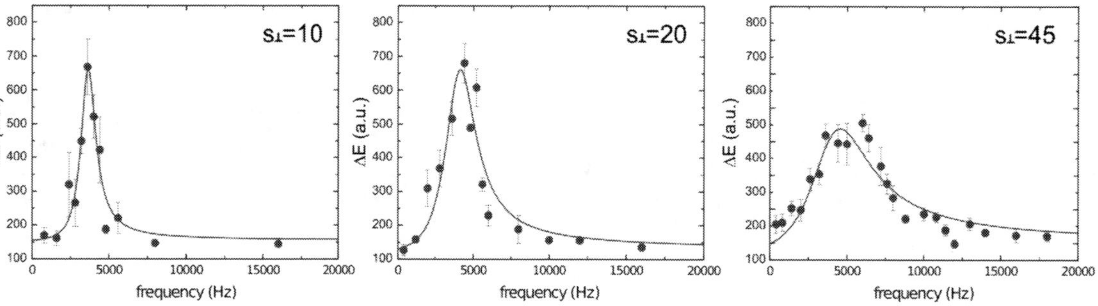

Figure 4. Bragg spectra for an array of 1D bosonic tubes obtained by loading a BEC into a two-dimensional optical lattice at different values of the optical potential barriers $s_\perp = 10$, 20 and 45, respectively. The continuos line is a fit of the data points with the function $\nu L(\nu)$ where $L(\nu)$ is a Lorentzian.

system gives $\gamma \simeq 0.3$ at $s_\perp = 10$ and $\gamma \simeq 0.9$ at $s_\perp = 45$ and thus we expect correlations to play a minor role in the observed spectra. We think that the broadening of the Bragg spectra in the range of parameters we have investigated would thus provide a measurement of the reduced coherence length of our sample caused by the increase in thermal fluctuations. An estimation of the temperature would be possible after a detailed modeling of the system taking into account the fact that we are investigating an inhomogenous array of 1D atomic tubes each of them having a different L_ϕ [41].

4. Conclusions

We have used inelastic light scattering to probe dynamical properties of strongly correlated phases of bosons loaded into optical lattices. In particular we monitored the formation of an energy gap across the superfluid to Mott insulator transition of an array of 1D atomic gases loaded in a longitudinal optical lattice s_y. Signatures of correlations were obtained also in the superfluid phase. In the Mott regime the spectra show different peaks corresponding to different phases present in our inhomogeneous sample. We also induced Bragg excitations to higher excited bands pointing out the possibility to use this method to access the single particle spectral function of the Mott phase. In the last section, we have presented some preliminary results obtained performing Bragg spectroscopy in an array of 1D atomic samples for different transverse confinement and discussed the perspective to use Bragg spectroscopy to investigate the effect of correlations and temperature fluctuations in interacting 1D Bose systems.

5. References

[1] Greiner M, Mandel O, Esslinger T, Hänsch TW and Bloch I 2002 *Nature* **415** 39
[2] Jördens R, Strohmaier N, Günter K, Moritz H and Esslinger T 2008 *Nature* **455** 204
[3] Schneider U, Hackermüller L, Will S, Best Th, Bloch I, Costi T A, Helmes R W, Rasch D and Rosch A 2008 *Science* **322** 1520
[4] Paredes B, Widera A, Murg V, Mandel O, Fölling S, Cirac I, Shlyapnikov G V, Hänsch T W and Bloch I 2004 *Nature* **429** 277
[5] Kinoshita T, Wenger T and Weiss D S 2004 *Science* **305** 1125
[6] van Oosten D, Dickerscheid D B M, Farid B, van der Straten P and Stoof H T C 2005 *Phys. Rev A* **71** 021601
[7] Rey A M, Blakie B, Pupillo G, Williams C J and Clark C W 2005 *Phys. Rev A* **72** 023407
[8] Henshaw D G and Woods A D B 1961 *Phys. Rev.* **121** 1266
[9] Brunello A, Dalfovo F, Pitaevskii L, Stringari S and Zambelli F 2001 *Phys. Rev A* **64** 063614
[10] Kozuma M, Deng L, Hagley E W, Wen J, Lutwak R, Helmerson K, Rolston S L and Phillips W D 1999 *Phys. Rev. Lett* **82** 871
[11] Stenger J, Inouye S, Chikkatur A P, Stamper-Kurn D M, Pritchard D E and Ketterle W 1999 *Phys. Rev. Lett* **82** 4569
[12] Steinhauer J, Ozeri R, Katz N and Davidson N 2002 *Phys. Rev. Lett* **88** 120407
[13] Richard S, Gerbier F, Thywissen J H, Hugbart M, Bouyer P and Aspect A 2003 *Phys. Rev. Lett* **91** 010405
[14] Papp S B, Pino J M, Wild R J, Ronen S, Wieman C E, Jin D S, and Cornell E A 2008 *Phys. Rev. Lett* **101** 135301
[15] Veeravalli G, Kuhnle E, Dyke P, Vale C J 2008 *Phys. Rev. Lett* **101** 250403
[16] Du X, Wan S, Yesilada E, Ryu C and Heinzen D J 2007 *arXiv:0704.2623v2*
[17] Fabbri N, Clément D, Fallani L, Fort C, Modugno M, van der Stam K M R and Inguscio M 2009 *Phys. Rev. A* **79** 043623
[18] Ernst P T, Götze S, Krauser J S, Pyka K, Lühmann D S, Pfannkuche D and Sengstock K 2009 *Nat. Phys.* **5** 1
[19] Clément D, Fabbri N, Fallani L, Fort C and Inguscio M 2009 *Phys. Rev. Lett* **102** 155301
[20] Jaksch D, Bruder C, Cirac J I, Gardiner C W, and Zoller P 1998 *Phys. Rev. Lett.* **81** 3108; Batrouni G G, Rousseau V, Scalettar R T, Rigol M, Muramatsu A and Denteneer P J H 2002 *Phys. Rev. Lett.* **89** 117203; Campbell G K, Mun J, Boyd M, Medley P, Leanhardt A E, Marcassa L G, Pritchard D E and Ketterle W 2006 *Science* **313** 649; Fölling S, Widera A, Müller T, Gerbier F and Bloch I 2006 *Phys. Rev. Lett.* **97** 060403
[21] Stöferle T, Moritz H, Schori C, Köhl M and Esslinger T 2004 *Phys. Rev. Lett.* **92** 130403
[22] Clément D, Fabbri N, Fallani F, Fort C and Inguscio M 2009 *New J. Phys.* **11** 103030
[23] Kühner T D and Monien H 1998 *Phys. Rev. B* **58** R14 742
[24] Kollath C, Iucci A, Giamarchi T, Hofstetter W and Schollwöck U 2006 *Phys. Rev. Lett.* **97** 050402
[25] Batrouni G G, Assaad F F, Scalettar R T and Denteneer P J H 2005 *Phys. Rev. A* **72** 031601R
[26] Pupillo G, Rey A M and Batrouni G G 2006 *Phys. Rev. A* **74** 013601
[27] Michele Modugno, private communication.
[28] Huber S D, Altman E, Büchler H P, and Blatter G 2007 *Phys. Rev. B* **75** 085106
[29] Menotti C and Trivedi N 2008 *Phys. Rev. B* **77** 235120
[30] Clément D, Fabbri N, Fallani L, Fort C, and Inguscio M 2010 *J. Low Temp. Phys.* **158** 5
[31] Work is in progress in collaboration with E. Altman and S. Huber
[32] Giamarchi T Quantum Physics in One Dimension Oxford Science Publications (Oxford, 2004)
[33] Petrov D S, Shlyapnikov G V and Walraven J T M 2000 *Phys. Rev. Lett.* **85** 3745
[34] Girardeau M 1960 *J. Math. Phys.* **1** 516
[35] Lieb E H and Liniger W 1963 *Phys. Rev.* **130** 1605
[36] Caux J S and Calabrese P 2006 *Phys. Rev. A* **74** 031605R
[37] Golovach V N, Minguzzi A and Glazman L I 2009 *Phys. Rev. A* **80** 043611
[38] Dunjko V, Lorent V and Olshanii M 2001*Phys. Rev. Lett.* **86** 5413
[39] Petrov D S, Shlyapnikov G V and Walraven J T M 2000 *Phys. Rev. Lett.* **87** 050404
[40] Note that our system consists of an array of inhomogeneous 1D atomic sample each of them characterized by a different density and thus a different γ. The values of γ given in the text are averaged over the whole array.
[41] Work is in progress to extract temperature estimation from the measured Bragg spectra.

Journal of Physics: Conference Series **264** (2011) 012019 doi:10.1088/1742-6596/264/1/012019

Hybrid quantum systems of atoms and ions

Christoph Zipkes, Lothar Ratschbacher, Stefan Palzer, Carlo Sias, Michael Köhl

Cavendish Laboratory, University of Cambridge, JJ Thomson Avenue, Cambridge CB3 0HE, United Kingdom

Abstract. In recent years, ultracold atoms have emerged as an exceptionally controllable experimental system to investigate fundamental physics, ranging from quantum information science to simulations of condensed matter models. Here we go one step further and explore how cold atoms can be combined with other quantum systems to create new quantum hybrids with tailored properties. Coupling atomic quantum many-body states to an independently controllable single-particle gives access to a wealth of novel physics and to completely new detection and manipulation techniques. We report on recent experiments in which we have for the first time deterministically placed a single ion into an atomic Bose Einstein condensate. A trapped ion, which currently constitutes the most pristine single particle quantum system, can be observed and manipulated at the single particle level. In this single-particle/many-body composite quantum system we show sympathetic cooling of the ion and observe chemical reactions of single particles in situ.

1. Introduction

Hybrid systems allow the exploration of physics far beyond that which can be studied in the individual components. This has been highlighted, for example, by the immersion of distinguishable particles into a quantum liquid which has contributed significantly to our understanding of many-body systems. In liquid helium vortex lattices have been observed using charged particles as markers for the vortex lines [1]. For future investigations with distinct single particles in quantum matter the degree of control achievable over these particles will play a decisive role. To this end, the excellent control over the external and internal degrees of freedom of a single ion trapped in a Paul trap appears highly promising. Immersed into a quantum gas a single trapped ion not just acts as a probe but could be used for local manipulation. Numerous phenomena have been foreseen, including sympathetic cooling [2], the nucleation of localized density fluctuations in a Bose gas [3, 4, 5], scanning probe microscopy with previously unattainable spatial resolution [6, 7], and hybrid atom-ion quantum processors [8].

2. Combining atom trapping and ion trapping

Ion trapping and atom trapping, even though both originating from the desire to control atoms with great experimental precision, have developed very much independently over the past 30 years. Even though there are some common grounds in experimental techniques, such as laser cooling, only little cross-pollination has been seen. Trapped ions, on one hand, have had a tremendous impact on quantum information processing and precision spectroscopy as the best controllable few particle systems. On the other hand, cold atoms and in particular quantum degenerate gases have profoundly stimulated research on quantum matter at ultralow temperatures. The experimental combination of ion trapping and atomic quantum gases has been demonstrated only very recently [9, 10].

In our experimental setup, the ion is confined in the ponderomotive potential (pseudo-potential) of a linear Paul trap which is created from a very rapidly oscillating electric quadrupole field. The quadrupole electric field is generated by four parallel electrodes oriented along the x-direction is $\vec{E}(x,y,z) = \frac{V_{RF}}{R^2}(0,y,-z)\cos(\Omega t)$. The trap operates at an RF frequency of $\Omega = 2\pi \times 42.7\,\mathrm{MHz}$, well above the typical frequencies used in evaporative cooling of the neutral atoms. The distance between the ion and the RF electrodes is $R = 0.4\,\mathrm{mm}$ and the applied voltage of $V_{RF} \approx 170\,\mathrm{V}$ gives rise to a radial confinement of $\omega_{\perp,ion} = 2\pi \times 1.5 \cdot 10^5\,\mathrm{Hz}$. The axial confinement $\omega_{x,ion} = 2\pi \times 4 \cdot 10^4\,\mathrm{Hz}$ is provided by two endcap electrodes charged to a few volts. A single $^{174}\mathrm{Yb}^+$ ion is loaded into the ion trap by isotope-selective Doppler-free two-photon ionization from a neutral ytterbium atomic beam using resonant light at 398 nm and 369.5 nm [11]. We pulse the atomic oven generating the atomic beam for 100 ms only which prevents deteriorating effects of the ion loading procedure on the ultrahigh vacuum ($< 2 \times 10^{-11}$ mbar) in our experimental setup. Subsequently, the ion is laser cooled on the $^2S_{1/2} \to {}^2P_{1/2}$ transition at 369.5 nm (see Figure 1). Repumping light is provided at 935 nm to clear population out of the metastable $^2D_{3/2}$ state. From time to time the ion may fall into the dark $F_{7/2}$ state in which case we discard the ion and load a new one. We continually cancel residual dc electric fields, which would result in excess micromotion of the ion, to better than 0.5 V/m using a photon-correlation fluorescence technique [12].

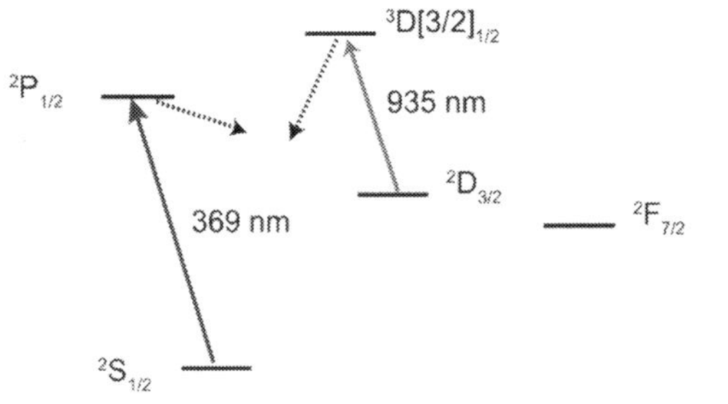

Figure 1. Level structure of singly ionized Ytterbium with the relevant energy levels shown.

Ultracold $^{87}\mathrm{Rb}$ atoms are prepared using standard techniques of laser-cooling and evaporative cooling in a magnetic trap [13], approximately 8 mm away from the ion's position (see Figure 2). After reaching a temperature of $\sim 1\,\mu\mathrm{K}$, which is slightly above the critical temperature for Bose-Einstein condensation, we transport the neutral atomic cloud from its initial position into the ion trap by displacing the potential minimum of the magnetic Ioffe trap using suitably timed changes of the electrical currents in the solenoids. The neutral atoms enter the ion trap through a 700 μm wide bore in the end cap electrode, moving along the symmetry axis of the linear ion trap. The final magnetic trap has trap frequencies of $\omega_x = 2\pi \times (7.8 \pm 0.1)\,\mathrm{Hz}$, $\omega_y = 2\pi \times (29.3 \pm 1.8)\,\mathrm{Hz}$, and $\omega_z = 2\pi \times (29.7 \pm 0.2)\,\mathrm{Hz}$.

The electric field $\vec{E}(\vec{r})$ of the ion trap provides also a potential for the neutral atoms via the quadratic dc-Stark effect (or polarisation potential): $V_{dc}(x,y,z) = -\frac{\alpha}{2}\langle|\vec{E}(\vec{r})|^2\rangle = -\frac{\alpha V_{RF}^2}{4R^4}(y^2+z^2)$. Here α denotes the dc-polarizability of the neutral atoms and $\langle . \rangle$ denotes the time average over one period of the ion trap drive frequency. We verify that the RF-potential of the ion trap is centred in the magnetic trap of the neutral atoms by measuring the electric field induced displacement of the neutral atom cloud in-situ. Since the dc-Stark potential is harmonic, the force on the atoms is linear in position and zero when the atoms are located exactly at the origin of the electric quadrupole. We calibrate the measurement by applying a homogeneous offset magnetic field and monitor the shift of the position of the atoms in situ without the ion trap switched on (see Figure 3). Repeating the same measurement with the ion

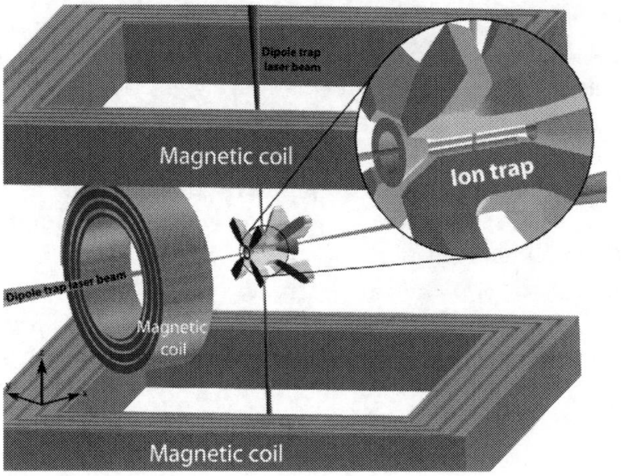

Figure 2. Drawing of the apparatus. Shown are the solenoids providing magnetic confinement to the atoms and the linear Paul trap sandwiched in between. The drawing is to scale and the spacing between the magnetic field coils is 36 mm. The inset shows the ion trap electrodes in detail and the laser beams. Adapted from reference [9].

trap switched on reveals the same zero-position but a different slope due to the dc-Stark effect indicating that the trap centre positions are well matched.

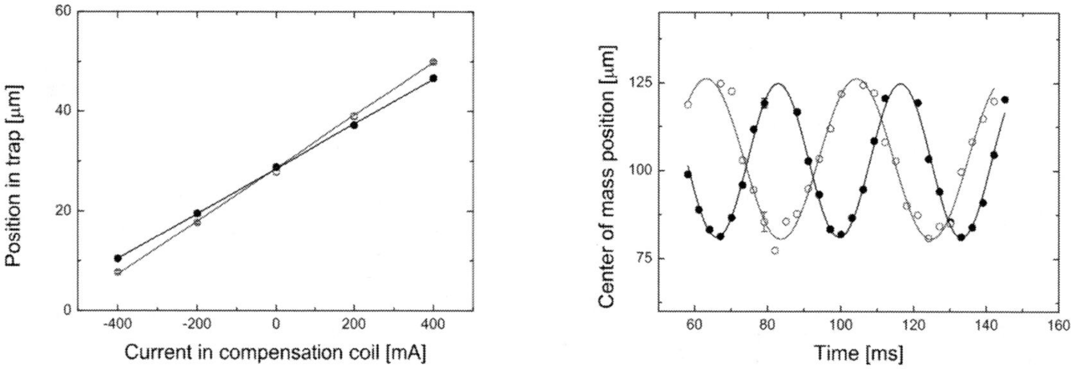

Figure 3. Left: Shift of the atom cloud in situ upon applying a homogeneous magnetic offset field. The filled symbols show the reference measurement in the absence of the ion trap potential. When the ion trap is switched on (open symbols), we observe that the zero position does not shift but only the slope of the displacement changes. This implies that at the equilibrium position the ion trap does not impose an additional force on the atoms and that the traps are well aligned. Right: Dipole oscillations of neutral atoms in the magnetic trap in the absence (filled symbols) and the presence (open symbols) of the ion trapping potential. Data are taken after 25 ms of ballistic expansion.

The dc-Stark effect gives rise to a weak anti-trapping harmonic potential. We have measured the anti-confining effect by observing a frequency shift of the collective dipole mode of the neutral atoms in the magnetic trap in the presence of the electric field (see Figure 3). The measured trapping frequency with the ion trap on is $\tilde{\omega}_z = 2\pi \times (24.4 \pm 0.2)$ Hz. This results in an effective anti-trapping frequency of $\omega_M = 2\pi \times 17$ Hz due to the ion trap.

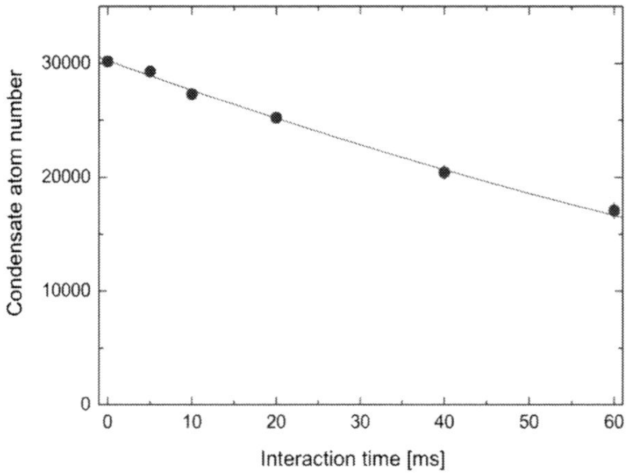

Figure 4. Atom number loss of a Bose-Einstein condensate due to collisions with a single ion. The solid line is a fit using a theoretical model (see [9]) and is used to determine the cross section σ_{al} for neutral atom loss. Each data point is averaged over approximately 40 realizations and the statistical error is given. The bare atom loss rate without the ion present has been subtracted. Data taken from reference [9].

3. Elastic collisions

Atom-ion scattering is dominated by the polarization potential, which asymptotically behaves as $V(r) = -\frac{C_4}{2r^4}$, and a hard-core repulsion at the length scale of the Bohr radius. Here r is the internuclear separation, $C_4 = \frac{\alpha q^2}{(4\pi\epsilon_0)^2}$, q denotes the charge of the ion and ϵ_0 is the vacuum permittivity. Atom-ion scattering was theoretically investigated as early as 1905 when Langevin calculated the drift velocity of ions in a buffer gas. The Langevin model is based on classical mechanics and the predicted cross section is $\sigma_L = \pi\sqrt{2C_4/E}$, leading to an energy independent collision rate constant. Langevin-type collisions occur in close encounters between an atom and ion when the impact parameter is below a critical value $b_c = (2C_4/E)^{1/4}$ [14] and exhibit a large momentum transfer between the atom and ion. Going beyond the Langevin model, the quantum mechanical description of scattering yields the angular dependence of the differential cross section in more detail. In the energy range under consideration, elastic scattering from the polarization potential involves several partial waves since the s-wave scattering regime of atom-ion collisions is only at sub-microkelvin temperatures [15, 16]. The semi-classical approximation of the full quantum mechanical elastic scattering cross section for a collision energy E is $\sigma(E) = \pi(1+\pi^2/16)\left(\frac{\mu C_4^2}{\hbar^2 E}\right)^{1/3}$ [17, 15] (μ denotes the reduced mass). This includes in particular a large contribution towards forward scattering under very small angles [18], mainly from the centrifugal barrier.

To study collisions between the ion and atoms in a Bose-Einstein condensate we first produce a Bose-Einstein condensate inside the ion trap. To this end we load the neutral atoms into an optical dipole trap formed by two crossed laser beams operating at 935 nm wavelength centred onto the ideal position of the ion while the ion is displaced by 140 μm along the x-direction. Each of the beams is focused to a waist of approximately 70 μm and powered by up to a few hundred milliwatts. In the dipole trap we perform evaporative cooling by lowering the laser intensity to reach Bose-Einstein condensates of 3×10^4 atoms. The final trap frequencies of the optical trap are $\omega_{\mathrm{x,opt}} = 2\pi \times 51$ Hz, $\omega_{\mathrm{y,opt}} = 2\pi \times 144$ Hz, $\omega_{\mathrm{z,opt}} = 2\pi \times 135$ Hz. We apply a homogeneous bias magnetic field of less than 1 G to prevent depolarization of the neutral atomic sample.

In order to observe interactions between an ion and a Bose-Einstein condensate we quickly move the pre-cooled ion over a distance of $\Delta x = 140\,\mu$m into the center of the Bose-Einstein condensate. The potential depth of the optical dipole trap is 1 μK which is less than the vibrational level spacing of the ion trap potential. Therefore, every collision will result in a neutral atom loss. We monitor this loss as a function of the interaction time. The data are displayed in Figure 4. From the decay of the atom number we estimate the cross section for atom loss to be $\sigma_{al} = (2.2 \pm 0.2) \times 10^{-13}\,\mathrm{m}^2$. We have time-resolved the dynamics of the immersion cooling process by immersing a hot ion directly into the Bose-Einstein condensate. For this measurement we partially suppress precooling of the ion in the neutral thermal atom cloud by displacing the ion along the diagonal (**x+y**)-direction. As a result, the initial temperature of the ion is approximately 4 K. Then we quickly move the ion into the Bose-Einstein condensate, wait for a variable interaction time, release the neutral atoms, and measure the ion's temperature using a fluorescence method [19, 20] (see Figure 3). The fluorescence method, even though limited in its temperature resolution, is a reliable and independent method in this temperature range. We observe sympathetic cooling of the ion in a Bose-Einstein condensate on a time scale of a few tens of milliseconds. After 60 ms the ion has reached a temperature of $T = (0.6 \pm 0.7)\,\mathrm{K}$, corresponding to a temperature as low as our resolution limit.

4. Charge exchange reactions

Charge exchange is one of the most fundamental reaction processes in atom-ion scattering. Most recently, with the progress into the regime of cold collisions, theory [15, 2, 16, 18] and experiments [21, 22] on charge exchange have regained significant attention. In our case, the Rb $(5s\,^2S_{1/2})$ atom in the hyperfine ground state $|F = 2, m_F = 2\rangle$ and the Yb^+ $(6s\,^2S_{1/2})$ ion in the electronic ground state $|J = 1/2, m_J = \pm 1/2\rangle$ collide in the excited $\mathrm{A}^1\Sigma^+$ singlet or the $\mathrm{a}^3\Sigma^+$ triplet state of the $(\mathrm{RbYb})^+$ molecular potential. The electronic ground state $X^1\Sigma^+$, which asymptotically corresponds to Rb^+ $(4p^6\,^1S_0)$ and neutral Yb $(6s^2\,^1S_0)$, lies 2.08 eV below. The most striking experimental observation of a charge exchange process is the disappearance of the Yb^+ fluorescence which we detect after the interaction with the neutral atoms. We have performed an energy-resolved study of the charge exchange reaction rate by tuning the kinetic energy of the ion [22]. To this end we apply a homogeneous electric offset field E_o to the ion trap and induce excess micromotion of the ion with an energy $\sim q^2 E_o^2/(m_{ion}\omega_\perp^2)$. We have calibrated this energy using the same laser fluorescence technique that we employed to monitor the ion's temperature during sympathetic cooling. For two different isotopes, $^{172}\mathrm{Yb}$ and $^{174}\mathrm{Yb}$, we find the rate constant to be independent of the collision energy, as predicted [23, 14] (see Figure 5). The average rate constants of $R_{172} = (2.8 \pm 0.3) \times 10^{-20}\,\mathrm{m}^3/\mathrm{s}$ and $R_{174} = (4.0 \pm 0.3) \times 10^{-20}\,\mathrm{m}^3/\mathrm{s}$ are five orders of magnitude smaller than in the homonuclear case [21] and of the same order of magnitude as predicted for the similarly asymmetric system $(\mathrm{NaCa})^+$ [2]. Because of an additional systematic error between the measurements for the two isotopes of 15%, due to uncertainty of the density determination of the thermal cloud, the difference of the rates for the two different isotopes is statistically not significant.

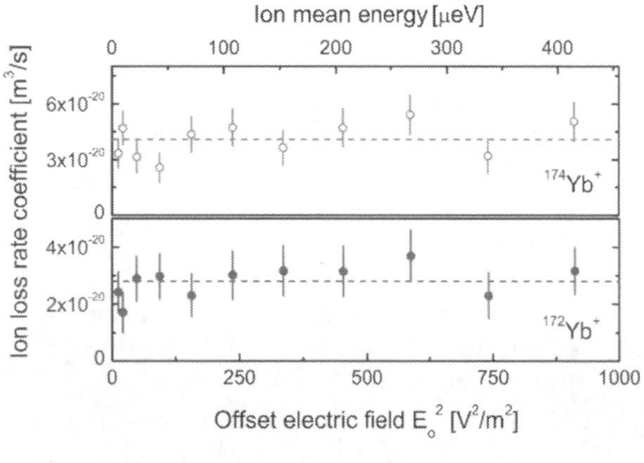

Figure 5. Measured ion loss probability as a function of offset electric field. Each data point is averaged over approximately 100 repetitions of the experiment. The dashed lines show the average of the data. Data taken from reference [22].

Charge exchange can occur by emission of a photon (radiative charge exchange) or as a nonadiabatic transition between molecular levels. For low temperatures, radiative charge exchange has been predicted to be the dominating process for $(NaCa)^+$ [2]. In order to investigate the reaction products of the charge exchange process we perform mass spectrometry. To this end, we load two Yb^+ ions into the ion trap and overlap them with the neutral cloud. In the cases in which only one of the two ions undergoes a reaction, the other ion serves to identify the reaction product by measuring a common vibrational mode in the ion trap [24, 25]. We excite the axial secular motion of Yb^+ by applying intensity modulated light at 370 nm at an angle of 60° relative to the symmetry axis of the Paul trap and monitor the fluorescence. The frequency of the intensity modulation is swept linearly from 30 kHz to 55 kHz in 2.5 s. When the intensity modulation coincides with a collective resonance of the secular motion, the ions are heated and the fluorescence reduces (see Fig. 6). The collective mode for a Rb^+ and a Yb^+ ion in the trap is 13% above the bare Yb^+ mode at 42 kHz, whereas the mode of $(RbYb)^+$ and Yb^+ would be 12% lower in frequency. In the 486 cases in which one Yb^+ ion is lost, we find approximately 30% probability for the production of cold Rb^+ and 70% for a complete loss. If the reaction products take up 2 eV as kinetic energy in a non-radiative charge exchange they would be lost due to the finite depth (~ 150 meV) of our ion trap. We have not observed the formation of a charged molecule in this process.

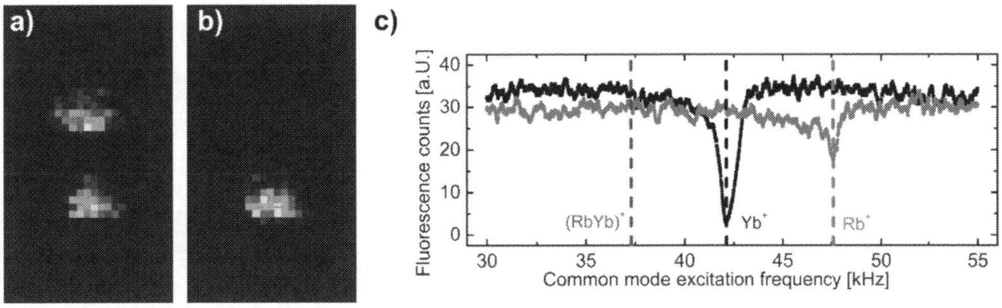

Figure 6. a) Ion crystal of two Yb^+ ions. **b)** A dark ion has been created. **c)** Mass spectrometry signal together with dashed lines indicating the expected frequencies. The black trace corresponds to two Yb^+ ions. For one Rb^+ ion and one Yb^+ ion (red curve) the collective mode frequency is higher. No signals at the resonance for Yb^+ and $(RbYb)^+$ were obtained. Mass spectrometry data taken from reference [22].

5. Conclusion

In conclusion, we have demonstrated the successful realization of a hybrid system of a single trapped ion and a quantum degenerate atomic gas. We have observed elastic collisions between the atoms and the ion leading to sympathetic cooling of the ion in an ultracold neutral environment. Moreover, we have observed charge exchange reactions between a single ion and ultracold atoms and analyzed the reaction products. Our results provide an excellent starting point for future experiments targeting local probing of atomic quantum fluids with a single particle and the full quantum control of chemical reactions at the single particle level.

Acknowledgments

We thank EPSRC (EP/F016379/1, EP/H005676/1), ERC (Grant number 240335), and the Herchel Smith Fund (C.S.) for support.

References

[1] Yarmchuk E J, Gordon M J V and Packard R E 1979 *Phys. Rev. Lett.* **43** 214

[2] Makarov O P, Côté R, Michels H and Smith W W 2003 *Phys. Rev. A* **67** 042705
[3] Côté R, Kharchenko V and Lukin M D 2002 *Phys. Rev. Lett.* **89** 093001
[4] Massignan P, Pethick C J and Smith H 2005 *Phys. Rev. A* **71** 023606
[5] Goold J, Doerk-Bending H, Calarco T and Busch T 2009 *Phys. Rev. A* **81** 041601
[6] Kollath C, Köhl M and Giamarchi T 2007 *Phys. Rev. A* **76** 063602
[7] Sherkunov Y, Muzykantskii B, d'Ambrumenil N and Simons B D 2009 *Phys. Rev. A* **79** 023604
[8] Idziaszek Z, Calarco T and Zoller P 2007 *Phys. Rev. A* **76** 033409
[9] Zipkes C, Palzer S, Sias C and Köhl M 2010 *Nature* **464** 388
[10] Schmid S, Härter A and Hecker-Denschlag J 2010 *Phys. Rev. Lett.* **105** 133202
[11] Balzer C, Braun A, Hannemann T, Paape C, Ettler M, Neuhauser W and Wunderlich C 2006 *Phys. Rev. A* **73** 041407
[12] Berkeland D J, Miller J D, Bergquist J C, Itano W M and Wineland D J 1998 *Journal of Applied Physics* **83** 5025–5033
[13] Palzer S, Zipkes C, Sias C and Köhl M 2009 *Phys. Rev. Lett.* **103** 150601
[14] Vogt E and Wannier G H 1954 *Phys. Rev.* **95** 1190
[15] Côté R and Dalgarno A 2000 *Phys. Rev. A* **62** 012709
[16] Idziaszek Z, Calarco T, Julienne P S and Simoni A 2009 *Phys. Rev. A* **79** 010702
[17] Massey H 1971 *Contemporary Physics* **12** 537
[18] Zhang P, Dalgarno A and Côté R 2009 *Phys. Rev. A* **80** 030703
[19] Wesenberg J H, Epstein R J, Leibfried D, Blakestad R B, Britton J, Home J P, Itano W M, Jost J D, Knill E, Langer C, Ozeri R, Seidelin S and Wineland D J 2007 *Phys. Rev. A* **76** 053416
[20] Epstein R J, Seidelin S, Leibfried D, Wesenberg J H, Bollinger J J, Amini J M, Blakestad R B, Britton J, Home J P, Itano W M, Jost J D, Knill E, Langer C, Ozeri R, Shiga N and Wineland D J 2007 *Phys. Rev. A* **76** 033411
[21] Grier A T, Cetina M, Oručević F and Vuletić V 2009 *Phys. Rev. Lett.* **102** 223201
[22] Zipkes C, Palzer S, Ratschbacher L, Sias C and Köhl M 2010 *Phys. Rev. Lett.* **105** 133201
[23] Langevin P 1905 *Ann. Chim. Phys.* **5** 245
[24] Drewsen M, Mortensen A, Martinussen R, Staanum P and Sørensen J L 2004 *Phys. Rev. Lett.* **93** 243201
[25] Willitsch S, Bell M T, Gingell A D, Procter S R and Softley T P 2008 *Phys. Rev. Lett.* **100** 043203

Quantum simulation of relativistic quantum physics with trapped ions

C. F. Roos[1,2], R. Gerritsma[1,2], G. Kirchmair[1,2], F. Zähringer[1,2], E. Solano[3,4] and R. Blatt[1,2]

[1] Institut für Quantenoptik und Quanteninformation, Österreichische Akademie der Wissenschaften, Otto-Hittmair-Platz 1, A-6020 Innsbruck, Austria

[2] Institut für Experimentalphysik, Universität Innsbruck, Technikerstr. 25, A-6020 Innsbruck, Austria

[3] Departamento de Química Física, Universidad del País Vasco - Euskal Herriko Unibertsitatea, Apartado 644, 48080 Bilbao, Spain

[4] IKERBASQUE, Basque Foundation for Science, Alameda Urquijo 36, 48011 Bilbao, Spain

E-mail: christian.roos@uibk.ac.at

Abstract. Coupling internal and vibrational states of a string of trapped ions has proven to be an effective way of entangling the ions' internal states. This mechanism can be used for high-fidelity quantum gates, QND measurements of spin correlations and creation of large entangled states. However, spin-motion interactions are also of interest for the purpose of quantum simulations where the motional state no longer acts as an auxiliary quantum system only. Here, we describe an experiment where a laser-cooled trapped ion is set to behave as a free relativistic quantum particle.

1. Introduction

Quantum simulation aims at simulating a quantum system of interest with a controllable laboratory system described by the same mathematical model. In this way, it might be possible to simulate quantum systems that can neither be efficiently simulated on a classical computer [1] nor easily accessed experimentally. For this, the laboratory system needs to be very well understood in terms of the Hamiltonian describing it. To turn it into a useful quantum simulator, it should allow for parameter tunability and for the measurement of observables that provide important insights into the physics of the system to be simulated. There are two types of quantum simulators currently discussed in the literature. Digital quantum simulators [2, 3, 4] try to translate the unitaries describing the system dynamics into quantum circuits consisting of elementary gate operations. In this approach, a universal quantum computer could be used for efficiently simulating all quantum systems with local interactions [2]. The second class, that one might call analogue quantum simulators, builds on the principle of engineering a system having exactly the Hamiltonian one is interested in. The main motivation behind these approaches is to find solutions of problems in quantum-many body physics that cannot be efficiently simulated on classical computers. Examples include quantum phase transitions, quantum magnetism or high-temperature superconductivity.

Many proposals and experiments in this area currently focus on neutral atoms trapped in optical lattices as a possible quantum simulator [5]. The parallelism that is naturally occurring in the interactions between atoms filling an optical lattice is a very promising starting point for simulating many-body Hamiltonians [6], and experiments demonstrating, for example, the transition from a Bose-Einstein condensate to a Mott insulator state [7] are exciting steps in this direction.

Laser-manipulated trapped ions provide another quantum optical system for which a number of proposals have been made [5, 8]. While it is more difficult to scale up ion trap systems to large numbers of particles, experiments with trapped ions offer superb possibilities for measuring observables and manipulating the system at the level of individual particles. Trapped ions have the potential of simulating quantum spin systems [9, 10, 11], spin-boson models [12] or strongly correlated phonons [13, 14]. But even systems of just a single or a few ions offer the possibility to simulate other quantum systems. The earliest example is probably the observation of C. Blockley and coworkers [15] that a resonant red-sideband interaction between two internal states of a single trapped ion and a vibrational mode is described by the Jaynes-Cummings-Hamiltonian which originally was meant to describe a two-level atom interacting with a single mode of the radiation field [16]. Other more recent examples include the suggestions to simulate quantum relativistic effects, quantum fields or particle generation [17, 18, 19, 20].

Even though the system dynamics can also be numerically simulated in experiments involving just a single ion, these experiments are nevertheless interesting as a test bed for a larger quantum simulator. For these systems, it is indeed possible to test the concept of a quantum simulator by a comparison with well-known solutions. In ref. [18], the proposal was made that a laser-manipulated trapped ion can be used to simulate the physics of a free Dirac particle. The Dirac equation for a spin-1/2 particle with rest mass m was put forward in 1928 to describe a quantum-relativistic particle. It is given by:

$$i\hbar\frac{\partial\psi}{\partial t} = (c\alpha\cdot\hat{\mathbf{p}} + \beta mc^2)\psi \qquad (1)$$

where c is the speed of light, $\hat{\mathbf{p}}$ is the momentum operator, α and β are the Dirac matrices, and the wave functions ψ are 4-component spinors. It provides a natural description of the electron spin and predicts the existence of antimatter [21]. When extended to the case of a particle moving in a $1/r$ potential, it accurately predicts the spectrum of the hydrogen atom. However, it also predicts peculiar effects like Klein's paradox [22] and *Zitterbewegung* [23], a trembling motion of relativistic particles in the absence of an external potential. For electrons, this motion would have a very small amplitude ($\approx 10^{-13}$ m) and an extremely high frequency ($\approx 10^{21}$ Hz). It arises as an interference effect between the positive and negative energy parts of the spinor and does not appear for spinors that consist entirely of positive (or negative) energy. The real existence of *Zitterbewegung*, in relativistic quantum mechanics and in quantum field theory, has been a recurrent subject of discussion in the last years [24, 25].

Lamata *et al.* suggested [18] that the physics of a free Dirac particle could be simulated using an ion with four internal states and a set of laser beams coupling these states to the three harmonic oscillators describing the ion motion in the trapping potential. Luckily, the simulation of a free Dirac particle in one spatial dimension is much simpler. In one dimension, there is no need for introducing a four-component spinor. Instead, only a two-component spinor appears in the 1+1 version of the Dirac equation

$$i\hbar\frac{\partial\psi}{\partial t} = H_D\,\psi = (c\,\hat{p}\,\sigma_x + mc^2\,\sigma_z)\psi, \qquad (2)$$

which couples only to a single motional degree of freedom. Here, σ_x and σ_z denote the usual Pauli matrices. In order to simulate the physics described by this equation with a trapped ion, the ion needs to be prepared in a desired input state and subjected to a non-relativistic quantum dynamics described by a Schrödinger Hamiltonian that has exactly the form of the

Dirac Hamiltonian H_D appearing in (2). Moreover, techniques need to be devised to in order to extract relevant physical observables such as the mean position of the particle $\langle x \rangle$ as a function of time.

2. Experimental realization

For our experiment [26], we work with a single trapped $^{40}Ca^+$ ion held in a linear trap with axial and transverse vibrational frequencies of about $\omega_{ax} = (2\pi)\,1.36$ MHz and $\omega_{tr} = (2\pi)\,3$ MHz, respectively. The two spinor components needed in Eq. (2) are encoded in the ion's $S_{1/2}$ ground state and the metastable $D_{5/2}$ state. A magnetic field of about 4 G is applied to lift the degeneracy of the Zeeman states. In the experiment described below, we chose to work with the states $S_{1/2}, m = 1/2$ and $D_{5/2}, m = 3/2$.

For a laser-cooled trapped ion, the first term appearing in H_D can be realized by a bichromatic laser field simultaneously coupling to the first red and blue vibrational sidebands of the ion's electronic transition of interest [27]:

$$H_{\mathrm{D}} = \hbar\eta\tilde{\Omega}(\sigma_x \cos\phi_+ - \sigma_y \sin\phi_+) \otimes ((a + a^\dagger)\cos\phi_- + i(a^\dagger - a)\sin\phi_-) \tag{3}$$

The phases $\phi_\pm$ are linear combinations of the phases ϕ_r, ϕ_b of the red- and blue-detuned light fields, $2\phi_\pm = \phi_b \pm \phi_r$, η is the Lamb Dicke parameter and $\tilde{\Omega}$ denotes the Rabi frequency. For the choice of $\phi_+ = 0$ and $\phi_- = \pi/2$, one obtains a spin-motion coupling proportional to $\sigma_x\hat{p}$. The second term of H_D could be realized by adding another light field off-resonantly coupling to an electronic transition in order to introduce a differential Stark shift described by a term proportional to σ_z. However, instead of shifting the ion's energy levels with respect to the frequency of the bichromatic light field creating the spin motion coupling, it is also possible to suitably detune the laser. By going into a reference frame where the atom-light interaction becomes time-independent, we find that the coupling is described by the Hamiltonian

$$H_D = 2\eta\Delta\tilde{\Omega}\sigma_x\hat{p} + \hbar\Omega\sigma_z, \tag{4}$$

where $\Delta = \sqrt{\hbar/(2\tilde{m}\omega_{ax})}$ is the size of the ion's ground state wave function and $\tilde{m}$ denotes the ion's mass. In this way, the Dirac Hamiltonian (2) is realized in our non-relativistic quantum system. In the ion trap system, the operators $\hat{x}$, $\hat{p}$ describing the motional state in (2) correspond to the position and the momentum of the particle in a frame rotating with the particle's oscillation frequency in the trap. By changing the strength of the off-resonant coupling term (or by changing the frequency of the bichromatic light field), the mass of the simulated particle can be varied from zero to finite values.

To determine the motional dynamics induced by the Hamiltonian (2), we need to measure the observables that characterize the ion's position as a function of time. As the motional wave packets are considerably smaller than an optical wavelength, there is no chance to determine the ion position from a simple light scattering experiment. Instead, we need to map the relevant information about the motional state onto an internal state of the ion which is then subsequently read out by a fluorescence measurement providing us with information about $\langle\sigma_z\rangle$. For this, we modify a proposal made in ref. [28] by using again a spin-motion coupling described by (3) prior to the fluorescence measurement. In this case, we make use of a state-dependent displacement operation $U_x = \exp(-ik\hat{x}\sigma_x/2)$ which is realized by applying the Hamiltonian (3) to the quantum state with $\phi_+ = \phi_- = 0$. The operation U_x followed by a measurement of σ_z is equivalent to measuring the observable

$$A(k) = U_p^\dagger\sigma_z U_p = \cos(k\hat{x})\sigma_z + \sin(k\hat{x})\sigma_y. \tag{5}$$

Here, $k = 2\eta\Omega_p t/\Delta$ is proportional to the interaction time t. If the ion's internal state is $|\uparrow\rangle$, the $+1$ eigenstate of σ_z, we have $\langle A(k)\rangle = \langle\cos(k\hat{x})\rangle$. For $|+\rangle_y$, the $+1$ eigenstate of

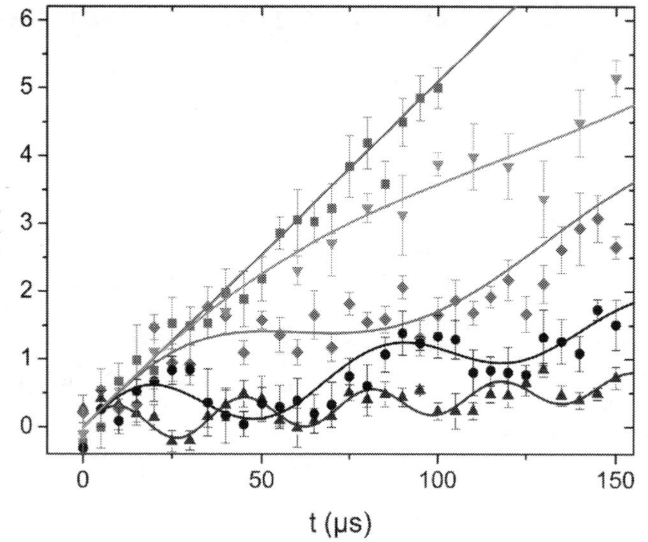

Figure 1. Expectation values $\langle \hat{x}(t) \rangle$ for particles with different masses. The linear curve (■) represents a massless particle ($\Omega = 0$) moving with the speed of light given by $c = 2\eta\tilde{\Omega}\Delta = 0.052\ \Delta/\mu s$ for all curves. The other curves are for particles with increasing mass moving down from the linear curve. Their Compton wavelengths are given by $\lambda_C := 2\eta\tilde{\Omega}\Delta/\Omega = 5.4\Delta$ (▼), 2.5Δ (♦), 1.2Δ (•) and 0.6Δ (▲), respectively. The solid curves represent numerical simulations. The figure shows *Zitterbewegung* for the crossover from the relativistic $2\eta\tilde{\Omega} \gg \Omega$ to the nonrelativistic limit $2\eta\tilde{\Omega} \ll \Omega$.

σ_y, $\langle A(k) \rangle = \langle \sin(k\hat{x}) \rangle$. It is also possible to measure the moments $\langle \hat{x}^k \rangle$, $k = 1, 2, \ldots$, of the distribution by taking derivatives of the expectation values of (5) with respect to k [28, 29]. For example, to measure $\langle \hat{x} \rangle$, we need to prepare $|+\rangle_y$ and calculate $\frac{d}{dk}\langle A(k) \rangle$ by applying U_x for short times (i.e., small values of k) and record the changing excitation. Moreover, a Fourier transformation of $\langle \cos(k\hat{x}) \rangle$ and $\langle \sin(k\hat{x}) \rangle$ yields the probability density $\langle \delta(\hat{x} - x) \rangle$ in position space.

Figure 1 demonstrates that we indeed observe *Zitterbewegung* in the motion of a relativistic particle described by the 1+1 Dirac equation. Here, we prepared the initial state $\psi(x; t = 0) = (\sqrt{2\pi}2\Delta)^{-\frac{1}{2}} e^{-\frac{x^2}{4\Delta^2}} \binom{1}{1}$ by sideband cooling and application of a $\pi/2$-pulse where $\binom{\alpha}{\beta} = \alpha|S_{1/2}, m = 1/2\rangle + \beta|D_{5/2}, m = 3/2\rangle$. The upper curve corresponds to the case of a massless particle ($\Omega = 0$) realized by making the bichromatic light field exactly resonant with the sideband transitions. Only in this case a uniform motion is observed. For a simulated particle with non-zero rest mass, obtained by tuning the centre frequency of the bichromatic laser field away from the atomic transition frequency, an oscillatory motion on top of the linear one is observed. In general, the Hamiltonian H_D entangles the spinor and motional states. For the measurement shown in Fig. 1, the spinor populations were incoherently recombined by optical pumping of the ion's $D_{5/2}$ state to $S_{1/2}, m = 1/2$ so that the ion could be prepared in an eigenstate of σ_y prior to the bichromatic laser pulse applied for measuring the particle's position. The resulting quantum states can be further characterized by reconstructing the spatial probability density $\langle \delta(\hat{x} - x) \rangle$ using again bichromatic light fields for the measurement process. Figure 2 shows another example for an input state with non-zero momentum. Here, the amplitude of the oscillatory motion vanishes after some time because positive and negative energy contributions of the state move in opposite direction and become unable to interfere after a while. The inset

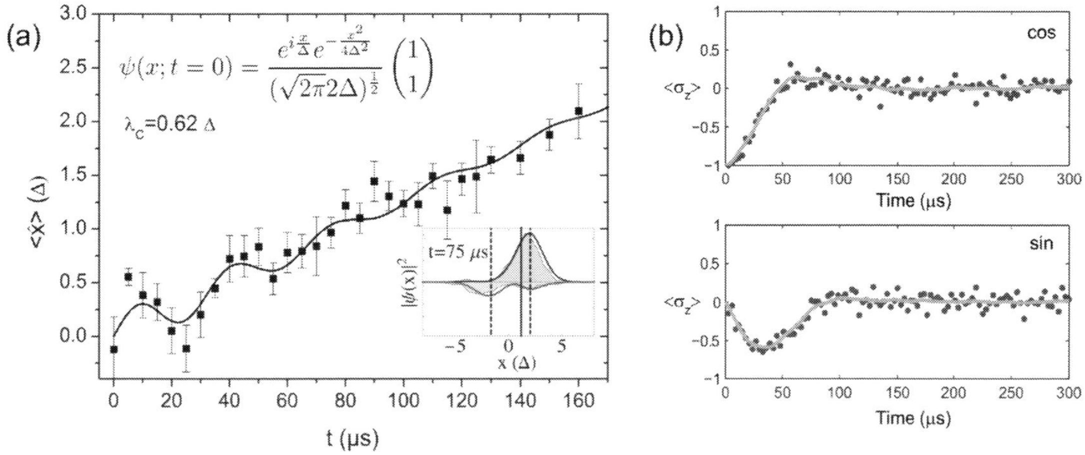

Figure 2. *Zitterbewegung* for a state with non-zero average momentum. (a) Initially, *Zitterbewegung* appears due to interference of positive and negative energy parts of the state. As these parts separate, the oscillatory motion fades away. The inset shows a position space reconstruction of the two spinor states at time $t = 75\ \mu s$. The solid curve represents a numerical simulation. (b) Measured expectation values $\langle \cos k\hat{x} \rangle$ and $\langle \sin k\hat{x} \rangle$ used for reconstructing the probability distribution $|\psi(x)|^2$ of the spinor state plotted above the horizontal line in the inset of (a). The solid line is a fit to the data based on a convex optimization routine.

shows a reconstruction of the probability distribution based on the two spinor states after a time $t = 75\ \mu s$. In order to reconstruct the spinor states individually, we replace the population recombination used in Fig. 1 by a projective measurement. Only if the ion scatters no light because it is projected into the state $D_{5/2}$, we use the subsequent measurement for reconstructing the probability distribution. By adding an additional π pulse to interchange the population of the spinor states, we are able to reconstruct both of them in two separate experiments. On the right-hand side of Fig. 2, the raw data $\langle \cos k\hat{x} \rangle$ and $\langle \sin k\hat{x} \rangle$ used for the reconstruction are shown. Here, $k = 2\eta\Omega_p t/\Delta$ is varied by changing the duration t of the bichromatic measurement pulse with Ω_p its Rabi frequency. To avoid unphysical negative densities, the reconstruction in position space is actually not carried out by a simple Fourier transformation. Instead, we use a convex optimization routine that realizes a least-squares fit to the data with the additional constraint of yielding non-negative densities in position space. The solid lines in Fig. 2 b are fits based on this convex optimization routine.

We consider this experiment to be an important first step that will pave the way towards more complex quantum simulations. Possible extensions include the simulation of relativistic particles in potentials that can be created by adding another ion to the setup [30]. In this extension of the work described here, the second ion interacts with another bichromatic laser field that induces a coupling proportional to $\hat{x}\sigma_x$. Now, the position operator $\hat{x}$ describes a centre-of-mass of the two-ion string. By preparing the second ion in an eigenstate of σ_x, one obtains an additional term in the Hamiltonian that looks like a linear potential for the Dirac particle encoded in the other ion. In this way, other relativistic phenomena like Klein tunneling could be observed in the setup.

Acknowledgments

We gratefully acknowledge support by the Austrian Science Fund (FWF), by the European Commission (Marie-Curie program) and by the Institut für Quanteninformation GmbH. E.S.

acknowledges support of UPV-EHU Grant GIU07/40 and EU project EuroSQIP. This material is based upon work supported in part by IARPA.

References

[1] Feynman R 1982 *Int. J. Theoret. Phys.* **21** 467–488
[2] Lloyd S 1996 *Science* **273** 1073–1078
[3] Abrams D S and Lloyd S 1997 *Phys. Rev. Lett.* **79** 2586–2589
[4] Zalka C 1998 *Proc. Roy. Soc. London A* **454** 313–322
[5] Jane E, Vidal G, Dür W, Zoller P and Cirac J I 2003 *Quantum Info. Comp.* **3** 15–37
[6] Bloch I, Dalibard J and Zwerger W 2008 *Rev. Mod. Phys.* **80** 885–964
[7] Greiner M, Mandel O, Esslinger T, Hänsch T W and Bloch I 2002 *Nature* **415** 39–44
[8] Johanning M, Varón A F and Wunderlich C 2009 *J. Phys. B* **42** 154009
[9] Porras D and Cirac J I 2004 *Phys. Rev. Lett.* **92** 207901
[10] Schmied R, Roscilde T, Murg V, Porras D and Cirac J I 2008 *New. J. Phys.* **10** 045017
[11] Friedenauer H, Schmitz H, Glueckert J, Porras D and Schaetz T 2008 *Nat. Phys.* **4** 757–761
[12] Porras D, Marquardt F, von Delft J and Cirac J I 2008 *Phys. Rev. A* **78** 010101
[13] Porras D and Cirac J I 2004 *Phys. Rev. Lett.* **93** 263602
[14] Deng X L, Porras D and Cirac J I 2008 *Phys. Rev. A* **77** 033403
[15] Blockley C A, Walls D F and Risken H 1992 *Europhys. Lett.* **17** 509–514
[16] Shore B W and Knight P L 1993 *J. Mod. Opt* **40** 1195–1238
[17] Alsing P M, Dowling J P and Milburn G J 2005 *Phys. Rev. Lett.* **94** 220401
[18] Lamata L, León J, Schätz T and Solano E 2007 *Phys. Rev. Lett.* **98** 253005
[19] Bermudez A, Martin-Delgado M A and Solano E 2007 *Phys. Rev. A* **76** 041801(R)
[20] Schützhold R, Uhlmann M, Petersen L, Schmitz H, Friedenauer A and Schätz T 2007 *Phys. Rev. Lett.* **99** 201301
[21] Anderson C D 1933 *Phys. Rev.* **43** 491–494
[22] Klein O 1929 *Z. Phys.* **53** 157–165
[23] Schrödinger E 1930 *Sitz. Preuss. Akad. Wiss. Phys.-Math. Kl.* **24** 418–428
[24] Krekora P, Su Q and Grobe R 2004 *Phys. Rev. Lett.* **93** 043004
[25] Wang Z Y and Xiong C D 2008 *Phys. Rev. A* **77** 045402
[26] Gerritsma R, Kirchmair G, Zähringer F, Solano E, Blatt R and Roos C F 2010 *Nature* **463** 68
[27] Zähringer F, Kirchmair G, Gerritsma R, Solano E, Blatt R and Roos C F 2010 *Phys. Rev. Lett.* **104** 100503
[28] Lougovski P, Walther H and Solano E 2006 *Eur. Phys. J. D* **38** 423–426
[29] Santos M F, Giedke G and Solano E 2007 *Phys. Rev. Lett.* **98** 020401
[30] Casanova J, Garcia-Ripoll J J, Gerritsma R, Roos C F and Solano E 2010 *Phys. Rev. A* **82** 020101(R)

Journal of Physics: Conference Series **264** (2011) 012021 doi:10.1088/1742-6596/264/1/012021

Coherently Walking, Rocking and Blinding Single Neutral Atoms

Artur Widera, Wolfgang Alt, and Dieter Meschede

Institut für Angewandte Physik, Universität Bonn, Wegelerstr. 8, 53115 Bonn, Germany

E-mail: widera@uni-bonn.de

Abstract. Advances in the preparation and detection, but most importantly in the coherent manipulation of single neutral atoms have allowed the observation of intriguing phenomena of quantum physics in recent years. We discuss developments to prepare and detect single neutral atoms in a one-dimensional optical lattice potential with single site resolution. Moreover, using two different experimental techniques, a state-dependent optical lattice potential on the one hand and a high-finesse optical cavity on the other hand, we have obtained coherent control over single neutral atoms. The former has enabled us to observe the quantum walk of atoms in position space, and to coherently control the motion of trapped atoms via microwave radiation. The latter offers a means to non-destructively detect the atomic spin state, thereby revealing quantum jumps of single atoms, or the altered optical properties of single atoms when subject to electromagnetically-induced transparency.

1. Introduction

Single neutral atoms have been, for a long time, envisioned to be candidates for a controlled assembly of interesting quantum states. Coherent control of internal and external atomic degrees of freedom lies at the heart of proposed quantum engineering procedures such as entanglement generation, quantum cellular automata, or single photon optical switching. Here we report on the implementation of two different tools for coherent control, a state dependent optical lattice and a high finesse optical resonator. Each system realises a text book example of quantum physics but at the same time yields novel insight into the feasibility and challenges for quantum state engineering with neutral atoms.

1.1. Preparation and Detection of Atoms with Single Site Resolution in an Optical Lattice

We capture and cool single and few neutral caesium (Cs) atoms in a high-gradient magneto-optical trap and subsequently transfer the atoms into a far detuned, standing wave optical dipole trap (1D optical lattice) at $\lambda = 866\,\text{nm}$, see Fig. 1(a). By means of optical pumping we prepare the atoms in the $|F = 4, m_F = +4\rangle$ state, where F is the total angular momentum and m_F its projection onto the quantisation axis along the lattice axis. This state is coherently coupled to $|F = 3, m_F = +3\rangle$ by microwave radiation around 9.2 GHz, with which we obtain bare Rabi frequencies of up to $\Omega_0/2\pi = 60\,\text{kHz}$. State selective detection is performed by first applying a resonant laser pulse which removes selectively atoms in $F = 4$. Then, standard fluorescence imaging is used to image the configuration of the remaining trapped atoms onto an EMCCD-camera. Numerical image processing finally allows us to determine the number and position of the atoms with nearest neighbour resolution [1].

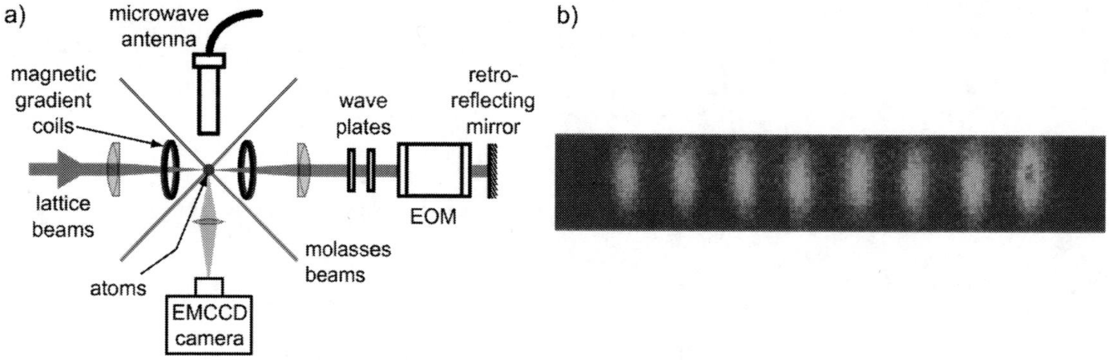

Figure 1. (a) Schematic experimental setup. (b) String of eight atoms with a distance of 17 lattice sites ($\approx 7\,\mu$m). The pattern has been prepared by magnetic resonance methods [2]. From the fluorescence images, the position of the atoms can be deduced with single site resolution [1], verifying the targeted inter-atomic distance. The picture is the sum of approximately 50 single fluorescence images.

We use this method to precisely prepare arbitrary patterns of atoms in our 1D lattice by magnetic resonance methods. The lattice is initially filled with many atoms ($10^2 \dots 10^3$). A magnetic field gradient along the lattice axis of 120 G/cm spatially detunes the microwave transition between the $|F = 3, m_F = +3\rangle$ and the $|F = 4, m_F = +4\rangle$ Zeeman substates by 30 kHz/μm. Application of a microwave pulse maps the frequency spectrum onto the spatial population distribution of atoms in $F = 3$. After the atoms in $F = 4$ having been removed from the trap, the spatial occupation distribution reflects the original spectrum of the pulse. For sufficiently narrow Gaussian pulses, single sites of the lattice can be addressed, and thus arbitrary strings of atoms can be designed by a proper choice of pulse frequencies, see Fig.1(b). For details see Ref. [2].

2. Atoms in Spin-Dependent Potentials

State-dependent potentials have been proposed originally in order to induce controlled coherent collisions between different trapped atoms [3], and they have been exploited for large scale entanglement in such systems [4]. A state-dependent optical lattice can be visualized as two initially overlapped lattices, where each lattice forms the trap for one of the internal atomic states. If the lattices are shifted with respect to each other, atoms in different internal states can be transported in opposite directions. This entangles the spin degree of freedom with the atomic position, which allows for instance the creation of spatially delocalized quantum states or the control of the atomic motion by microwaves.

2.1. Experimental Realisation

It can be shown [3] that at our lattice wavelength of $\lambda = 866$ nm the two internal states $|F = 4, m_F = +4\rangle$ and $|F = 3, m_F = +3\rangle$ couple to a good approximation to only the left- or right-handed circularly polarized components of the light field. Experimentally, the one-dimensional lattice is formed by retro-reflecting a linearly polarized Gaussian laser beam. Its plane of polarization can be rotated by an angle ϑ by a combination of an electro-optical modulator (EOM) and passive wave plates, see Fig.1(a). If the lin-ϑ-lin light field is decomposed into a circular basis, two standing waves emerge which are spatially separated by a distance $\Delta x = \vartheta\,\lambda/(2\pi)$, each trapping atoms in one of the internal states.

The angle ϑ is typically set to zero for the MOT-cooling and spin polarisation phases. It can

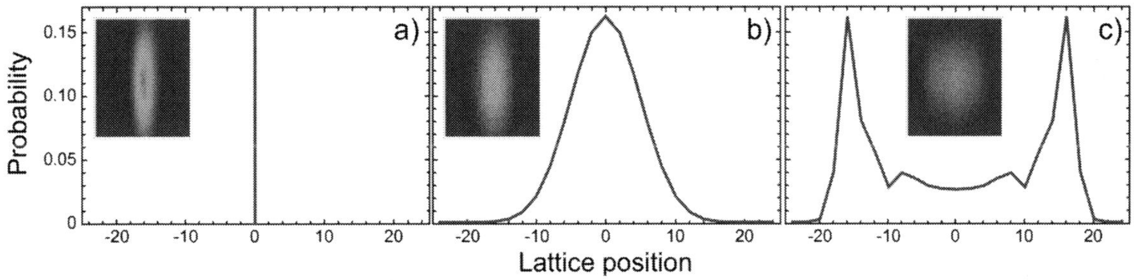

Figure 2. Calculated theoretical probability distributions for (a) a localized atom, (b) a random walk, and (c) a quantum walk. The insets show the corresponding atomic wavefunction after experimentally performing a random and a quantum walk with N=24 steps in (b) and (c), respectively. The images are direct superpositions of more than 500 unprocessed individual fluorescence images, thus blurred by our optical resolution of about $2\,\mu$m . The faster spreading due to ballistic expansion of the quantum walk can be clearly seen in comparison to the diffusive spread of the classical random walk.

be continuously rotated in the range between $[0, \pi]$ by applying a voltage to the EOM. Typical times of voltage ramps are on the order of $10\,\mu$s, implying that such ramps can in principle excite higher vibrational levels [5]. For our usual experiments this is avoided by proper choice of the ramp parameters.

2.2. Quantum Walks

The classical random walk is a well known model throughout science, describing a large variety of random processes such as Brownian motion. In a simple case, a particle is walking along a one-dimensional, discrete line. In every time step a coin is tossed, and depending on the random outcome, heads or tails, the particle moves one step to the right or to the left. After N steps, the probability to find the walker at a certain distance d away from the initial position follows the well-known binomial distribution, see Fig. 2(b). In particular, the width of this probability distribution scales proportional to $\sqrt{N}$.

For a quantum particle, two differences strongly alter this probability distribution: First, a quantum particle can be in a coherent superposition of moving to the left *and* to the right instead of going left *or* right. Thus its wave packet becomes coherently delocalized over several sites of the line. If this procedure is continued, different partial wavepackets of the particle are re-combined at a common site after having gone different paths. Here, matter wave interference occurs, which is the second difference to the classical case. The resulting multi-path matter wave interference strongly alters the probability distribution to find the particle at a certain distance from its initial position, see Fig. 2(c). In particular, the width of this distribution scales linearly with the number of steps N. This scaling lies at the heart of proposals to exploit this quantum walk phenomenon for quantum search algorithms.

In our experiment, the spin dependent lattice together with coherent microwave radiation to couple the two internal states provides the necessary tool to induce such a quantum walk: An atom is trapped in the lattice and its position is determined by fluorescence imaging. Then, a microwave $\pi/2$-pulse creates a coherent superposition of the two internal states. Subsequently, the two states are transported into opposite directions by a shift of the spin dependent lattice. This procedure defines one "step" of the walk, which can be repeated for a given number of steps N. To detect the distribution of walked distances, fluorescence images of the final positions of many walks are superposed (corrected for the initial positions), see Fig. 2(a-c). If, after every step, the internal state coherence in the system is intentionally destroyed by suitable waiting time intervals, the classical random walk is recovered. Details can be found in Ref. [6], including the characterization of the quantum walk state by local quantum state tomography, or the coherent reversal of the entire walk by application of a time-inverted sequence. Future applications include single atom interferometry, the generation of atom-atom entanglement by controlled coherent collisions and interacting quantum walks or quantum cellular automata.

2.3. Coherent Motional Dynamics

Beyond the ability to create coherent spatially delocalized states, the spin dependent lattice allows tight coherent control of the atomic motion. Traditionally, the quantized motional state of trapped atoms is manipulated by laser light-induced motional sideband transitions. There, the light frequency (photon energy) allows selecting the change of vibrational quanta, and the photon momentum determines the strength of the transition, characterized by the Lamb-Dicke factor. With the microwave fields applied in our experiment, the corresponding photon momentum is roughly five orders of magnitude smaller than in the optical case, and the resulting sideband transition strength is negligible.

The spin dependent lattice, however, offers an alternative way to enable sideband transitions, as different vibrational states in two mutually displaced potentials are no longer orthogonal. Formally this can be cast into the form of an effective Lamb-Dicke parameter, which in our case is not fixed by the choice of photon wavelength, but can be adjusted via the lattice displacement Δx, see Fig. 3(a).

Experimentally, we deduce information about the motional state of the atoms by microwave spectroscopy. While in the case where no lattice shift is applied ($\vartheta = \Delta x = 0$) only the carrier transition can be observed, sidebands emerge if this lattice shift is non-zero, see Fig. 3(b). Based on these sidebands, we have developed a cooling scheme using microwave radiation without the need of phase-locked laser sources, which yields up to 97% ground state population, see Fig. 3(c). Moreover, our method allows us to directly and coherently access all vibrational states in our trap by a single microwave pulse, and to map out the wave function overlap of our trapped atoms with nanometer resolution. For details see Ref. [7]. These techniques may be extended to optical lattices of other wavelengths and to three dimensions, or may be applied to increase the available Hilbert space in quantum information processing by storing an additional qubit in the vibrational states.

3. Atoms in a High-Finesse Optical Cavity

Single photons interacting with single atoms form a paradigm of quantum mechanics, where on the one hand quantum effects can be studied in a relatively simple system, and on the other hand novel applications for, e.g., quantum communication can be envisioned.

We realise this paradigm experimentally by inserting single neutral atoms into a high-finesse optical cavity as shown in Fig. 4(a). For details of the experimental setup, see Ref. [8]. In short, the relevant parameters for the coupled system are the coherent atom-field coupling strength g, the photon decay rate κ, and the atomic decay rate γ, which in our case amount to $\{g, \kappa, \gamma\}/(2\pi) = \{12, 0.4, 2.6\}$ MHz, where for g the maximal possible coupling strength for our system has been given.

3.1. Quantum Jumps

An intriguing prospect is the application of the cavity to measure the atomic state without destroying it. We approach such a quantum non-demolition measurement by exploiting the normal mode splitting due to the strong coupling of the atom to the cavity field in only one hyperfine state. We tune the probe laser field into resonance with the cavity, and both are slightly blue detuned from the $F = 4 \rightarrow F' = 5$ transition on the Cs D2-line. While for an empty cavity the resonant probe laser beam would be fully transmitted, the strong coupling

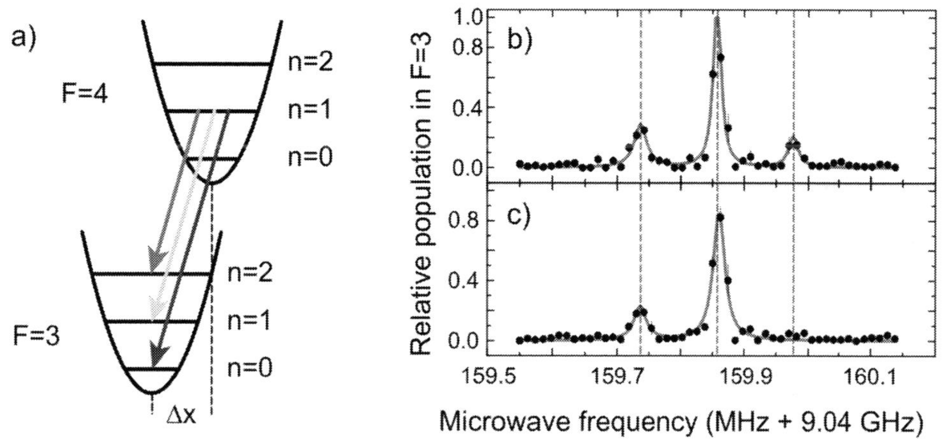

Figure 3. (a) If the trapping potentials of the initial and final states are spatially displaced by Δx, the vibrational quantum state can be changed in microwave transitions, even though the photon momentum is negligible. Microwave spectra in a displaced state dependent optical lattice ($\Delta x = 8\,\mathrm{nm}$) for (b) molasses cooled atoms, and (c) after microwave cooling on the blue sideband. In the latter case, the absence of the blue sideband indicates an axial ground state population of approximately 98%. In our case, the role of red and blue sidebands are exchanged with respect to the typical case of sideband cooling, because we start from the energetically higher $F = 4$ state.

of a single atom in the $F = 4$ hyperfine state leads to the formation of new, frequency-shifted eigenstates according to the well known Jaynes-Cummings model. Hence, the probe laser beam is rather reflected and the transmission level becomes low. If the atom is in the $F = 3$ hyperfine state, however, it is sufficiently far detuned so that interaction with the cavity field can be neglected and effectively the empty cavity transmission is recovered, i.e., the transmission level of the probe laser beam is high. This method provides a fast means to determine the hyperfine state non-destructively, i.e., neither losing the atom, nor, to good approximation, changing its state by more than a quantum-mechanical projection.

Our quantum non-demolition measurement is not perfect, however, because the atom can be transferred from the $F = 4$ to the $F = 3$ state by off-resonant excitation through the cavity field to the $F' = 4$ state and subsequent decay to $F = 3$. By adding a weak additional repumping laser to bring the atom back to $F = 4$, the probe laser transmission resembles a random telegraph signal as depicted in Fig. 4, reflecting the quantum jumps of the atomic hyperfine state between the strongly coupled $F = 4$ and the uncoupled $F = 3$ state.

An interesting situation arises for two atoms in the resonator. If both atoms have been initially prepared in the same hyperfine state and one atom undergoes a quantum jump, the final state resembles a quantum correlated state, provided no information about the respective state of any atom present can be obtained. It turns out, however, that the two atom telegraph signals do not have sufficient signal-to-noise ratio in order to identify the transmission level where only one atom is in either spin state. Instead of the simple threshold analysis given above, we have analysed the data by a Bayesian analysis using independently obtained information about the system to estimate a conditional probability of the atom to be in a certain spin state at a specific time, given the measured probe laser transmission at this time and the estimated probability at the previous time. This allows us to reconstruct the spin state of the two-atom system with improved fidelity, for details see Ref. [9]. An interesting extension of this method will be a real time analysis by a fast digital signal processor, which opens the possibility for

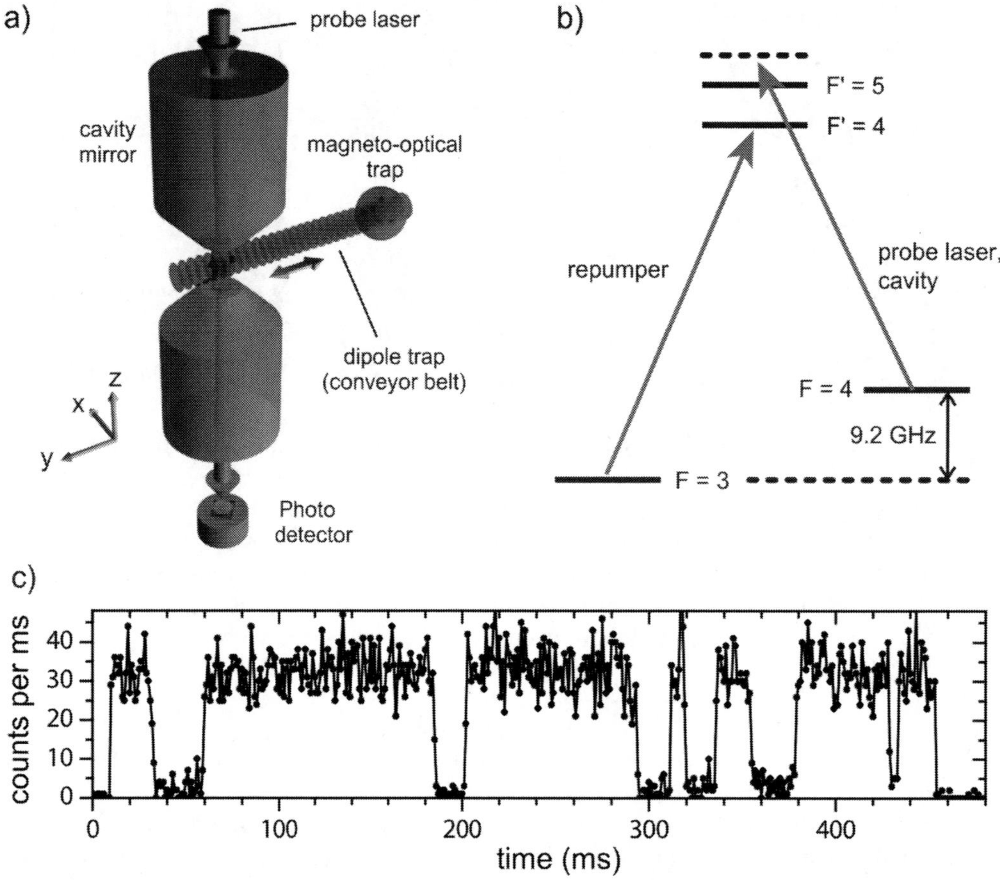

Figure 4. (a) Schematics of the experimental setup to couple single and few atoms to a high finesse optical resonator. (b) Relevant level scheme for Cs. The $F = 4$ ground state is strongly coupled to the resonator, while the $F = 3$ ground state is far detuned. Excitations to $F' = 4$ due to the cavity field or an external repumping laser and subsequent decay can change the ground state. (c) Resulting random telegraph signal for the spin state jumping between $F = 4$ (low transmission level) and $F = 3$ (high transmission level). For details see Ref. [9].

feedback onto the quantum state of several atoms simultaneously coupled to the cavity.

3.2. Controlling the Optical Properties of Single Atoms

Electromagnetically induced transparency (EIT) is a widely known effect in three-level Λ-type systems, where a dense atomic medium becomes transparent for a probe light field when the system is also illuminated by an additional control light field such that the two ground states are coherently coupled [10]. For single atoms the challenge lies in detecting these altered optical properties, as the atom medium has to be initially optically thick to detect a strong change in transmission. Here we have used our high finesse optical resonator to enhance this detection: Due to the strong atom-cavity field coupling, already a single atom can render the system optically thick for the probe laser beam as explained above. If the system is illuminated from the side with a control laser beam phase locked to the probe laser, we observe an increased transmission of the probe laser at the two-photon resonance. A peculiarity of our system is that both laser beams are detuned from the atomic resonance in order to be sensitive to both absorption and dispersive effects, for details see Ref. [11].

A surprising application of the EIT condition in our system arises from the observation of a strong cooling effect right on the two-photon resonance. We have observed that the storage time of atoms in the dipole trap is almost twenty-fold increased if the two-photon resonance is fulfilled, compared to the case without a control laser beam. In future the Cavity-EIT could be used for storing light in single or few atoms to realize quantum memories, or for performing single-photon-induced optical switching.

Acknowledgments

Many people have contributed significantly to the experiments outlined in this paper. We gratefully acknowledge enthusiastic work by Andrea Alberti, Noomen Belmechri, Stefan Brakhane, Jai-Min Choi, Martin Eckstein, Leonid Förster, Tobias Kampschulte, Michał Karski, Mkrtych Khudaverdyan, Lingbo Kong, Miguel Martinez Dorantes, Sebastian Reick, René Reimann, Andreas Steffen, and Alexander Thobe. We also thank Poul Jessen and Klaus Mølmer for fruitful collaborations and valuable discussions. We acknowledge financial support from the EU through the IP SCALA and ACQUTE and from the DFG through research unit 635.

References

[1] Karski M, Förster L, Choi J M, Alt W, Widera A and Meschede D 2009 *Phys. Rev. Lett.* **102** 053001
[2] Karski M, Förster L, Choi J, Steffen A, Belmechri N, Alt W, Meschede D and Widera A 2010 *New J. Phys.* **12** 065027
[3] Jaksch D, Briegel H, Cirac J I, Gardiner C W and Zoller P 1999 *Phys. Rev. Lett.* **82** 1975
[4] Mandel O, Greiner M, Widera A, Rom T, Hänsch T W and Bloch I 2003 *Nature* **425** 937
[5] Mandel O, Greiner M, Widera A, Rom T, Hänsch T W and Bloch I 2003 *Phys. Rev. Lett.* **91** 010407
[6] Karski M, Förster L, Choi J, Steffen A, Alt W, Meschede D and Widera A 2009 *Science* **325** 174
[7] Förster L, Karski M, Choi J, Steffen A, Alt W, Meschede D, Widera A, Montano E, Lee J, Rakreungdet W and Jessen P S 2009 *Phys. Rev. Lett.* **103** 233001
[8] Khudaverdyan M, Alt W, Kampschulte T, Reick S, Thobe A, Widera A and Meschede D 2009 *Phys. Rev. Lett.* **103** 123006
[9] Reick S, Mølmer K, Alt W, Eckstein M, Kampschulte T, Kong L, Reimann R, Thobe A, Widera A and Meschede D 2010 *J. Opt. Soc. Am. B* **27** A152
[10] Fleischhauer M, Imamoglu A and Marangos J P 2005 *Rev. Mod. Phys.* **77** 633
[11] Kampschulte T, Alt W, Brakhane S, Eckstein M, Reimann R, Widera A and Meschede D 2010 *Phys. Rev. Lett.* **105** 153603

Journal of Physics: Conference Series **264** (2011) 012022 doi:10.1088/1742-6596/264/1/012022

Quantum memory, entanglement and sensing with room temperature atoms

K. Jensen[1], W. Wasilewski[1], H. Krauter[1], T. Fernholz[1], B. M. Nielsen[1], J. M. Petersen[1], J. J. Renema[1], M. V. Balabas[1], M. Owari[2], M. B. Plenio[2], A. Serafini[3], M. M. Wolf[1], C. A. Muschik[4], J. I. Cirac[4], J. H. Müller[1] and E. S. Polzik[1,*]

[1] Niels Bohr Institute, Danish Quantum Optics Center QUANTOP, Copenhagen University, Blegdamsvej 17, 2100 Copenhagen, Denmark.

[2] Institut für Theoretische Physik, Universität Ulm, Albert-Einstein Allee 11, D-89069 Ulm, Germany.

[3] University College London, Department of Physics & Astronomy, Gower Street, London WC1E 6BT.

[4] Max-Planck-Institut für Quantenoptik, Hans-Kopfermann-Str. 1, D-85748 Garching, Germany.

E-mail: *polzik@nbi.dk

Abstract. Room temperature atomic ensembles in a spin-protected environment are useful systems both for quantum information science and metrology. Here we utilize a setup consisting of two atomic ensembles as a memory for quantum information initially encoded in the polarization state of two entangled light modes. We also use the ensembles as a radio frequency entanglement-assisted magnetometer with projection noise limited sensitivity below femtoTesla/$\sqrt{\text{Hz}}$. The performance of the quantum memory as well as the magnetometer was improved by spin-squeezed or entangled atomic states generated by quantum non demolition measurements. Finally, we present preliminary results of long lived entangled atomic states generated by dissipation. With the method presented, one should be able to generate an entangled steady state.

1. Quantum Memory

A quantum memory for light should have two properties (i) it should be able to store a wide range of input states, and (ii) it should be able to store entanglement such as one part of an Einstein-Podolski-Rosen (EPR) pair. We demonstrate a quantum memory for a range of input states (an alphabet) which are squeezed by 6 dB and displaced in phase space [1]. A classical benchmark fidelity is calculated, and we find that the experimental fidelity is higher than the benchmark which shows that the memory is truly a quantum memory. This is an advance with respect to earlier experiments where the storage of coherent states was demonstrated in the same setup [2]. Although a memory for squeezed vacua had been realised in other setups [3, 4], the classical fidelity benchmark was never beaten in such previous attempts [5]. Before discussing our experimental results we will first define canonical variables for atoms and light which is the common language used to describe information transfer between atoms and light.

In the experiment we utilize two ensembles of cesium atoms kept in separate paraffin-coated glass cells. At room-temperature each cell contains around $N_A \approx 10^{11} - 10^{12}$ atoms. The atoms are located in a bias magnetic field $B = 0.92$ Gauss pointing in the x-direction. Each ensemble can be described by the total spin vector $\hat{\vec{J}} = (\hat{J}_x, \hat{J}_y, \hat{J}_z)$. The two ensembles (1 and 2) are optically pumped into opposite directions with macroscopic spin projections $J_{x1} = -J_{x2} = |J_x|$ which can be considered classical. The transverse spin components $\hat{J}_y$ and $\hat{J}_z$ of each ensemble are quantum variables and will rotate at the Larmor frequency Ω_L. We define two sets (c and s) of non-local canonical variables by

$$\hat{x}_{Ac} = \frac{\hat{J}_{y1}^{\rm rot} - \hat{J}_{y2}^{\rm rot}}{\sqrt{2\,|J_x|}}, \quad \hat{p}_{Ac} = \frac{\hat{J}_{z1}^{\rm rot} + \hat{J}_{z2}^{\rm rot}}{\sqrt{2\,|J_x|}} \quad \text{and} \quad \hat{x}_{As} = -\frac{\hat{J}_{z1}^{\rm rot} - \hat{J}_{z2}^{\rm rot}}{\sqrt{2\,|J_x|}}, \quad \hat{p}_{As} = \frac{\hat{J}_{y1}^{\rm rot} + \hat{J}_{y2}^{\rm rot}}{\sqrt{2\,|J_x|}}, \tag{1}$$

where *rot* refers to the fact that the spins are defined in a frame rotating at the Larmor frequency. Based on the commutation relations for the spin, $\left[\hat{J}_y, \hat{J}_z\right] = iJ_x$, we find the commutation relations for the canonical variables $[\hat{x}_A, \hat{p}_A] = i$ and the Heisenberg uncertainty relation $\mathrm{Var}\,(\hat{x}_A) \cdot \mathrm{Var}\,(\hat{p}_A) \geq 1/4$. The two sets c and s of canonical variables commute (for instance $[\hat{x}_{Ac}, \hat{p}_{As}] = 0$). Coherent spin states, such as states which are fully pumped along the x-direction, are minimal uncertainty states with $\mathrm{Var}\,(\hat{x}_A) = \mathrm{Var}\,(\hat{p}_A) = 1/2$.

The polarization of light can be described by the Stokes vector $\hat{\vec{S}} = (\hat{S}_1, \hat{S}_2, \hat{S}_3)$. For light propagating along the z-direction and strongly polarized along the y-direction, the $\hat{S}_1$ operator can be replaced by a classical number. In this situation $\hat{S}_2$ and $\hat{S}_3$ describe the quantum fluctuations in the x-polarization mode. For a pulse of light with duration T we define canonical variables for a cosine mode with frequency Ω and exponentially rising modefunction with time constant $1/\gamma$ by

$$\hat{x}_{Lc}^{\rm in} \propto \int_0^T \frac{\hat{S}_2(t)}{\sqrt{|S_1|}} \cos(\Omega t) e^{+\gamma t} dt \quad \text{and} \quad \hat{p}_{Lc}^{\rm in} \propto \int_0^T \frac{\hat{S}_3(t)}{\sqrt{|S_1|}} \cos(\Omega t) e^{+\gamma t} dt, \tag{2}$$

and similarly for the sine mode with canonical variables $\hat{x}_{Ls}$ and $\hat{p}_{Ls}$. The cosine and sine modes are orthogonal and each have the commutation relation $[\hat{x}_L, \hat{p}_L] = i$. Coherent light states (such as vacuum) have $\mathrm{Var}\,(\hat{x}_L) = \mathrm{Var}\,(\hat{p}_L) = 1/2$, and states squeezed for instance along the x-direction have $\mathrm{Var}\,(\hat{x}_L) < 1/2$.

The interaction between off-resonant light and two highly polarized and oppositely oriented ensembles of cesium atoms can be described as a swap and squeezing of the quantum states [6]. We can write input-output equations describing the operators before and after the interaction

$$\begin{aligned} \hat{x}_A^{\rm out} &= \sqrt{1 - \frac{\kappa^2}{Z^2}}\hat{x}_A^{\rm in} + \kappa \hat{p}_L^{\rm in}, & \hat{p}_A^{\rm out} &= \sqrt{1 - \frac{\kappa^2}{Z^2}}\hat{p}_A^{\rm in} - \frac{\kappa}{Z^2}\hat{x}_L^{\rm in}, \\ \hat{x}_L^{\rm out} &= \sqrt{1 - \frac{\kappa^2}{Z^2}}\hat{x}_L^{\rm in} + \kappa \hat{p}_A^{\rm in}, & \hat{p}_L^{\rm out} &= \sqrt{1 - \frac{\kappa^2}{Z^2}}\hat{p}_L^{\rm in} - \frac{\kappa}{Z^2}\hat{x}_A^{\rm in}. \end{aligned} \tag{3}$$

These equations are valid both for the cosine mode and the sine mode and the index c, s has therefore been omitted. Input (output) light operators are defined with exponentially rising (falling) modefunctions with a time constant $1/\gamma_{\rm sw}$ where $\gamma_{\rm sw}$ is the swap rate proportional to number of photons and number of atoms. The coupling constant $\kappa = Z\sqrt{1 - e^{-2\gamma_{\rm sw}T}}$, $\sqrt{1 - \kappa^2/Z^2} = e^{-\gamma_{\rm sw}T}$, and $1/Z^2 = 14a_2/a_1$ is a measure of departure from the quantum non demolition (QND) interaction [7]. a_1 and a_2 are the vector and tensor polarizabilities. For cesium atoms in the $F = 4$ ground state hyperfine manifold and a probe blue-detuning of $\Delta = 850$ MHz from the $F = 4 \to F' = 5$ component of the D_2 line we have $Z^2 = 6.3$. In the QND limit $a_2(\Delta) \to 0$, Eq. (3) reduces to the QND-interaction where $\hat{p}_A$ and $\hat{p}_L$ are conserved. The interaction given by Eq. (3) can be used to measure the atomic state since the output light operator $\hat{x}_L^{\mathrm{out}}$ contains an atomic input operator. For long times $\gamma_{\mathrm{sw}} T \gg 1$ the atoms and light swap their quantum states and are squeezed by a factor Z^2 in the variance

$$\hat{x}_A^{\mathrm{out}} = Z\hat{p}_L^{\mathrm{in}}, \quad \hat{p}_A^{\mathrm{out}} = -\frac{1}{Z}\hat{x}_L^{\mathrm{in}} \quad \text{and} \quad \hat{x}_L^{\mathrm{out}} = Z\hat{p}_A^{\mathrm{in}}, \quad \hat{p}_L^{\mathrm{out}} = -\frac{1}{Z}\hat{x}_A^{\mathrm{in}}. \tag{4}$$

This swap and squeezing interaction could be used directly as a memory; however, the swap time $1/\gamma_{\mathrm{sw}}$ is too long compared to the atomic decoherence time. Instead our memory protocol consists of a light atom interaction with strength $\kappa = 1$ followed by a feedback of the measured $\hat{x}_L^{\mathrm{out}}$ to the atomic $\hat{p}_A$ operator. The final atomic operators after optimal feedback is

$$\hat{x}_A^{\mathrm{fin}} = \sqrt{1 - \frac{1}{Z^2}}\hat{x}_A^{\mathrm{in}} + \hat{p}_L^{\mathrm{in}} \quad \text{and} \quad \hat{p}_A^{\mathrm{fin}} = -\hat{x}_L^{\mathrm{in}}. \tag{5}$$

We see that $\hat{x}_L^{\mathrm{in}}$ is stored perfectly in $\hat{p}_A^{\mathrm{fin}}$ (up to a sign). Likewise $\hat{p}_L^{\mathrm{in}}$ is stored into $\hat{x}_A^{\mathrm{in}}$; however, the initial atomic $\hat{x}_A^{\mathrm{in}}$ is not cancelled, only slightly reduced. This will lead to additional noise in the final atomic state. This noise can be reduced if one squeezes $\hat{x}_A^{\mathrm{in}}$ prior to the storage. This can be done with a near QND-measurement followed by feedback to the atoms [8]. In the experiment, we reduced the intial variance in $\hat{x}_A^{\mathrm{in}}$ from $1.10(8) \cdot \frac{1}{2}$ to $0.86(6) \cdot \frac{1}{2}$. Anti-squeezing in $\hat{p}_A^{\mathrm{in}}$ was measured to be $2.1(1) \cdot \frac{1}{2}$.

The states which are stored in the memory are two entangled sidebands of light with creation operators $\hat{a}_+$ and $\hat{a}_-$ and frequencies $\omega_\pm = \omega_0 \pm \Omega_L$ where ω_0 is the optical carrier frequency (see Fig. 4). Entanglement of the sidebands is equivalent to squeezing in the *cosine* and *sine* modes which are described by the two sets of canonical operators $\hat{x}_{Ac}$, $\hat{p}_{Ac}$ and $\hat{x}_{As}$, $\hat{p}_{As}$. The Gaussian input states are characterized by the mean values and variances of these canonical operators, and in the experiment [1] we store 18 different states with mean values $\{\langle\hat{x}_L\rangle ; \langle\hat{p}_L\rangle = 0, 3.8, 7.6; 0, 3.8, 7.6\}$ and variances squeezed by 6 dB either along the $\hat{x}_L$- or $\hat{p}_L$-direction (Fig. 1). The mean values are created by phase and amplitude modulation with electro-optical modulators, and the squeezed light is produced using an optical parametric oscillator. The fidelity of a state transfer is the overlap between the intial light state and the final atomic state. For a given distribution of input states one can calculate the average fidelity. We calculated the average fidelity for three different experimental input distrubutions parametrized by $d_{\mathrm{max}} = \max\{\langle\hat{x}_L\rangle, \langle\hat{p}_L\rangle\} = 0, 3.8$ and 7.6, see Fig. 1. For these three input distributions we obtain average experimental fidelities of 0.58(2), 0.52(2), and 0.41(4), respectively. The choice of $\kappa = 1$ (which minimizes the noise added to the stored states) together with optical losses before the memory had the consequence that the stored atomic mean values were smaller by a factor 0.85 than the input mean values. This is the main reason for the reduction of the experimental fidelity with larger displacements.

In order for a memory to be truly quantum, it should be able to outperform any classical memory. For a single known quantum state the classical memory works with fidelity one. For an unknown state selected randomly from an input distribution of states similar to the ones in Fig. 1, an upper limit (benchmark) for the fidelity was calculated by semi-definite programming methods [9]. For $d_{\mathrm{max}} = 3.8$, the classical benchmark fidelity is 0.44, below the measured fidelity of 0.52(2). The memory presented here is therefore a true quantum memory for this input set. This means that if we choose to input one part (for instance, the upper sideband mode $\hat{a}_+$) of an EPR-pair into the memory, the memory would be entangled with the other part (the lower sideband mode $\hat{a}_-$). In the actual experiment, both modes of a two mode EPR entangled state were stored. One part of the EPR pair ($\hat{a}_+$) was stored in the first ensemble and one part ($\hat{a}_-$) of the EPR pair was stored in the second ensemble (see also Fig. 4). The two modes could in principle have been separated spatially using narrowband filter cavities. In this case, only one part of the EPR pair would be stored and the other would be left as a propagating mode. Based on the performance of our memory, we would expect an EPR variance of $\Sigma_{\mathrm{EPR}} = 0.76 < 1$ between the stored mode and the travelling mode ($\Sigma_{\mathrm{EPR}} = 1$ is the limit for entanglement).

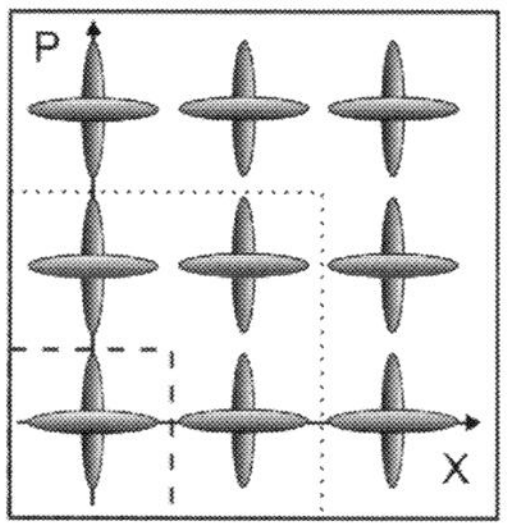

Figure 1. Input States. In total we store 18 different input states represented as ellipses in phase space. We divide the states up into three alphabets. The first alphabet (dashed line) contains 2 states (not displaced, squeezing phase $\phi = 0$ or 90 deg). The second alphabet (dotted line) contains 8 states with displacements up to $d_{\max} = 3.8$. The third alphabet (solid line) contains all 18 states with a maximum displacement $d_{\max} = 7.6$.

2. Radio frequency magnetometry

An atomic ensemble can be used as a sensor for magnetic fields [10]. Below we will describe how an atomic radio frequency (RF) magnetometer works in terms of the total spin vector $\hat{\vec{J}}$ and the Stokes vector $\hat{\vec{S}}$. The atomic ensemble is initially optically pumped such that the mean spin points in the x-direction. The transverse spin components have zero mean and variances given by the Heisenberg uncertainty relation for a minimal uncertainty state $\mathrm{Var}\left(\hat{J}_y\right) = \mathrm{Var}\left(\hat{J}_z\right) = J_x/2$. The spin can be envisioned as a long vector with an uncertainty disk on top (see Fig. 2a). The atomic ensemble is again located in a bias field leading to Larmor precession of the transverse spin components. If an RF magnetic field with amplitude B_{RF} oscillating at the Larmor frequency in the transverse directions is applied for a certain duration τ, the spin vector will tilt and a mean transverse component is created (see Fig. 2a)

$$\left\langle \hat{J}_{\perp} \right\rangle = \Gamma B_{\mathrm{RF}} J_x T_2 \left(1 - e^{-\tau/T_2}\right), \tag{6}$$

where $\Gamma = 2\pi\Omega_L/B = 2.2 \cdot 10^{10}$ rad/Tesla for cesium, and T_2 is the transverse spin decay time. The projection noise limited minimal detectable magnetic field amplitude is given by equating $\left\langle \hat{J}_{\perp} \right\rangle$ to the projection noise $\sqrt{J_x/2}$, and we find

$$B_{\min} = \left[\Gamma \sqrt{2J_x} T_2 \left(1 - e^{-\tau/T_2}\right)\right]^{-1}. \tag{7}$$

The projection noise limited sensitivity is then $B_{\min}\sqrt{T_{\mathrm{RF}}}$, has the units Tesla/$\sqrt{\mathrm{Hz}}$, and is the minimal detectable field using one second of measurement time.

The precessing transverse spin component $\hat{J}_{\perp}$ can be measured using linearly polarized light. The output Stokes operator is

$$\hat{S}_2^{\mathrm{out}}(t) = \hat{S}_2^{\mathrm{in}}(t) + aS_1 \left(\hat{J}_y^{\mathrm{rot}} \sin\Omega t + \hat{J}_z^{\mathrm{rot}} \cos\Omega t\right), \tag{8}$$

and consists of shot noise and an atomic signal. The factor a is a coupling constant. For a single atomic ensemble located in a magnetic field, back-action noise from the measurement will pile up in both atomic operators $\hat{J}_y^{\mathrm{rot}}$ and $\hat{J}_z^{\mathrm{rot}}$. We, therefore, choose to implement the magnetometer with two atomic ensembles where a near QND measurement of the collective operators is used to cancel the backaction noise. The measured signals are

$$\hat{S}_{2c}^{\mathrm{out}} = \sqrt{1 - \kappa^2/Z^2}\,\hat{S}_{2c}^{\mathrm{in}} + \kappa\sqrt{\frac{\Phi}{FN_A}} \left(\hat{J}_{y1}^{\mathrm{rot}} + \hat{J}_{y2}^{\mathrm{rot}}\right), \tag{9}$$

and $\hat{S}_{2s}^{\mathrm{out}}$ which is given by a similar expression with $\hat{J}_y$ replaced by $\hat{J}_z$. Φ is the photon flux, N_A the number of atoms, and $F = 4$ for the specific hyperfine ground state of cesium. This equation

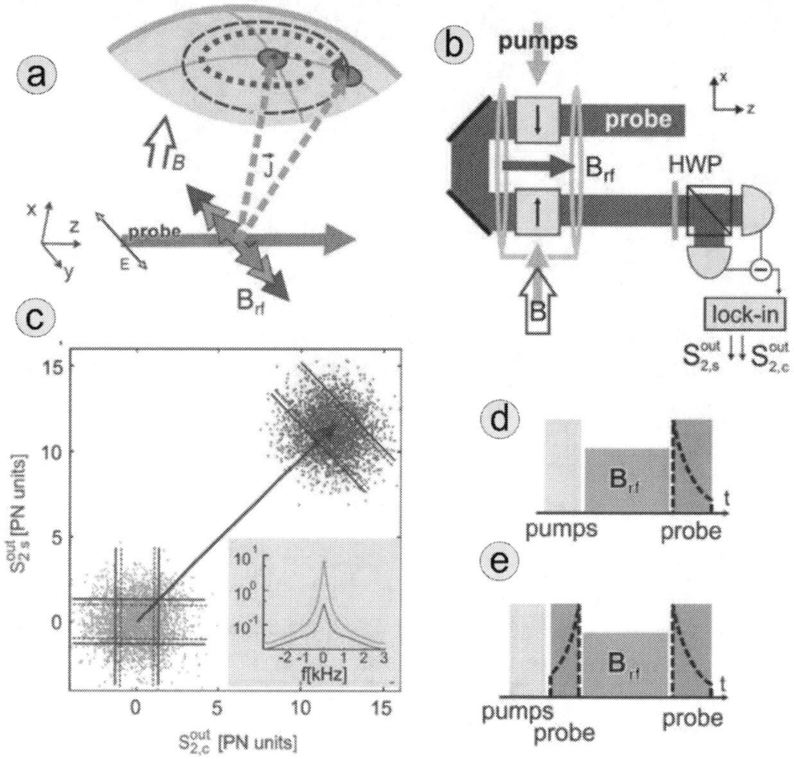

Figure 2. (a) An illustration of the RF magnetometer. The atomic ensemble is located in a bias static field B and an oscillating magnetic field B_{RF}. (b) Setup. The atomic ensembles are probed by off resonant light which is measured by polarization homodyning. HWP: half wave plate. (c) Results of a series of measurements with and without B_{RF} applied. Each dot represents a measurement point. Dashed lines are the projection noise limit and solid lines are the measured standard deviations. Inset: Power spectrum of the signal in arbitrary units. (d) Pulse sequence for the magnetometry experiment. Firstly, atoms are polarized by optical pumping, then the magnetic RF pulse is applied and, finally, the atomic spin is probed with off-resonant light. (e) Pulse sequence for entanglement assisted magnetometry. An extra probe pulse prior to the RF pulse is used to entangle the two atomic ensembles. Figures are from [11]. ©2010 The American Physical Society

is identical to Eq. (3) but is written in terms of Stokes and spin operators instead of canonical operators. The displacement given by Eq. (6) gives a mean value in the signals $\hat{S}_{2c}^{\mathrm{out}}$ and $\hat{S}_{2s}^{\mathrm{out}}$ proportional to the RF field amplitude B_{RF}. The variance of the signal will be

$$\mathrm{Var}\left(\hat{S}_{2c}^{\mathrm{out}}\right) = \left(1-\kappa^2/Z^2\right)\mathrm{Var}\left(\hat{S}_{2c}^{\mathrm{in}}\right) + \kappa^2\left(\frac{\Phi}{FN_A}\right)\mathrm{Var}\left(\hat{J}_{y1}^{\mathrm{rot}} + \hat{J}_{y2}^{\mathrm{rot}}\right). \quad (10)$$

For shot noise limited light $\mathrm{Var}\left(\hat{S}_{2c}^{\mathrm{in}}\right) = S_x/2$ and for projection noise limited atoms in the coherent spin state $\mathrm{Var}\left(\hat{J}_y\right) = \mathrm{Var}\left(\hat{J}_z\right) = J_x/2$.

In the experiment a 36(3) femtoTesla RF field was applied to the atoms for a duration of 15 ms. The measured signals $\hat{S}_{2c}^{\mathrm{out}}$ and $\hat{S}_{2s}^{\mathrm{out}}$ are plotted in Fig. 2c from which we can deduce a signal to noise ratio $SNR = 12.3$. We then calculate the sensitivity $B_{\mathrm{RF}}\sqrt{\tau}/SNR = 3.6(4)\cdot 10^{-16}$

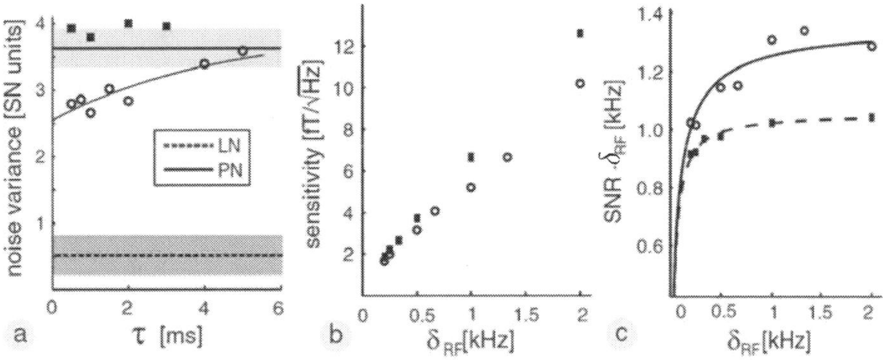

Figure 3. (a) Noise variance of $\hat{S}_{2c}^{\text{out}}$ as a function of RF magnetic field duration. Filled squares: without entanglement. Circles: with entangled atoms. (b) Magnetic field sensitivity as a function of RF bandwidth $\delta_{\text{RF}} = 1/\tau$. (c) The product $SNR \cdot \delta_{\text{RF}}$ reaches a constant value for large RF bandwidths. Figure is from [11]. ©2010 The American Physical Society

Tesla/$\sqrt{\text{Hz}}$ which is close to the theoretical value $2.7(4) \cdot 10^{-16}$ Tesla/$\sqrt{\text{Hz}}$ for the projection noise limited sensitivity calculated from Eq. (7) with $N_A = 2 \cdot 7.2(7) \cdot 10^{11}$ atoms and $T_2 = 32$ ms. The best experimental sensitivity obtained was $B_{\text{RF}}\sqrt{T_{\text{tot}}}/SNR = 4.2(8) \cdot 10^{-16}$ Tesla/$\sqrt{\text{Hz}}$ calculated using the duty cycle time T_{tot} (optical pumping pulse of 6 ms, probe pulse of 1.5 ms and RF pulse duration of 22 ms). This sensitivity is comparable to the sensitivity of the state-of-the-art atomic magnetometer [12] which operates with 10^4 times more atoms. This is expected since the projection noise limited measurement of the magnetic field yields the best sensitivity per atom without the use of entanglement.

The signal to noise ratio can be improved by reducing the uncertainty in the variables $\hat{J}_{y1}^{\text{rot}} + \hat{J}_{y2}^{\text{rot}}$ and $\hat{J}_{z1}^{\text{rot}} + \hat{J}_{z2}^{\text{rot}}$. This can be achieved by entangling the two atomic ensembles. The criterion for entanglement between the ensembles is that the Einstein-Podolsky-Rosen variance

$$\Sigma_{\text{EPR}} = \frac{\text{Var}\left(\hat{J}_{y1}^{\text{rot}} + \hat{J}_{y2}^{\text{rot}}\right) + \text{Var}\left(\hat{J}_{z1}^{\text{rot}} + \hat{J}_{z2}^{\text{rot}}\right)}{2J_x} < 1. \tag{11}$$

Such entanglement can be created with a near QND measurement followed by feedback. In the magnetometry experiment the atomic projection noise was initially $\Sigma_{\text{EPR}} = 1.10(8)$ and was reduced to -1.5dB (30%) below the projection noise level (see Fig. 3a). The entanglement decays with time constant $T_{2E} = 4$ ms. Since the B_{RF} is applied after the entanglement creation, entanglement only improves the magnetometer for short RF pulses or, equivalent, broadband pulses. Experimentally we obtain a 15% improvement in sensitivity for RF pulses with bandwidth greater than 1 kHz (Fig. 3b).

3. Entanglement generated by dissipation

A novel method for generating entanglement of atomic systems in a steady state has been recently proposed in [13]. Dissipation is used as the entangling mechanism, such that the atoms reach an entangled steady state independent of the initial state. This kind of steady state entanglement is highly desirable since it provides a robust entanglement source which is available on demand at any instance in time. In order to describe the entanglement generation, we implement a master equation approach detailed below.

We consider two oppositely polarized atomic ensembles with level structure as shown in Fig. 4. Atoms in ensemble 1 (2) initially occupy the level $F = 4, m_F = 4$, ($F = 4, m_F = -4$). The

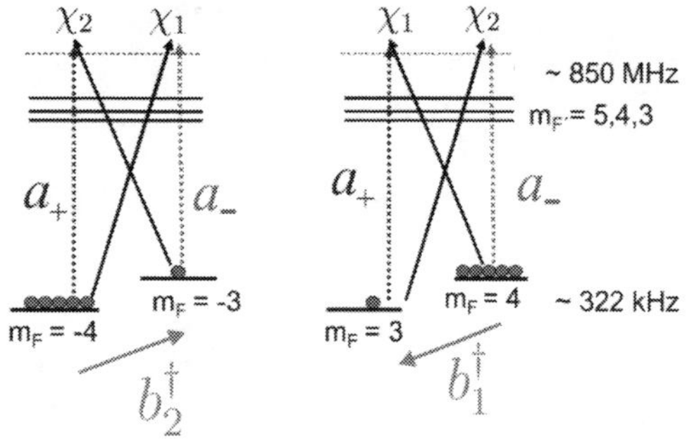

Figure 4. Level structure and light fields for the two atomic ensembles. $\hat{a}_+$ and $\hat{a}_-$ are annihilation operators for the upper and lower sidebands. $\hat{b}_1^\dagger$ and $\hat{b}_2^\dagger$ are creation operators for the collective atomic excitation for the first and second ensemble. χ_1 and χ_2 are coupling strengths.

ensembles interact with strong linearly y-polarized light represented in the x-quantization axis as a sum of right hand and left hand circularly polarized light. Situated in this light field, the atoms can be excited and spontaneously emit photons in all directions. The photons are described by modes $\hat{a}_{\vec{k}}$ where $\vec{k}$ is the wavevector. Since spontaneous emission is a resonant process and the atomic ground state levels are separated by the energy $\hbar\Omega_L$ the frequency of the spontaneously emitted photons will be ω_+ or ω_- where $\omega_\pm = \omega_0 \pm \Omega_L$. The atoms in ensemble 1 (2) couple to σ_--polarized light with coefficient χ_1 (χ_2) and to σ_+-polarized light with coefficient χ_2 (χ_1). Upon absorption of a σ_--polarized photon, a collective spin excitation $\hat{b}_1^\dagger$ can be created in the first ensemble together with creation of a photon in the upper sideband (see Fig. 4). This leads to a term proportional to $\int_{\Delta\omega_{us}} d\vec{k}\chi_1\hat{a}_{\vec{k}}\hat{b}_1 + h.c.$ in the Hamiltonian. The integration (over $\vec{k}$) runs over modes within a narrow frequency interval $\Delta\omega_{us}$ centred around ω_+ (us stands for upper sideband.) In total, we find four such terms which give rise to the total Hamiltonian

$$\hat{\mathcal{H}} \propto \int_{\Delta\omega_{ls}} d\vec{k}\left(\hat{A}\hat{a}_{\vec{k}}^\dagger + \hat{A}^\dagger\hat{a}_{\vec{k}}\right) + \int_{\Delta\omega_{us}} d\vec{k}\left(\hat{B}\hat{a}_{\vec{k}}^\dagger + \hat{B}^\dagger\hat{a}_{\vec{k}}\right), \tag{12}$$

where Δ_{ls} is a narrow frequency interval centred around ω_- and where we have defined the variables

$$\hat{A} = \chi_2\hat{b}_1 + \chi_1\hat{b}_2^\dagger \quad \text{and} \quad \hat{B} = \chi_2\hat{b}_2 + \chi_1\hat{b}_1^\dagger, \tag{13}$$

This Hamiltonian leads to a master equation for the density matrix ρ of the ensembles [13]

$$\frac{\partial\hat{\rho}}{\partial t} = \frac{d\Gamma}{2}\left(\hat{A}\hat{\rho}\hat{A}^\dagger - \hat{A}^\dagger\hat{A}\hat{\rho} + \hat{B}\hat{\rho}\hat{B}^\dagger - \hat{B}^\dagger\hat{B}\hat{\rho} + h.c.\right) + \hat{\mathcal{L}}_{\text{noise}}\rho, \tag{14}$$

where Γ is the single atom radiative decay rate and $\hat{\mathcal{L}}_{\text{noise}}$ represents undesired noise such as single atom spontaneous emission. Emission in the forward propagating modes is enhanced by the optical depth d as compared to the modes into other directions and give rises to the terms in Eq. (14) which are proportional to d. With such a master equation, the collective state of the atomic ensembles will be driven into an entangled steady state with $\Sigma_{\text{EPR}} < 1$ [13]. The emission of a forward propagating photon could arise from both the first and second ensemble, the two paths are indistingishable, and this leads to entanglement between the two atomic ensembles. The other modes act as in the usual spontaneous emission decoherence mechanism. The generated entanglement can also be explained by the simpler model given by Eq. (4) which does not include spontaneous emission in the non-forward propagating modes. For long interaction times $\gamma_{\text{sw}}T \to \infty$ the collective state of the two ensembles reaches a steady state where the operators $\hat{p}_{Ac}$ and $\hat{p}_{As}$ are squeezed by a factor Z^2 in the variances which means that the two atomic ensembles are entangled.

In the preliminary experiments two important things were observed. Firstly, starting from a coherent state with $\Sigma_{\text{EPR}} = 1$, entanglement is created with $\Sigma_{\text{EPR}} < 1$ for a period of more than

20 ms. Then the noise grows, $\Sigma_{\rm EPR} > 1$, and the atomic ensembles are no longer entangled. Secondly, the mean spin J_x of each ensemble decreases with time. Both the increase in noise and decrease in J_x for long times are expected due to atomic depolarization. The main depolarization mechanisms are spontaneous emission into the other modes and atomic collisions. Also, when J_x decreases, the collective enhancement $\propto J_x$ of the emission into the forward propagating modes is reduced and the entanglement mechanism loses strength.

In order to keep the atoms in the $F = 4$ manifold well polarized, we shine resonant σ_+ (σ_-) polarized light onto the atoms in ensemble 1 (2). This significantly increases the polarization $p = \frac{1}{4}\sum_{m=-4}^{4} m\rho_{mm}$ of the atoms in the $F = 4$ manifold. Preliminary results show that also the time with entanglement could be extended beyond the 20 ms. Still, atoms are lost to the $F = 3$ manifold with time and, unfortunately, we do not reach entanglement in a steady state. A full account of the results can be found in [14]. Possible improvements could be to add repump light which can recycle atoms lost to the $F = 3$ manifold. According to our model, the combination of repumping and higher optical depths would yield steady state entanglement. Another possibility which may lead to steady state "forever entangled" atomic systems is a combination of the dissipation mechanism with the continuous measurement.

4. Outlook

Optically thick room temperature alkali atoms in a spin protected environment have the potential to become a robust platform for quantum technologies. Recently, a lifetime of the ground spin state exceeding several seconds with an improved spin protecting coating of the cell walls has been demonstrated in [15, 16]. Combination of improved coating with a fibre-coupled microcell design and low finesse optical cavities yields a scalable and versatile system for quantum information processing and sensing.

Acknowledgments

This research was supported by EU projects Q-ESSENCE, HIDEAS, CORNER, COMPAS, EMALI, COQUIT, QAP and QUEVADIS and the Elite Network of Bavaria (ENB) project QCCC.

References

[1] Jensen K, Wasilewski W, Krauter H, Fernholz T, Nielsen B M, Owari M, Plenio M B, Serafini A, Wolf M M and Polzik E S *Nature Physics advance online publication, 07 November 2010 (DOI 10.1038/nphys1819)*
[2] Julsgaard B, Sherson J, Cirac J I, Fiurášek J and Polzik E S 2004 *Nature* **432** 482–486
[3] Appel J, Figueroa E, Korystov D, Lobino M and Lvovsky A I 2008 *Phys. Rev. Lett.* **100** 093602
[4] Honda K, Akamatsu D, Arikawa M, Yokoi Y, Akiba K, Nagatsuka S, Tanimura T, Furusawa A and Kozuma M 2008 *Phys. Rev. Lett.* **100** 093601
[5] Adesso G and Chiribella G 2008 *Phys. Rev. Lett.* **100** 170503
[6] Wasilewski W, Fernholz T, Jensen K, Madsen L S, Krauter H, Muschik C and Polzik E S 2009 *Opt. Express* **17** 14444–14457
[7] Hammerer K, Sørensen A S and Polzik E S 2010 *Rev. Mod. Phys.* **82** 1041–1093
[8] Julsgaard B, Kozhekin A and Polzik E S 2001 *Nature* **413** 400–403
[9] Owari M, Plenio M B, Polzik E S, Serafini A and Wolf M M 2008 *New J. Phys.* **10** 113014
[10] Budker D and Romalis M 2007 *Nat. Phys.* **3** 227–234
[11] Wasilewski W, Jensen K, Krauter H, Renema J J, Balabas M V and Polzik E S 2010 *Phys. Rev. Lett.* **104** 133601
[12] Lee S K, Sauer K L, Seltzer S J, Alem O and Romalis M V 2006 *Appl. Phys. Lett.* **89** 214106
[13] Muschik C A, Polzik E S and Cirac J I 2010 (*Preprint* arXiv:1007.2209v1)
[14] Krauter H, Muschik C A, Jensen K, Wasilewski W, Petersen J M, Cirac J I and Polzik E S 2010 (*Preprint* arXiv:1006.4344v1)
[15] Balabas M V, Jensen K, Wasilewski W, Krauter H, Madsen L S, Müller J H, Fernholz T and Polzik E S 2010 *Opt. Express* **18** 5825–5830
[16] Balabas M V, Karaulanov T, Ledbetter M P and Budker D 2010 *Phys. Rev. Lett.* **105** 070801

Rydberg state mediated quantum gates and entanglement of pairs of neutral atoms

M Saffman, X L Zhang, A T Gill, L Isenhower, and T G Walker

Department of Physics, University of Wisconsin, 1150 University Avenue, Madison, Wisconsin 53706 USA

E-mail: msaffman@wisc.edu

Abstract. Experiments performed within the last year have demonstrated Rydberg state mediated quantum gates and deterministic entanglement between pairs of trapped neutral atoms. These experiments validate ten year old proposals for Rydberg mediated quantum logic, but are only the beginning of ongoing efforts to improve the fidelity of the results obtained and scale the experiments to larger numbers of qubits. We present here a summary of the results to date, along with a critical evaluation of the prospects for higher fidelity Rydberg gates.

1. Introduction

One of the central challenges in building devices suitable for quantum logic operations is the need to isolate quantum information from environmental decoherence while simultaneously allowing for strong and controllable interactions between pairs of qubits. Neutral atoms in optical traps formed by far off-resonance laser beams represent one of the best isolated systems known. Quantum information can be stored in long-lived hyperfine states, with decoherence times that have been measured to be as long as several seconds[1]. The challenge for neutral atom based quantum computing has centered around the difficulty of introducing a strong and controllable two-atom interaction. A range of ideas have been proposed [2, 3, 4, 5, 6, 7] and collisional entanglement[3] has been demonstrated in many atom optical lattice experiments[8, 9].

In 2000 Jaksch, et al.[10] introduced the idea of using strong dipolar interactions between Rydberg excited atoms for quantum gates. If a "control" atom is Rydberg excited with a resonant laser pulse, subsequent excitation of a second proximal "target" atom will be off-resonant, and therefore blocked, due to dipolar interactions between the Rydberg atoms. This Rydberg blockade form of gate operation provides a straightforward way of implementing a controlled π phase shift on the target atom using a three pulse sequence (read from right to left): $U_Z = U^{(c)}_{r\to g}(\pi)U^{(t)}_{g\to r\to g}(2\pi)U^{(c)}_{g\to r}(\pi)$. Here $U^{(c/t)}(\theta)$ is a Rabi pulse of area θ applied between ground ($|g\rangle$) and Rydberg ($|r\rangle$) states of the control (c) or target (t) atom. This sequence gives a U_Z controlled phase gate, which can be readily converted into a controlled-NOT (CNOT) gate by adding single qubit operations on the target atom before and after the pulse sequence[11]. We note that Rydberg states are also suitable for engineering entanglement between non-trapped atoms as was demonstrated in 1997 in cavity QED experiments[12].

The Rydberg blockade gate has a number of desirable characteristics. As long as the dipolar blockade interaction B is very large compared to the excitation Rabi frequency Ω the gate works well. This is true even if the exact value of B is not known. This means that the separation of the atoms does not have to be controlled very precisely which adds to the robustness of the gate. Furthermore when $\mathsf{B} \gg \Omega$ the target atom is never excited to a Rydberg state during the gate. This implies that the strong dipolar interaction does not lead to any energy shifts or forces between the atoms during the gate operation. Remarkably we can have a very strong two-atom interaction without the atoms involved ever experiencing any forces.

In addition to these desirable features what really sets the Rydberg gate apart from other approaches is the strength of the interaction. Consider a pair of alkali atoms at $R = 10\ \mu$m separation. The interaction energy for ground state atoms at this length scale is dominated by the magnetostatic dipole-dipole interaction with a strength $U_{gg}/h \sim 10^{-5}$ Hz. The same atom pair excited to $n = 100$ Rydberg states has $U_{rr} \sim 10^7$ Hz, an increase by 12 orders of magnitude[13]! The ability to coherently turn the interaction on and off with a contrast of 12 orders of magnitude is possibly unique to the Rydberg system, and is a superb figure of merit for Rydberg mediated quantum logic. The intrinsic properties of the Rydberg blockade approach suggest that entangling gates should be possible with errors at the 10^{-3} level or better[14, 13].

Initial observations of Rydberg mediated excitation suppression were made in 2004[15, 16]. Those experiments were done in many atom samples and it was not possible to verify that excitation was limited to a single atom. Rydberg blockade at the level of single atoms was demonstrated by groups from the University of Wisconsin and Institut d'Optique in 2009[17, 18]. In addition to blockade the Institut d'Optique experiment showed that the effective Rabi frequency when simultaneously addressing two atoms was increased by a factor of two, compared to the single atom rate. The reason for this is as follows. In an N atom ensemble subject to strong blockade an effective two-level system is formed between the states $|g\rangle = |g_1 g_2 ... g_N\rangle$ and $|\mathsf{s}\rangle = \frac{1}{\sqrt{N}} \sum_{=1,N} |g...r_j...g\rangle$. The state $|\mathsf{s}\rangle$ has one unit of excitation symmetrically shared between all N atoms. This entangled state is coupled to $|g\rangle$ with the Rabi frequency $\Omega_N = \sqrt{N}\Omega$ which is $\sqrt{N}$ larger than the one-atom coupling[19]. Observation of a $\sqrt{2}$ speed up by Gaëtan, et al.[18] implied preparation of the entangled state $|\psi\rangle = \frac{1}{\sqrt{2}}(|gr\rangle + e^{\imath\phi'}|rg\rangle))$ and was thus the first tantalizing observation of Rydberg blockade mediated entanglement. However, the entanglement in this case involves short lived Rydberg states which are not well suited for scalable quantum logic.

The next step was to map the Rydberg entanglement onto long lived ground states with an additional laser pulse, $|r\rangle \to |g_1\rangle$. One thereby creates the state

$$|\psi\rangle = \frac{1}{\sqrt{2}}(|01\rangle + e^{\imath\varphi}|10\rangle) \tag{1}$$

where 0, 1 correspond to different long lived hyperfine ground states of ^{87}Rb $(F = 1, 2)$. This was done by Wilk, et al. [20] in 2010. The fidelity of the expected entangled Bell state was extracted from measurements of parity oscillations[21] and was found to be under the entanglement threshold of $F = 0.5$. Correction of the data for loss of atoms during the preparation sequence revealed a fidelity of $F = 0.74$ for the 62% of the trials that had two atoms remaining after the preparation sequence. Note that the phase φ appearing in Eq. (1) may be stochastic due to several experimental issues. This limits the Bell state fidelity, a point which we will return to below.

Independently of the Institut d'Optique experiments the Wisconsin group implemented the U_Z operation between two trapped atoms and added $\pi/2$ pulses on the target atom to generate a full CNOT gate matrix. The CNOT gate was also demonstrated using a different protocol referred to as a controlled amplitude swap[22]. The CNOT gate was then used with the control atom in the superposition state $\frac{1}{\sqrt{2}}(|0\rangle + i|1\rangle)$ and the target atom in either of the two hyperfine states to create the Bell states $|B_1\rangle = \frac{1}{\sqrt{2}}(|00\rangle + |11\rangle)$ or $|B_2\rangle = \frac{1}{\sqrt{2}}(|01\rangle + |10\rangle)$[22]. The fidelity of the Bell state $|B_1\rangle$ was evaluated with data extracted from parity oscillation measurements and was under $F = 0.5$. Correcting for the fraction of atom pairs remaining after the preparation steps revealed post-selected entanglement with fidelity $F = 0.58$.

These data were subsequently improved upon using an upgraded apparatus which resulted in Bell state fidelity of $F = 0.58 \pm 0.04$ without applying any corrections for atom loss or trace loss[23], and a loss corrected fidelity of $F = 0.71$. These latest results demonstrate that Rydberg blockade can be used for deterministic entanglement of neutral atom pairs which is a prerequisite for larger scale quantum information processing.

The results obtained in the past year are a significant step forwards in neutral atom quantum logic. Nevertheless it is also apparent that a large gap exists between what has been achieved and what should be possible using Rydberg blockade. In Sec. 2 of this contribution we will take a closer look at the technical and intrinsic errors associated with Rydberg mediated entanglement. The results of this analysis suggest that it should indeed be possible to reach 10^{-3} gate errors, or better, but that doing so will not be particularly simple and will require a carefully designed experimental apparatus. In Sec. 3 we conclude with some remarks on the potential for scaling to a many qubit experiment.

2. Entangling gate errors

Imperfections in quantum gates arise from two categories of errors. Those we term *intrinsic* errors are due to the choice of gate protocol and the characteristics of the physical system used to encode the qubits. In addition there are *technical* errors which arise from imperfections and noise in the applied pulses, fluctuations of background magnetic or electric fields, and motional effects at finite atomic temperature. A rigorous treatment would be based on including all of these error sources in the Bloch equations, solving them for the C_Z pulse sequence, and then averaging the results over the computational input states. In the limit where all errors are small, which is the situation we are most interested in, a simpler and more physically transparent approach is based on separately estimating the various error sources and then adding them together. We follow this latter approach here.

Furthermore we will treat only the errors associated with the basic two-qubit entangling operation. A quantum computer requires additional capabilities including state preparation, single qubit operations, and measurements. A full discussion of those errors for a neutral atom implementation is outside the scope of this paper.

2.1. *Intrinsic errors*

The intrinsic error of a Rydberg blockade controlled phase gate can be exceedingly low. When the hyperfine qubit states have a frequency separation $\omega_{\rm hf}$ that is large compared to both Ω and B the optimum excitation Rabi frequency is[13] $\Omega_{\rm opt} = (7\pi)^{1/3}\frac{\mathsf{B}^{2/3}}{\tau^{1/3}}$ and the corresponding minimum gate error is $E_{\rm min} = \frac{3(7\pi)^{2/3}}{8}\frac{1}{(\mathsf{B}\tau)^{2/3}}$, with τ the Rydberg radiative lifetime at room temperature. Very low gate errors are obtained when both B and Ω become large and are no longer completely negligible compared to $\omega_{\rm hf}$. In this case we must use the full expression[13]

$$E \simeq \frac{7\pi}{4\Omega\tau}\left(1+\frac{\Omega^2}{\omega_{\rm hf}^2}+\frac{\Omega^2}{7\mathsf{B}^2}\right)+\frac{\Omega^2}{8\mathsf{B}^2}\left(1+6\frac{\mathsf{B}^2}{\omega_{\rm hf}^2}\right). \quad (2)$$

Figure 1 shows E evaluated from Eq. (2) at $\Omega = \Omega_{\rm opt}$ as a function of atom separation R for ^{87}Rb atoms excited to $ns_{1/2}$ states. The blockade shift $\mathsf{B}(R)$ was calculated using the methods described in [24]. We see that the intrinsic error can fall under 10^{-4} for $n > 100$ although this requires rather high Rabi excitation frequencies. For example the $150s_{1/2}$ state has an intrinsic error of $E = 5.5 \times 10^{-5}$ at $R = 5.$ μm using $\Omega/2\pi \simeq 30$ MHz.

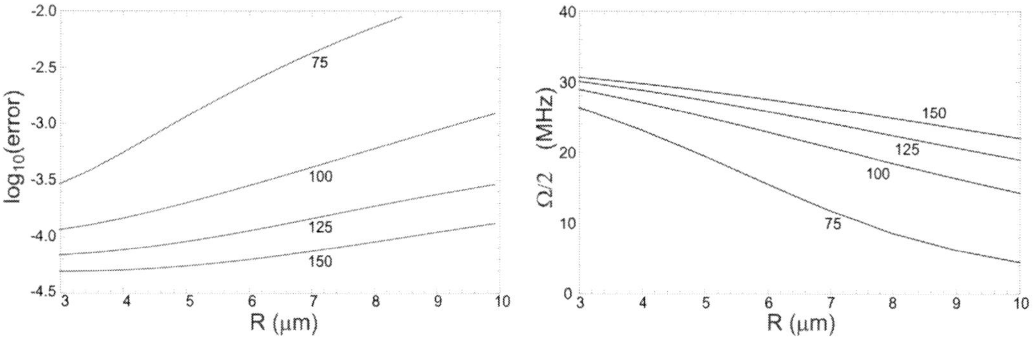

Figure 1. Intrinsic error of the C_Z gate (left) and corresponding Rabi excitation frequency (right) for Rb $ns_{1/2}$ states with $n = 75, 100, 125, 150$. Calculations used radiative lifetimes of $\tau = 180, 340, 570, 860\ \mu$s.

2.2. Technical errors

The main technical errors that affect the C_Z operation are spontaneous emission during Rydberg excitation, magnetic noise and Doppler effects. The spontaneous emission probability can always be made sufficiently small provided sufficient laser power is available for the Rydberg excitation step, although this may be a technical challenge. For example let us consider the feasibility of achieving a 30 MHz excitation frequency to the $150s_{1/2}$ state while maintaining $< 10^{-3}$ probability of spontaneous emission in a Rydberg π pulse. In a two-photon excitation scheme spontaneous emission is minimized by setting the one-photon Rabi frequencies equal, in which case the probability of emission from the intermediate p level during a π excitation pulse is related to the two-photon Rabi frequency by [13]

$$\Omega = \frac{P_{\rm se}}{\pi}\frac{|\Omega_1|^2}{\gamma_p},$$

with Ω_1 the one-photon Rabi frequency and γ_p the radiative decay rate of the p level.

Some sample numbers for excitation through $5p_{1/2}$ are 20 μW of 795 nm light, 10 W of 474 nm light focused to beam waists of 1 μm using a detuning of 37 GHz, which gives $\Omega/2\pi = 30$ MHz and $P_{\rm se} = 3 \times 10^{-4}$. Generating such high single frequency power at 474 nm would probably require using a pulsed amplifier that is then frequency doubled, as in[25]. This would be technically challenging in comparison to the cw lasers used in current Rydberg Rabi oscillation experiments[26]. An alternative is to perform excitation through the second resonance level, $5s_{1/2} \rightarrow 6p_{1/2} \rightarrow ns_{1/2}$ using 422 and 1004 nm lasers, as has been demonstrated in [27]. The smaller value of $\gamma_{6p_{1/2}} = 1./125$ ns, and larger matrix element to the Rydberg level leads to excitation of $150s_{1/2}$ with $\Omega/2\pi = 30$ MHz and $P_{\rm se} = 1 \times 10^{-4}$ using 5 μW of 422 nm light and 2 W of 1004 nm light. The high power requirement is thereby moved to the infrared which is technically more convenient.

Doppler broadening due to atomic motion leads to imperfect Rydberg excitation. This is a small effect provided $\Delta_D \ll \Omega$ where $\Delta_D = \mathbf{k}_{2\nu} \cdot \mathbf{v}$ is the Doppler shift, $\mathbf{v}$ is the atomic velocity and $\mathbf{k}_{2\nu} = \mathbf{k}_1 - \mathbf{k}_2$ is the two-photon excitation wavenumber. For example at $\Omega/2\pi = 30$ MHz and $T = 100\ \mu$K atoms the probability of an error in exciting a Rydberg atom is under 10^{-5}.

A more serious technical error arises from dephasing of the Rydberg state relative to the ground state during gate operation. We have observed this dephasing by performing a Ramsey interference experiment between the ground state $|g\rangle = |5s_{1/2}, f = 2, m_f = 0\rangle$ and the Rydberg state $|r\rangle = |97d_{5/2}, m_j = 5/2\rangle$. The results are shown in Fig. 2. A short analysis shows that the envelope of the Ramsey oscillations is given by $\langle e^{\imath\varphi_{\rm st}}\rangle$ with $\varphi_{\rm st}$ any stochastic phase that is accumulated during the Ramsey gap time t.

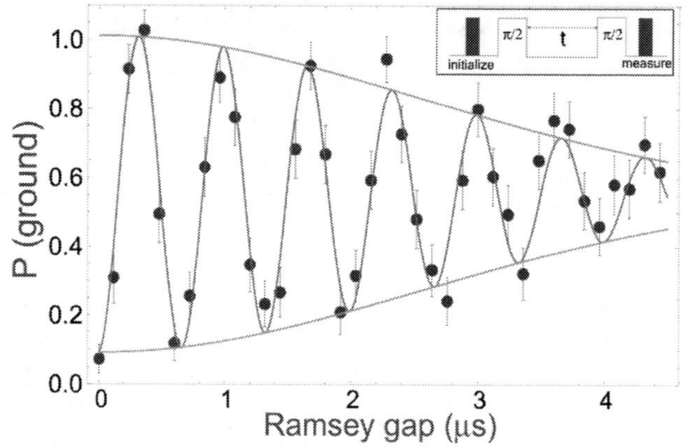

Figure 2. Ramsey experiment between ground and Rydberg states. A fit to the envelope gives a decay time of $T_2 = 3.6\ \mu$s.

The stochastic phase is due to two main sources, magnetic field fluctuations and motional effects. Modeling magnetic field fluctuations as a random Gaussian process with standard deviation σ we find

$$\langle e^{\imath\varphi_{\rm st}}\rangle = e^{-t^2/T_{2,B}^2}$$

where

$$T_{2,B} = \frac{2^{3/2}\pi\hbar}{|g_R m_{jR} - g_g m_{fg}|\mu_B\sigma}$$

where $|g_R m_{jR} - g_g m_{fg}|\mu_B$ is the difference in Zeeman shift per unit field strength between Rydberg fine structure and ground hyperfine states with μ_B a Bohr magneton. This definition of T_2 corresponds to the $1/e$ decay time of the Ramsey fringes.

Doppler broadening, in addition to imperfect Rydberg excitation, leads to motional decoherence of the Rydberg state according to

$$\langle e^{\imath\varphi_{\rm st}}\rangle = e^{-t^2/T_{2,D}^2}$$

where

$$T_{2,D} = \left(\frac{2m}{k_BT}\right)^{1/2}\frac{1}{k_{2\nu}}$$

with m the atomic mass, T the temperature, and k_B Boltzmann's constant. A physical picture of this decoherence mechanism is that motion of the Rydberg excited atom causes it to see a different phase of the two-photon field which gets imprinted onto the wavefunction upon deexcitation.

In the presence of both magnetic and Doppler dephasing the effective T_2 is

$$T_2 = \frac{T_{2,B}T_{2,D}}{(T_{2,B}^2 + T_{2,D}^2)^{1/2}}.$$

The decoherence due to magnetic and Doppler effects can then be extracted from Fig. 2 if we know the magnetic field fluctuations σ. The field fluctuations were estimated to be $\sigma \sim 2.5\times10^{-6}$ T by measuring T_2 of the hyperfine qubit at two different bias magnetic field strengths. Figure 2 then implies that $T \sim 60\ \mu$K, $T_{2,B} = 13\ \mu$s and $T_{2,D} = 3.7\ \mu$s. We see that the Rydberg decoherence is dominated by Doppler dephasing in these experiments.

Dephasing of the Ramsey interference signal also limits the fidelity of the entanglement that is created using Rydberg interactions. Averaging over the phase φ in Eq. (1) leads to

$$F = \frac{1}{2}\left(1 + \langle e^{\imath\varphi_{\rm st}}\rangle\right). \qquad (3)$$

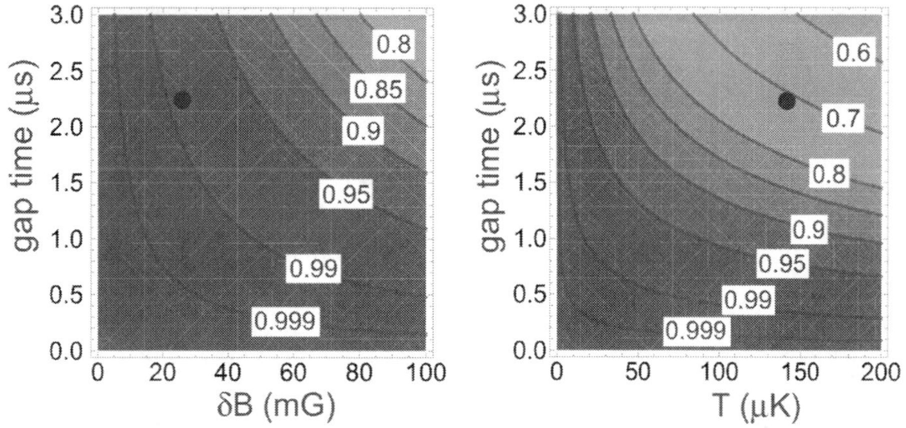

Figure 3. Bell state fidelity limits set by magnetic and Doppler dephasing. Calculations are for ^{87}Rb with $k_{2\nu} = k_{480\ \mathrm{nm}} - k_{780\ \mathrm{nm}}$. The blue dots correspond to the experiments of [23].

This limit was first pointed out in [20]. Figure 3 shows the fidelity limits set by magnetic and Doppler dephasing. These limits are in agreement with the atom loss corrected fidelity of $F = 0.71$ reported in [23].

2.3. Discussion

The most accurate quantum gate experiments to date are those of the Innsbruck ion trap group who have reported two-qubit entanglement fidelities better than 99%[28]. Can neutral atom Rydberg gates hope to reach similar fidelities?

The discussion in the previous sections has identified the leading error sources as being due to intrinsic gate errors, Doppler excitation error, spontaneous emission, magnetic dephasing, and Doppler dephasing. Assuming excitation of $150s_{1/2}$ states with $\Omega/2\pi = 30$ MHz the intrinsic gate error, Doppler excitation error, and spontaneous emission error can all be under 10^{-4} as discussed above. The minimum gap time during which the control atom is Rydberg excited is $2\pi/\Omega = .03\ \mu$s. Looking at Fig. 3 we see that such short gap times would lead to dephasing errors well below 10^{-3} even with our current values of magnetic noise and atomic temperature.

The remaining technical challenges concern accurate timing with 30 MHz Rabi frequencies as well as sufficient laser power as discussed above. A 10^{-3} error in the population of a target state on a π pulse at 30 MHz requires timing accuracy at the level of about 1 ns, which is feasible with available technology. An interim conclusion is that gate errors at the 10^{-3} level are possible, but will require careful attention to technical issues.

3. Outlook

The utility of neutral atom qubits for quantum information processing depends on more than high gate fidelities. For circuit model quantum computing it is also necessary to develop architectures which support site selective gates, initialization, and measurements for a many-qubit array. Although techniques for loading millions of atoms into optical lattices are well developed there are specific challenges associated with Rydberg blockade quantum computing.

One of the foremost issues is that red detuned dipole traps can store ground state atoms with low decoherence but they do not trap Rydberg states. The polarizability of a highly excited Rydberg atom is essentially that of a free electron, and is negative. Thus a red detuned trap presents repulsive mechanical potential for the Rydberg atom. Furthermore the trapping light rapidly photoionizes the Rydberg atom[14]. This problem is handled in current experiments[20, 22, 23] by turning off the trapping light for a few μs during Rydberg excitation. Although this leads to a small amount of mechanical heating and decoherence due to entanglement between the qubit state and the center of mass motion, this has not been a significant issue until now.

In a multi-qubit array turning off the traps whenever a Rydberg gate is performed could lead to a significant heating and decoherence problem. One solution would be to work with an optical system where each trap could be turned on and off individually on μs time scales. This is not compatible with multi-trap architectures based on multi-beam optical lattices[29], or fixed diffraction gratings[30], but can be achieved at few ms time scales using spatial light modulator based approaches[31, 32].

A solution which negates the need to turn the traps off is to work at a wavelength where the trap is attractive for both ground state and Rydberg atoms. This can be achieved in a blue detuned optical lattice[29] or a blue detuned bottle beam trap[33]. Bottle beam traps specifically designed for "magic" trapping of ground and Rydberg atoms are currently under development at UW Madison. It is also necessary to consider Rydberg state photoionization rates due to the trapping light. In a blue detuned trap the nucleus is in a region of low light intensity, and since the photoionization matrix element is dominated by the region near the core, the rate tends to be strongly suppressed[34].

All of these issues are challenging, but also susceptible to solution. We believe that the combination of potentially high gate fidelity, and suitable architectures for many-qubit implementations, bode well for the future development of Rydberg blockade based quantum computing.

This work was supported by the NSF under grant PHY-1005550, IARPA through ARO contract W911NF-10-1-0347, and DARPA under award FA8750-09-1-0230.

References

[1] Treutlein P, Hommelhoff P, Steinmetz T, Hänsch T W and Reichel J 2004 *Phys. Rev. Lett.* **92** 203005
[2] Brennen G K, Caves C M, Jessen P S and Deutsch I H 1999 *Phys. Rev. Lett.* **82** 1060
[3] Jaksch D, Briegel H J, Cirac J I, Gardiner C W and Zoller P 1999 *Phys. Rev. Lett.* **82** 1975
[4] Pellizzari T, Gardiner S A, Cirac J I and Zoller P 1995 *Phys. Rev. Lett.* **75** 3788
[5] You L and Chapman M S 2000 *Phys. Rev. A* **62** 052302
[6] Lukin M D and Hemmer P R 2000 *Phys. Rev. Lett.* **84** 2818–2821
[7] Mompart J, Eckert K, Ertmer W, Birkl G and Lewenstein M 2003 *Phys. Rev. Lett.* **90** 147901
[8] Mandel O, Greiner M, Widera A, Rom T, Hänsch T W and Bloch I 2003 *Nature (London)* **425** 937–940
[9] Anderlini M, Lee P J, Brown B L, Sebby-Strabley J, Phillips W D and Porto J V 2007 *Nature (London)* **448** 452–456
[10] Jaksch D, Cirac J I, Zoller P, Rolston S L, Côté R and Lukin M D 2000 *Phys. Rev. Lett.* **85** 2208–2211
[11] Nielsen M A and Chuang I L 2000 *Quantum computation and quantum information* (Cambridge University Press, Cambridge)
[12] Hagley E, Maître X, Nogues G, Wunderlich C, Brune M, Raimond J M and Haroche S 1997 *Phys. Rev. Lett.* **79** 1
[13] Saffman M, Walker T G and Mølmer K 2010 *Rev. Mod. Phys.* **82** 2313
[14] Saffman M and Walker T G 2005 *Phys. Rev. A* **72** 022347
[15] Singer K, Reetz-Lamour M, Amthor T, Marcassa L G and Weidemüller M 2004 *Phys. Rev. Lett.* **93** 163001
[16] Tong D, Farooqi S M, Stanojevic J, Krishnan S, Zhang Y P, Côté R, Eyler E E and Gould P L 2004 *Phys. Rev. Lett.* **93** 063001
[17] Urban E, Johnson T A, Henage T, Isenhower L, Yavuz D D, Walker T G and Saffman M 2009 *Nature Phys.* **5** 110
[18] Gaëtan A, Miroshnychenko Y, Wilk T, Chotia A, Viteau M, Comparat D, Pillet P, Browaeys A and Grangier P 2009 *Nature Phys.* **5** 115
[19] Lukin M D, Fleischhauer M, Cote R, Duan L M, Jaksch D, Cirac J I and Zoller P 2001 *Phys. Rev. Lett.* **87** 037901
[20] Wilk T, Gaëtan A, Evellin C, Wolters J, Miroshnychenko Y, Grangier P and Browaeys A 2010 *Phys. Rev. Lett.* **104** 010502
[21] Gaëtan A, Evellin C, Wolters J, Grangier P, Wilk T and Browaeys A 2010 *New J. Phys.* **12** 065040
[22] Isenhower L, Urban E, Zhang X L, Gill A T, Henage T, Johnson T A, Walker T G and Saffman M 2010 *Phys. Rev. Lett.* **104** 010503
[23] Zhang X L, Isenhower L, Gill A T, Walker T G and Saffman M 2010 *Phys. Rev. A* **82** 030306(R)

[24] Walker T G and Saffman M 2008 *Phys. Rev. A* **77** 032723
[25] Mamaev A V, Lodahl P and Saffman M 2003 *Opt. Lett.* **28** 31
[26] Johnson T A, Urban E, Henage T, Isenhower L, Yavuz D D, Walker T G and Saffman M 2008 *Phys. Rev. Lett.* **100** 113003
[27] Viteau M, Radogostowicz J, Chotia A, Bason M G, Malossi N, Fuso F, Ciampini D, Morsch O, Ryabtsev I I and Arimondo E 2010 *J. of Phys. B: At., Mol. and Opt. Phys.* **43** 155301
[28] Benhelm J, Kirchmair G, Roos C F and Blatt R 2008 *Nat. Phys.* **4** 463
[29] Nelson K D, Li X and Weiss D S 2007 *Nat. Phys.* **3** 556–560
[30] Knoernschild C, Zhang X L, Isenhower L, Gill A T, Lu F P, Saffman M and Kim J 2010 *Appl. Phys. Lett.* **97** 134101
[31] Bergamini S, Darquié B, Jones M, Jacubowiez L, Browaeys A and Grangier P 2004 *J. Opt. Soc. Am. B* **21** 1889–1894
[32] Kruse J, Gierl C, Schlosser M and Birkl G 2010 *Phys. Rev. A* **81** 060308
[33] Isenhower L, Williams W, Dally A and Saffman M 2009 *Opt. Lett.* **34** 1159
[34] F. Robicheaux, private communication.

22nd International Conference on Atomic Physics IOP Publishing
Journal of Physics: Conference Series **264** (2011) 012024 doi:10.1088/1742-6596/264/1/012024

Correlation measurements of individual microwave photons emitted from a symmetric cavity

D. Bozyigit, C. Lang, L. Steffen, J. M. Fink, C. Eichler, M. Baur, R. Bianchetti, P. J. Leek, S. Filipp and A. Wallraff

Department of Physics, ETH Zürich, CH-8093, Zürich, Switzerland.

E-mail: andreas.wallraff@phys.ethz.ch

M. P. da Silva, A. Blais

Département de Physique, Université de Sherbrooke, Sherbrooke, Québec, J1K 2R1 Canada.

Abstract. Superconducting circuits have been successfully established as systems to prepare and investigate microwave light fields at the quantum level. In contrast to optical experiments where light is detected using photon counters, microwaves are usually measured with well developed linear amplifiers. This makes measurements of correlation functions - one of the important tools in optics - harder to achieve because they traditionally rely on photon counters and beam splitters. Here, we demonstrate a system where we can prepare on demand single microwave photons in a cavity and detect them at the two outputs of the cavity using linear amplifiers. Together with efficient data processing, this allows us to measure different observables of the cavity photons, including the first-order correlation function. Using these techniques we demonstrate cooling of a thermal background field in the cavity.

Correlation function measurements are one of the most important tools in quantum optics to investigate the properties of radiation [1]. In such measurements one observes the ability of light to interfere with itself as quantified by first and second-order correlation functions [2, 3]. The radiation to be characterized is usually passed through a beam splitter and subsequently its interference is observed in Mach-Zehnder interferometer or Hanbury Brown-Twiss type setups [4]. At optical frequencies the light is commonly detected using readily available photon counters. In the microwave regime, however, no efficient photon counters exist to date, as photons at these much lower frequencies carry an energy that is orders of magnitude less than at optical frequencies. Instead, most experiments rely on a detection of electric field quadratures using linear amplifiers and oscilloscopes. Although linear microwave amplifiers can be used to reliably detect fields on the single photon level using averaging techniques [5] they necessarily add noise to the signal [6]. Despite the lack of photon detectors, it was shown recently [7] that correlation functions are also accessible in electric field quadratures detection schemes.

In the microwave regime beam splitters are typically implemented as so called hybrids [8]. Similar to the optical case, these are devices that rely on the interference of light propagating along different paths. Therefore, their size is determined by the radiation wavelength which is on the order of 1 cm at microwave frequencies. This makes an on-chip beam splitter a sizable device, which is not yet commonly used in quantum microwave experimental setups. While first realizations in superconducting microwave electronics are being demonstrated [9–12] they still

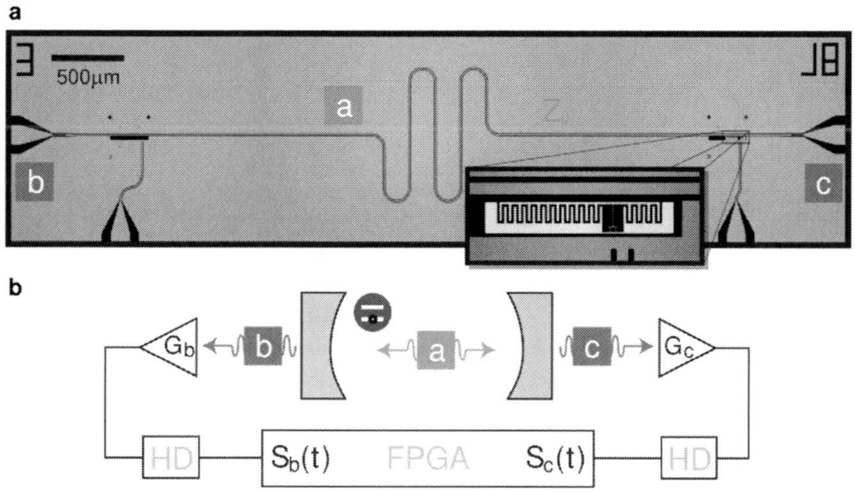

Figure 1. Experimental setup. **a**, Microscope image of our sample with superconducting microwave transmission line resonator and transmon qubit in the resonator gap (see inset). Two output lines are coupled symmetrically on the left and right. **b**, Schematic of the experimental setup. A two-level system emits a photon into the cavity mode a defined by two mirrors. Linear amplifiers with gain $G_{b,c}$ couple symmetrically to the cavity outputs amplifying the radiation in output modes b and c. Heterodyne detectors (HD) extract both quadratures of the output fields and feed results to field programmable gate array (FPGA) based digital correlation electronics.

require additional effort in the fabrication process.

Here, we present measurements of the first-order correlation function of single microwave photons. We explore an architecture which features two channel detection without employing a beam splitter by using the two outputs of a symmetric microwave cavity. Furthermore we demonstrate that by recording the time series of the detected quadrature signal – instead of its time average – and using efficient digital signal processing, we can extract the first-order correlation function of radiation generated by microwave frequency emitters. Interestingly they allow us to observe the cooling of a thermal background field in the resonator through the interaction with the qubit. The measurements presented here were performed before samples with integrated beam splitters were available to us. Nevertheless, they demonstrate the feasibility of our detection technique. Based on these results we have performed measurements of the second-order correlation function in an optimized setup which allows the observation of strictly quantum mechanical effects such as single photon antibunching [12].

For our experiments we have realized a superconducting electronic circuit (Fig. 1a), similar to the one presented in [5]. In our circuit, we coherently and controllably couple a single qubit to a high quality resonator to create an individual photon on demand. The superconducting transmon qubit [13] used in this experiment is characterized by its maximum Josephson energy $E_{\mathrm{J,max}} \approx 14.4\,\mathrm{GHz}$, its charging energy $E_{\mathrm{C}} \approx 500\,\mathrm{MHz}$ and its energy relaxation and dephasing times in excess of a few hundred nanoseconds. The transition frequency of the qubit is flux tunable using both a quasi-static magnetic field generated with a miniature coil and an on chip transmission line to generate nanosecond time scale flux pulses. By integrating our qubit into a superconducting coplanar transmission line resonator of frequency $\nu_{\mathrm{r}} = 6.433\,\mathrm{GHz}$ and quality factor $Q = 2060$ we couple it strongly to a single mode a of the radiation field stored in the resonator. This approach is known as circuit quantum electrodynamics [14] and allows to study

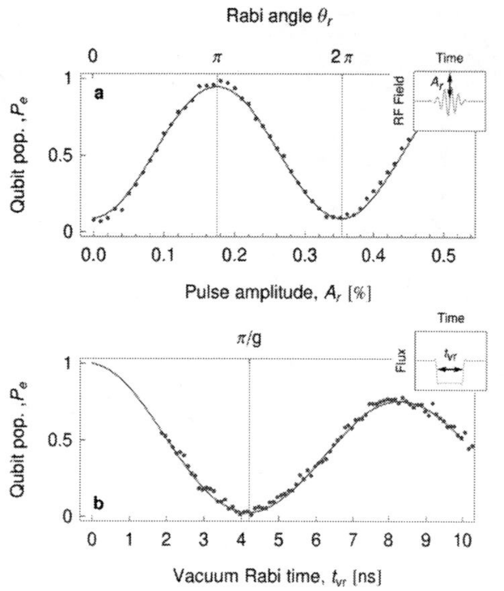

Figure 2. Control of qubit state and qubit/resonator interaction. a, Qubit excited state population P_e versus Rabi pulse amplitude A_r. The Rabi rotation angle θ_r is indicated on the top axis. The used pulse shape is shown in the inset. **b**, P_e versus qubit-resonator interaction time t_{vr} with the used flux pulse shape in the inset. Note that the interaction time can experimentally not be made shorter than 1.9 ns due to the finite rise time of the pulses. Dots are data and lines are fits to an analytic model of an exponentially decaying sinusoid.

in exquisite detail the interaction of quantum two-level systems with quantized radiation fields.

In this system we implement a single photon source using the following procedure. Applying a phase controlled truncated Gaussian microwave pulse of variable amplitude A_r and total duration $t_r = 18\,\text{ns}$ to the qubit biased at a transition frequency of $\nu_q = 6.933\,\text{GHz}$, we prepare a superposition state $|\psi_q\rangle = \alpha|g\rangle + \beta|e\rangle$ between the qubit ground $|g\rangle$ and excited states $|e\rangle$. The superposition is characterized by the two complex probability amplitudes $\alpha = \cos(\theta_r/2)$ and $\beta = \sin(\theta_r/2)e^{i\phi}$ which are parameterized by the polar (Rabi) angle θ_r and the phase angle ϕ. The phase angle ϕ is expressed relative to the phase of the local oscillator used in the heterodyne detection scheme and therefore can always be chosen to be zero. We characterize the prepared qubit state using a pulsed dispersive measurement of the resonator transmission [15] and clearly observe Rabi oscillations in the qubit population $P_e = |\beta|^2$ versus the amplitude $A_r \propto \theta_r$ (Fig. 2a). After qubit state preparation, we apply a current pulse of controlled amplitude and duration to the flux bias line to tune the qubit transition frequency into resonance with the resonator frequency ν_r. We time-resolve the resonant vacuum Rabi oscillations of the coupled system at a frequency of $2g/(2\pi) = 118\,\text{MHz}$ by dispersively measuring the qubit state after it has been tuned back to the frequency ν_q strongly detuned from the resonator (Fig. 2b). Adjusting the qubit-resonator interaction time to half a vacuum Rabi period $t_{vr} = \pi/g = 4.2\,\text{ns}$, we coherently map the qubit state $|\psi_q\rangle$ to an equivalent superposition state $|\psi_c\rangle = \alpha|0\rangle + \beta|1\rangle$ of the $|0\rangle$ and $|1\rangle$ photon Fock states stored in the resonator mode a, while the qubit returns to the ground state. Similar techniques have been used to prepare and measure a wide range of intra-cavity photon superposition states in recent experiments both with superconducting circuits [16] and with Rydberg atoms [17]. It is important to note that this technique relies on a reliable preparation of the vacuum state in the resonator prior to the controlled vacuum Rabi oscillations. The choice of the interaction time t_{vr} is based on the assumption that the coupled system carries maximally one excitation for which the vacuum Rabi frequency is $2g/(2\pi)$. If the resonator is in a weak thermal state the final state will not be $|\psi_c\rangle$ but generally an entangled state between qubit and resonator as discussed at the end of this report.

In the following we discuss the characterization of the photon states in the cavity by measuring the quadrature amplitudes of the microwave fields emitted from *both* ends of the cavity, see schematic in Fig. 1b. For this purpose we have realized a symmetric circuit QED setup in which the output fields are detected independently and simultaneously at both ports of the cavity. Our scheme comprises two independent detection chains similar to the one pioneered by Gabelli *et al.* [9]. Each chain consists of a cold amplifier with gain $G_{b,c} \approx 33\,\mathrm{dB}$ and noise temperature $T_{\mathrm{N(b,c)}} \approx 4.5\,\mathrm{K}$ followed by a two stage heterodyne detector in which the signal is down converted from the resonator frequency to 25 MHz in an analog stage and to d.c. in a digital homodyne stage. This allows measurement of the electric field at both ends of the resonator characterizing the radiation emitted into the modes labeled b and c. In this way, we extract the complex envelope [18] $S_{b,c}(t)$ of the amplified electric field $E_{b,c}^{(+)}$ described by

$$E_{b,c}^{(+)}(t) + N_{b,c}(t) = S_{b,c}(t)e^{-2\pi i\nu_r t} \tag{1}$$

where the real and imaginary part of $S_{b,c}(t)$ are the two field quadratures in the frame rotating at the resonator frequency and $N_{b,c}(t)$ is the noise added by the amplifier. Using input-output theory [19], one can show that the full information about the intra-cavity mode a can be extracted from a measurement of the propagating modes b and c [20]. It is interesting to note that our setup is essentially equivalent to one in which the radiation emitted by a source is investigated at the two output ports of a beam splitter [20]. Considering the expressions for all field operators in the Heisenberg picture we find a formal equivalence, where the resonator mode a is equivalent to the beam splitter input mode. Furthermore, the modes b and c are equivalent to the beam splitter output modes and the cavity input modes are equivalent to the vacuum port of the beam splitter. This allows us to realize a Mach-Zehnder interferometer or a Hanbury Brown-Twiss type setup to investigate the radiation in the mode a without explicitly implementing a microwave beam splitter.

As a first step, we characterize zero and one photon superposition states by measuring the time dependence of the average quadrature amplitudes of the electric field of the output mode b at one end of the resonator. This gives us access to the expectation value of the annihilation operator of the cavity field $\langle S_b(t)\rangle \propto \langle a(t)\rangle$ [20]. Similar measurements were presented in Ref. [5], where the cavity photon was created by Purcell limited spontaneous emission. Figure 3a shows the real part of $\langle S_b(t)\rangle$ versus time t after the preparation of the photon superposition state $|\psi_c\rangle$ characterized by the qubit Rabi angle θ_r used for its preparation. We find excellent agreement with the expected average field quadrature amplitude $\langle a\rangle \propto \sin(\theta_r)/2$ (Fig. 3c), where we get the largest signals for the superposition states $|\psi_c^+\rangle = (|0\rangle + |1\rangle)/\sqrt{2}$ and $|\psi_c^-\rangle = (|0\rangle - |1\rangle)/\sqrt{2}$ prepared using $\theta_r = \pi/2$ and $3\pi/2$, respectively. As expected from the uncertainty principle, the Fock states $|0\rangle$ and $|1\rangle$ prepared with $\theta_r = 0$ and π, respectively, do not show any quadrature amplitude signals (Fig. 3a) since their phase is completely uncertain. For all of the above measurements, the overall global phase of the signals is adjusted such that the imaginary part of $\langle S_b(t)\rangle$ is equal to zero which therefore is not displayed. We also note that the amplifier noise averages to zero in the quadrature amplitude measurement. Moreover, the time dependence of all measurement traces is well understood. For the state $|\psi_c^+\rangle$ (Fig. 3b), for example, the characteristic decay time is given by twice the cavity decay time $2T_\kappa = Q/\pi\nu_r = 102\,\mathrm{ns}$. The rise time, which should ideally be $t_{\mathrm{vr}} = 4.2\,\mathrm{ns}$, is limited by the bandwidth $BW = 15\,\mathrm{MHz}$ of our detection scheme.

In our measurement scheme, we simultaneously record the time dependent quadrature amplitudes $S_{b,c}(t)$ detected at both ports of the cavity for *each* single photon that we generate. This is realized using a two channel analog-to-digital converter (ADC) with a time resolution of 10 ns. Based on the measurement record of each event, we can then calculate *any* expectation value, such as averages, products or correlations, that can be expressed in terms of the detected output signals $S_{b,c}(t)$. Processing this data in real time using field programmable gate array (FPGA) electronics allows us to efficiently extract information even in the presence of substantial noise added by the amplifier.

As a first experiment taking advantage of this scheme, we have measured the expectation value of the instantaneous power $\langle S_b^*(t)S_b(t)\rangle$ emitted into the output mode b with the cavity

Journal of Physics: Conference Series **264** (2011) 012024 doi:10.1088/1742-6596/264/1/012024

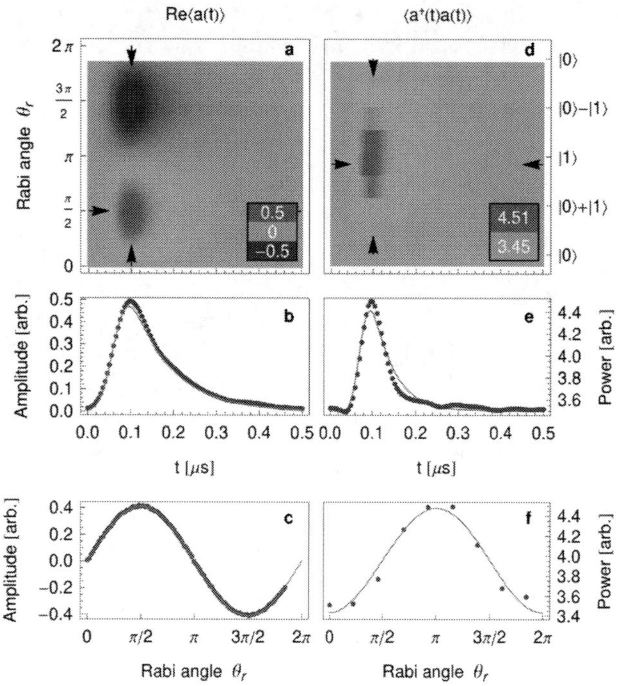

Figure 3. Quadrature amplitude and cross-power measurements. **a**, Measured time dependence of cavity field quadrature amplitude of mode b for zero and one photon superposition states characterized by the Rabi angle θ_r (left axis) and generated state (right axis). **b**, Single quadrature trace at $\theta_\mathrm{r} = \pi/2$ in **a** corresponding to $(|0\rangle + |1\rangle)/\sqrt{2}$. **c**, Dependence of maximum quadrature amplitude on θ_r. **d**, Measured time dependence of cross-power between modes b and c for the same preparation as in **a**. **e**, Single power trace at $\theta_\mathrm{r} = \pi$ in **c** corresponding to $|1\rangle$. **f**, Dependence of maximum cross-power on θ_r. Dots are data and lines are theory, see text.

mode a prepared in the Fock state $|1\rangle$ (not shown). It is important to note that we have digitally calculated the power by multiplying the quadrature amplitudes $S_b(t)$ for each single photon pulse instead of using a diode as a power meter in which the detection and the averaging is realized within the detector [5]. In this power measurement, the detected noise power of the amplifier dominates by a factor of about 1600 over the single photon power which is still observed using sufficient averaging (similar to that in Fig. 3e). By identifying the measured single photon pulse with the energy $h\nu_\mathrm{r}$ we can compare the measured average power $\bar{P}$ to the power of the noise background $\bar{P}(N_b)$ to determine the system noise temperature as

$$T_{\mathrm{N(sys)}} \approx \frac{h\nu_\mathrm{r}}{k_\mathrm{B}} \cdot \frac{\eta_\mathrm{filter}}{BWt_\mathrm{p}} \cdot \frac{1}{2}\frac{\bar{P}(N_b)}{\bar{P} - \bar{P}(N_b)} \approx 25\,\mathrm{K}, \tag{2}$$

with Planck constant h, Boltzmann constant k_B, and the time between photon preparations $t_\mathrm{p} = 512\,\mathrm{ns}$. Finally we account for the fact that the finite bandwidth of the detection chain filters out a part of the single photon pulse by introducing the efficiency $\eta_\mathrm{filter} = 0.78$ given by the overlap of the photon spectrum and the detection bandwidth. The factor $1/2$ reflects the fact that we only detect half of all emitted photons on one side of the cavity. We find that $T_{\mathrm{N(sys)}}$ is substantially higher than the noise temperature of the amplifiers because of absorption in the cables and insertion loss of components in the detection chain.

Calculating the cross-power $\langle S_b^*(t)S_c(t)\rangle$ between the two output modes instead of the direct power emitted into just a single output mode, we can reject most of the noise added by the amplifiers, demonstrating the versatility of our digital scheme. The detected cross-power is related to the average photon number in the cavity as [20]

$$\langle S_b^*(t)S_c(t)\rangle \propto \langle a^\dagger(t)a(t)\rangle + P(N_{bc}), \tag{3}$$

where $P(N_{bc})$ is the power of correlated noise between channels b and c. In these measurements, the detected noise cross-power is significantly smaller than the direct noise power of each amplifier as the two detection chains add predominantly uncorrelated noise. The equivalent temperature of the residual correlations is found to be 520 mK in these particular measurements, which can only be partly explained by the physical temperature of the sample. Therefore we conclude that a substantial part of these residual correlations are of technical origin, such as insufficient isolation of the two amplifier chains, correlated digitizer noise and residual resonator thermal noise due to incomplete thermalization of the resonator inputs. In more recent measurements where we use a similar two channel detection setup with improved thermalization and better insolation we find correlated noise with an equivalent temperature of about 80 mK which corresponds to a correlated noise power much smaller than the output power of the single photon source.

We have characterized the measured cross-power of our single photon source for the same set of cavity superposition states as used for the quadrature amplitude measurements (Fig. 3d). We find excellent agreement of the temporal evolution of the cavity photon number in dependence on the preparation angle of the photon state $\langle a^\dagger a\rangle \propto \sin^2(\theta_r/2)$ (Fig. 3f). The maximum cross-power is measured for the Fock state $|1\rangle$ ($\theta_r = \pi$) and the minimum power for the $|0\rangle$ state ($\theta_r = 0$ or 2π) (Fig. 3d).

Finally, we have characterized our single photon source using time-dependent first-order cross-correlation measurements of the two output modes of the resonator

$$\Gamma^{(1)}(\tau) = \int \langle S_b^*(t) S_c(t+\tau)\rangle dt\,. \tag{4}$$

For this purpose, we generate a train of 40 single photon pulses, each created using the procedure described above, with a pulse separation of $t_\mathrm{p} = 512\,\mathrm{ns}$ which is much greater than the qubit and cavity decay times. To remove the background of the correlated noise, we subtract the measured correlation function $\Gamma^{(1)}_{ss}(\tau)$ in the resonator steady-state from the signal acquired when performing the photon state preparation sequence $\Gamma^{(1)}(\tau)$. From the recorded quadrature amplitude data, we calculate in real-time

$$\Gamma^{(1)}(\tau) - \Gamma^{(1)}_{ss}(\tau) \propto G^{(1)}(\tau)\,, \tag{5}$$

which gives us access to the first-order correlation function $G^{(1)}(\tau) = \int\langle a^\dagger(t)a(t+\tau)\rangle dt$ of the resonator field [20]. To measure each trace in Fig. 4a, 128×10^6 trains of 40 photons were prepared in a specific state and $G^{(1)}(\tau)$ was calculated in real time using our FPGA based electronics, corresponding to more than 1 terabyte of data that have been evaluated in approximately 1 hour.

The correlation function data $G^{(1)}(\tau)$ (Fig. 4a) is characterized by a set of peaks that are separated by the repetition time t_p of the single photon source. The amplitude of $G^{(1)}$ at $\tau = 0$ and $\tau = nt_\mathrm{p}$, representing the correlation between a pulse i and $i+n$, depends in a characteristic fashion on θ_r. For the Fock state $|1\rangle$ (at $\theta_r = \pi$), the correlation function $G^{(1)}(0)$ is at a maximum and vanishes at $G^{(1)}(nt_\mathrm{p})$ as there is no coherence between photons emitted from the source at different times. In fact, $G^{(1)}(0) \propto \langle a^\dagger a\rangle \propto |\beta|^2 = \sin(\theta_r/2)^2$ oscillates sinusoidally with the preparation angle, as it essentially measures the average photon number of the generated field (Fig. 4b). For photon superposition states, the expectation values of $\langle a^\dagger\rangle$ and $\langle a\rangle$ of subsequently generated photon states have non-vanishing values, as discussed before. Since photons from different repetitions of the experiments are uncorrelated, $G^{(1)}(nt_\mathrm{p}) \propto \langle a^\dagger\rangle\langle a\rangle$ which has a finite value and oscillates at half the period. Thus, $G^{(1)}(nt_\mathrm{p}) \propto |\alpha\beta|^2 = \sin^2(\theta_r)/4$ is maximized for the states $|\psi_\mathrm{c}^+\rangle$ and $|\psi_\mathrm{c}^-\rangle$ (Fig. 4b).

The observed features of the first-order correlation function together with the results presented in Fig. 3 confirm that we have implemented a deterministic single photon source. Considering the well controlled procedure implemented for generating the single photon pulses,

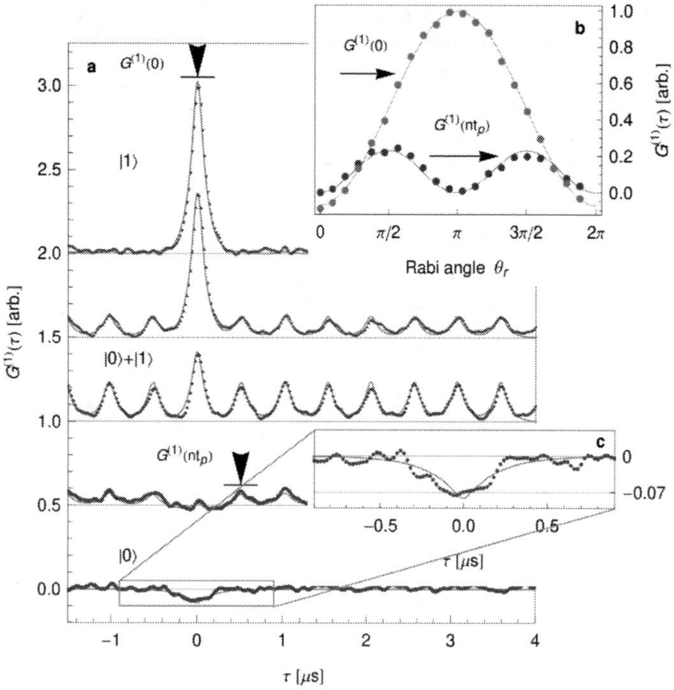

Figure 4. Correlation function measurements. a, Time dependence of first-order correlation function of the cavity field $G^{(1)}$ for indicated states. Data is offset for improved visibility. **b**, Correlation function at $\tau = 0$ and $\tau = nt_p$ versus θ_r. Dots are data, lines are predictions. **c**, Enlarged trace for $\theta_r = 0$ from **a** displaying cavity field cooling.

any other pulsed coherent or thermal field can essentially be ruled out as being the source of the measured correlations. This indicates for our scheme that the quantum mechanical properties of the radiation are not lost in the detection chain but are fully retained in the form of statistical information of the data and can be extracted by appropriate single-shot analysis. Measurements of the second-order correlation function $G^{(2)}$ provide unambiguous proof of the quantum character of the field, independent of any prior knowledge about its source [1, 12].

Interestingly, our time-dependent first-order correlation function measurements allow us to observe cooling of the resonator field through its interaction with the qubit in its ground state. If the thermal occupation of the qubit excited state is substantially smaller than the thermal occupation of the resonator, the qubit can absorb a photon from the resonator during the interaction time t_{vr}. The qubit then emits the photon into the environment at a different frequency thereby cooling the resonator field. If we take the theoretical values for $G^{(1)}$ and correct them to first order for a small thermal state in the resonator with mean photon number $n_{bg} \ll 1$ during the state preparation, we find for the first-order correlation measurements with background subtraction

$$G^{(1)}(nt_p) \propto (\sin^2(\theta_r)/4)(1 - n_{bg}(1 + \cos\sqrt{2}\pi))^2 \tag{6}$$

$$G^{(1)}(0) \propto \sin(\theta_r/2)^2(1 + n_{bg}\sin^2\sqrt{2}\pi) - n_{bg}. \tag{7}$$

where the factors $\sqrt{2}$ indicate the faster vacuum Rabi oscillation of two excitations in the system. When we try to prepare the vacuum state in the cavity ($\theta_r = 0$), the value of $G^{(1)}(0) = -n_{bg}$ is a direct measure of the background field photon number which is cooled by the qubit interaction. In our measurement in Fig. 4c this is observable as a pronounced dip in the measured $G^{(1)}(\tau)$ around $\tau = 0$, which allows us to extract the thermal background population of the resonator $n_{bg} = 0.07$ corresponding to a field temperature of $T_{bg} = 115\,\text{mK}$. This temperature is in good agreement with independent measurements of T_{bg} from the vacuum Rabi mode splitting spectrum [21]. Analyzing the time dependent correlation function measurements in the presence of a weak thermal background field, we find excellent agreement between our data and Eq. (6) for all prepared field states (see solid lines in Fig. 4).

Our experiments clearly demonstrate that correlation function measurements based on quadrature amplitude measurements can be used to characterize quantum properties of propagating microwave frequency radiation fields. Even in the presence of noise added by the amplifier, efficient data processing techniques and two channel detection allow measurements of higher statistical moments of the fields. We have also demonstrated that the cavity field can equivalently be characterized measuring both cavity outputs using a two channel detection method instead of using a beam splitter. When better, possibly quantum limited, amplifiers [22] become available the demonstrated techniques may also help to enable the full tomography of propagating radiation fields. Furthermore, the flexibility of circuit design and the high level of control achievable in circuit QED will enable a variety of future experiments with quantum microwave fields for basic research and applications.

Acknowledgments

We thank C. M. Caves, D. Esteve and especially K. W. Lehnert for very fruitful discussions. This work was supported by the European Research Council (ERC) through a Starting Grant and by ETHZ. M. P. S. was supported by a NSERC postdoctoral fellowship. A. B. was supported by NSERC, CIFAR, and the Alfred P. Sloan Foundation.

References

[1] Walls D and Milburn G 1994 *Quantum Optics* (Berlin: Springer Verlag)
[2] Glauber R J 1963 *Phys. Rev.* **130** 2529–2539
[3] Glauber R J 1963 *Phys. Rev. Lett.* **10** 84–86
[4] Brown R H and Twiss R Q 1957 *Proc. R. Soc. (London) A* **242** 300–324
[5] Houck A, Schuster D, Gambetta J, Schreier J, Johnson B, Chow J, Frunzio L, Majer J, Devoret M, Girvin S and Schoelkopf R 2007 *Nature* **449** 328–331
[6] Caves C M 1982 *Phys. Rev. D* **26** 1817–1839
[7] Grosse N B, Symul T, Stobinska M, Ralph T C and Lam P K 2007 *Phys. Rev. Lett.* **98** 153603
[8] Pozar D M 1993 *Microwave Engineering* (Addison-Wesley Publishing Company)
[9] Gabelli J, Reydellet L H, Feve G, Berroir J M, Placais B, Roche P and Glattli D C 2004 *Phys. Rev. Lett.* **93** 056801
[10] Mariantoni M, Menzel E P, Deppe F, Araque Caballero M A, Baust A, Niemczyk T, Hoffmann E, Solano E, Marx A and Gross R 2010 *Phys. Rev. Lett.* **105** 133601
[11] Menzel E P, Deppe F, Mariantoni M, Araque Caballero M A, Baust A, Niemczyk T, Hoffmann E, Marx A, Solano E and Gross R 2010 *Phys. Rev. Lett.* **105** 100401
[12] Bozyigit D, Lang C, Steffen L, Fink J M, Eichler C, Baur M, Bianchetti R, Leek P J, Filipp S, da Silva M P, Blais A and Wallraff A 2010 *Nat. Phys.*, in press
[13] Koch J, Yu T M, Gambetta J, Houck A A, Schuster D I, Majer J, Blais A, Devoret M H, Girvin S M and Schoelkopf R J 2007 *Phys. Rev. A* **76** 042319
[14] Wallraff A, Schuster D I, Blais A, Frunzio L, Huang R S, Majer J, Kumar S, Girvin S M and Schoelkopf R J 2004 *Nature* **431** 162–167
[15] Bianchetti R, Filipp S, Baur M, Fink J M, Göppl M, Leek P J, Steffen L, Blais A and Wallraff A 2009 *Phys. Rev. A* **80** 043840
[16] Hofheinz M, Wang H, Ansmann M, Bialczak R C, Lucero E, Neeley M, O'Connell A D, Sank D, Wenner J, Martinis J M and Cleland A N 2009 *Nature* **459** 546–549
[17] Deleglise S, Dotsenko I, Sayrin C, Bernu J, Brune M, Raimond J M and Haroche S 2008 *Nature* **455** 510–514
[18] Mandel L and Wolf E 1995 *Optical Coherence and Quantum Optics* (Cambridge University Press)
[19] Gardiner C W and Collett M J 1985 *Phys. Rev. A* **31** 3761–3774
[20] da Silva M P, Bozyigit D, Wallraff A and Blais A 2010 *Phys. Rev. A* **82** 043804
[21] Fink J M, Baur M, Bianchetti R, Filipp S, Göppl M, Leek P J, Steffen L, Blais A and Wallraff A 2009 *Phys. Scr.* **T137** 014013
[22] Castellanos-Beltran M A, Irwin K D, Hilton G C, Vale L R and Lehnert K W 2008 *Nat. Phys.* **4** 929–931

From cavity electromechanics to cavity optomechanics

C. A. Regal and K. W. Lehnert

JILA, National Institute of Standards and Technology and the University of Colorado, and Department of Physics, University of Colorado, Boulder, Colorado 80309, USA

E-mail: `regal@colorado.edu`

Abstract. We present an overview of experimental work to embed high-Q mesoscopic mechanical oscillators in microwave and optical cavities. Based upon recent progress, the prospect for a broad field of "cavity quantum mechanics" is very real. These systems introduce mesoscopic mechanical oscillators as a new quantum resource and also inherently couple their motion to photons throughout the electromagnetic spectrum.

Mechanical oscillators coupled to the electromagnetic mode of a cavity have emerged as an important new frontier in quantum optics. By utilizing low-optical-loss and high-Q nano- and micromechanical elements, researchers can now achieve significant coupling to the cavity mode compared to mechanical decoherence. This coupling presents an opportunity for ground-state cooling of the mechanical oscillator and even strong many-photon coupling between the mechanical oscillator and cavity mode. A unique feature of cavity mechanics is that the coupling does not rely on resonant behavior of the intracavity element. Thus, in principle, the concept works for any frequency of light where a mechanical element can be integrated into a high-finesse cavity. Indeed, experiments are proceeding in parallel throughout the electromagnetic spectrum.

Near ground-state cooling has now been achieved with both metallic oscillators coupled to microwave superconducting resonators [1, 2] and mirrors coupled to high-finesse optical cavities [3-5]. While in one case the electromagnetic resonance is defined by an LC oscillator and in the other case by the distance between two mirrors (Fig. 1), both employ the same parametric coupling. The motion of the mechanical oscillator changes the cavity resonance according to a coupling coefficient g; the term in the Hamiltonian describing this radiation pressure interaction is $H_{\rm int} = \hbar g x_{\rm zp} a^\dagger a(d^\dagger + d)$. Here a represents the cavity mode, and the position operator of the mechanical membrane is $\hat{x} = x_{\rm zp}(d^\dagger + d)$, where $x_{\rm zp}$ is the mechanical oscillator's zero point motion.

In this article, we review some of the progress towards quantum control of cavity mechanical systems, addressing both electromechanical and optomechanical realizations. Similar themes are found for the mechanical oscillators used for both electromechanics and optomechanics. Thus, we can propose that a mechanical oscillator could be physically common to two cavities, one at an optical frequency and one at a microwave frequency. Such an interface, in combination with strong coupling and slow decoherence, could allow for protocols such as quantum-state transfer between microwave and optical photons.

One of the first goals within the cavity mechanics community has been ground-state cooling of a mechanical oscillator by harnessing the radiation pressure within a cavity. This cooling has

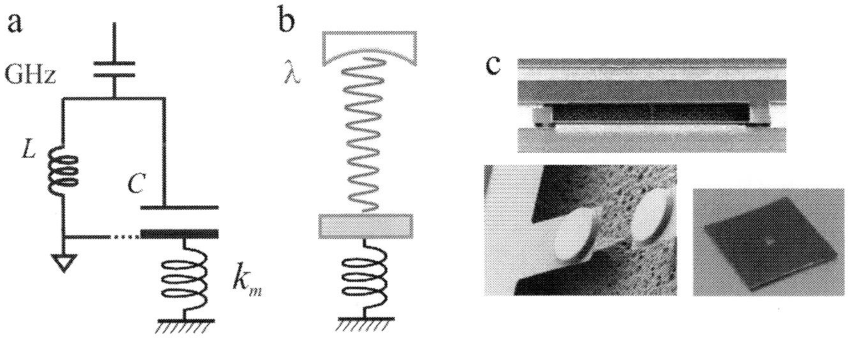

Figure 1. (a) Schematic of an electromagnetic cavity (LC oscillator) linearly coupled to the motion of a mechanical oscillator with spring constant $k_m = m\omega_m^2$. (b) The optical analog. A Fabry-Perot optical cavity with a moving end mirror. (c) Examples of mechanical elements used in these experiments. A metallic nanostring [6], a mirror on a cantilever [7], a stoichiometric silicon nitride membrane [8].

been discussed in a number of review articles (see, for example, [9]) and would be an important demonstration of the quantum capabilities of cavity mechanics. The idea is analogous to resolved sideband cooling used for many years in atomic physics experiments [10], with the difference that the dissipation is provided by photons leaking from the cavity rather than by spontaneous emission. This idea of cavity cooling was also demonstrated in experiments with single atoms and atomic gases in cavities [11-14].

As illustrated in Fig. 2, a sufficiently long cavity lifetime allows one to resolve individual motional states, i.e., achieve the resolved sideband limit where the mechanical frequency is larger than the cavity linewidth $\omega_m \gg \gamma$ [15, 16]. To achieve ground-state cooling, one must overcome the heating from the environment. The mechanical oscillator is physically coupled to a solid-state object at a temperature $T_{\rm bath}$ and only isolated via its mechanical quality factor $Q_m = \omega_m/\gamma_m$. Its heating rate $\gamma_m n_{\rm bath}$, where $n_{\rm bath} = k_B T_{\rm bath}/\hbar\omega_m$, must be overcome by the cooling rate. At the optimum red detuning of $\omega_c - \omega_0 = \Delta = \omega_m$ and in the fully resolved sideband limit, the cooling rate is $\Gamma_{\rm cool} = 4\Gamma^2/\gamma$, where $\Gamma = g x_{\rm zp}|\alpha|$ is the effective coupling with $|\alpha|^2$ photons in the cavity. The ratio of these parameters $\gamma_m n_{\rm bath}/\Gamma_{\rm cool}$ is the final phonon occupation. This analysis is valid in the limit where Γ and $\gamma_m n_{\rm bath}$ remain small compared to other parameters in the problem. For example, an important minimum condition is $\gamma_m n_{\rm bath} \ll \omega_m$, which is equivalent to a requirement on the Q-frequency product of $Q_m\omega_m \gg k_B T_{\rm bath}/\hbar$. This puts a stringent condition on the mechanical oscillators and, for the most part, has led to a focus on cryogenic precooling of cavity mechanics experiments.

Microwave and optical systems have different advantages and disadvantages for realizing cavity cooling to the ground state. Optical systems appear at first glance to be the most natural for harnessing radiation pressure in cavities. The coupling can naturally be quite large; for a moving end mirror the coupling is given by $g = \omega_c/L$. Hence, for 1 μm light and a mm-length cavity, $g/2\pi \sim 300$ kHz/pm. Although the mirrors typically used in optical cavities are too massive to be strongly influenced by radiation pressure, advances in the microfabrication of optical quality elements have allowed rapid progress. For example, experimenters have used microcavity technology [4, 5] and placed a nm-thick mechanical membrane inside of a traditional high-finesse Fabry-Perot cavity [17]. The challenge of optomechanics, however, is its compatibility with cryogenic operation. As mentioned above, with typical Q-frequency products of mechanical oscillators, a "quantum-enabled" system with $Q_m\omega_m > k_B T_{\rm bath}/\hbar$ is most likely to

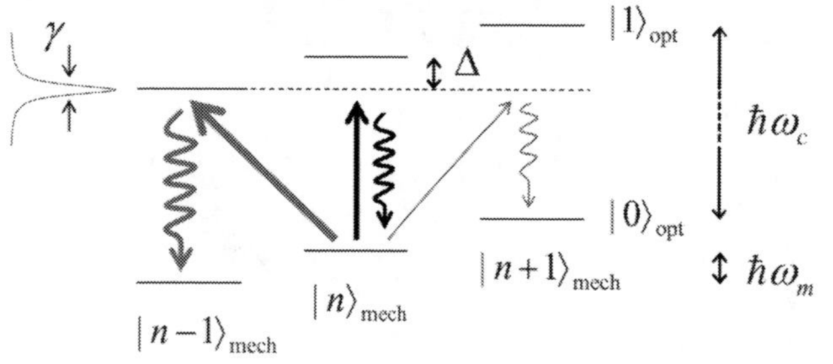

Figure 2. Ladder of mechanical oscillator motional states demonstrating resolved-sideband cavity cooling. The excitation laser is red-detuned of the cavity transition, ω_c, and the dominant transition is indicated by the largest arrows on the left. This absorption followed by emission of a higher-energy photon gives rise to mechanical cooling.

be realized with cryogenic precooling. However, alignment of high-finesse cavities at cryogenic temperatures is tricky, and absorption of optical power that heats $T_{\rm bath}$ must be avoided by use of low-loss optical materials. Despite these challenges, a number of experiments using a combination of cryogenic and cavity cooling have now reached the level of tens of phonons occupation in a mechanical oscillator.

Superconducting microwave cavities, on the other hand, are naturally compatible with a dilution refrigerator environment. This compatibility provides a large advantage for the amount of cavity cooling required. For example, starting at 50 mK with a 10 MHz mechanical oscillator leaves only $n_{\rm bath} = 100$. (In fact by using a GHz mechanical oscillator, ground-state cooling has recently been achieved with only conventional refrigeration techniques [18].) With a traditional microwave cavity at the cm scale, however, the coupling to a micromechanical object would be very small. Utilizing the one-dimensional (1D) geometry of a superconducting stripline cavity helped realize sufficient coupling to demonstrate microwave cavity cooling of cold oscillators [19, 1, 2], but increasing the coupling still remained a challenge. The most straightforward, but difficult, way to improve the coupling is to squeeze the capacitor plates incorporating the mechanical element closer and closer together. Another route is engineering the impedance $Z = \sqrt{L/C}$. This route allows the capacitance from the mechanical object to be the dominant capacitance while maintaining GHz resonant frequencies [20]. Recent experiments advancing both of these goals have allowed for a very large coupling $g/2\pi \sim 50$ kHz/pm [21] that is surprisingly comparable to optomechanical systems and should allow cooling to very low occupation [22].

Another important difference lies in detection. An optical interferometer offers a well-tested way of measuring the mechanical motion near the quantum limit, while in electromechanical systems, the technology is just being developed. Figure 3(a) illustrates a typical setup for interferometric measurement of the mechanical motion in the microwave domain. A key component is the cold microwave amplifier; the best available commercial amplifiers used in early experiments added 30 quanta of noise [6]. With these added quanta, the detection is equivalent to a square-law detector with a quantum efficiency of only $\eta = (1 + 2n_{\rm add})^{-1} \sim 2\%$. To allow measurement of motion near the quantum ground state in a feasible amount of time, new amplifier technology needed to be developed. In fact tunable Josephson parametric amplifiers (JPA) capable of detecting microwave fields with an added noise of less than half a quantum have recently been implemented [23, 20].

The mechanical elements used in both electromechanical and optomechanical experiments draw on a large variety of high-Q oscillators. Their masses range from the gram-scale mirrors that have been utilized by the LIGO collaboration [24] to picogram masses of 100 nm cross-section beams [6, 25]. For the micro- and nanomechanical elements, our field draws heavily on the success of micro- and nanoelectromechanical systems (MEMS and NEMS) [26]. Extensive

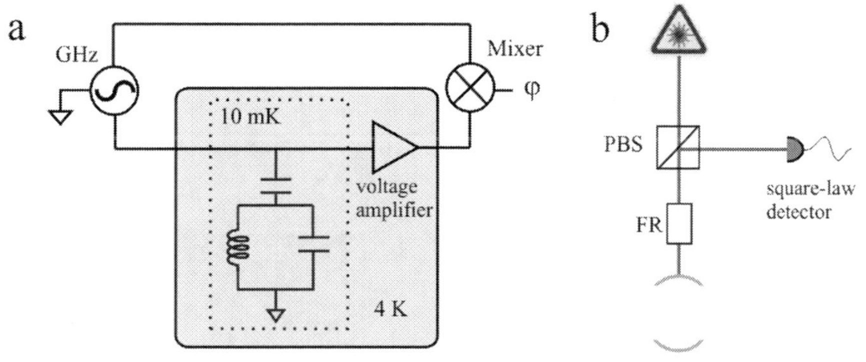

Figure 3. Sketches of interferometers for motion detection. In experiments thus far, the cavity is used as a transducer to measure the mechanical fluctuations. (a) Microwave interferometer. A typical cryogenic voltage amplifier is a high electron mobility transistor (HEMT) amplifier. Recent experiments have incorporated a chain of both a Josephson parametric amplifier (JPA) and a HEMT amplifier (see text). (b) Optical interferometer incorporating a Fabry-Perot cavity and a Pound-Drever-Hall scheme. Only the most basic elements are depicted.

work to understand and develop high-quality factor, low-mass oscillators in the 1990s resulted in a variety of sensitive mass and force detectors. In the quest to realize quantum cavity mechanics, we similarly want low mass and high quality factor, but we also want to optimize upon large absolute mechanical frequency compared to T_{bath}. This optimization on large frequency is not required for a force detector where the ultimate sensitivity is limited by the thermal force noise of spectral density $S_F = 4m\gamma_m k_B T_{\mathrm{bath}}$, which depends on the properties of the mechanical oscillator only through m and γ_m. Thus, in both optomechanics and electromechanics we have gravitated towards certain classes of mechanical objects that suit our particular needs.

One example is tensioned mechanical oscillators (Fig. 4). Much of the MEMS and NEMS field is based upon flexural modes of nanobeams, but if one adds tension to a 1D or 2D oscillator to make a string or drum the mechanical frequency can be significantly increased. In addition the quality factor increases, allowing for a double benefit in the Q-frequency product. Tensioned oscillators have been used in electromechanical devices in the form of polycrystalline metallic nanostrings [6], metallized silicon nitride (SiN) strings [27], and metallic drums [21]. SiN has been used for optomechanics in the form of both drums and strings [17, 8, 25]. The optical spring effect used in the LIGO collaboration also uses a similar idea of increasing stiffness without increasing loss [24]. When the stress of SiN film is pushed to its limit (approaching 1 GPa) by using the stoichiometric form, Si_3N_4, the Q-frequency products, even at room temperature, can be quite spectacular [28, 29, 30]. In fact, they can exceed $k_B T_{\mathrm{room}}/h = 6 \times 10^{12}$ Hz and be incorporated into low-loss cavity mechanics systems [8]. This extends the prospects of quantum cavity mechanics to the room temperature domain.

Improving upon this encouraging result requires that we understand the dependence of the quality factor on tension in these resonators. The gross behavior of a significant increase in Q_m is apparent; mainly this effect can be understood simply as increasing the resonance frequency while keeping the dissipation, as expressed through the linewidth γ_m, approximately constant. For the case of SiN strings with loss dominated by coupling to uniformly-distributed localized defects, this was verified quantitatively in the experiments of Ref. [31]. The bending energy, associated with the loss, was calculated and observed to change much less than the total energy. However, the model showed some change in the linewidth with tension. This work used SiN strings of varying lengths and mode index, but it will be interesting to investigate the loss mechanisms of tensioned resonators of different materials and different geometries. A variety of loss mechanisms considered extensively for flexural modes require reanalysis for tensioned resonators. For example, Q limits due to thermoelastic loss are expected to be surprisingly large [32], and anchor or clamping losses will need to be considered quantitatively in the presence of tension.

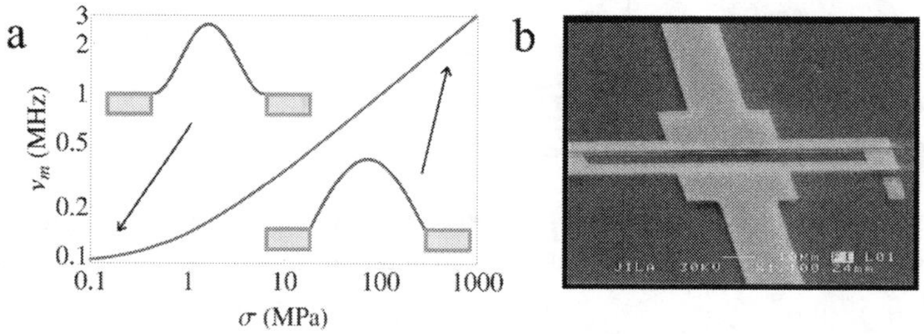

Figure 4. (a) Variation of resonant frequency with film stress σ for a beam with clamped-clamped boundary conditions [33, 31]. The typical string considered here has cross-sectional dimensions of 100 nm and is 100 μm long. Insets: Illustration of how the fundamental mode shape changes from that of a flexural mode to near a string mode. (b) Example of a metallic nanostring placed between two electrodes [picture credit P.-L. Yu].

Thus far we have considered the goal of initializing a mechanical oscillator in the ground state via cavity cooling. While this is an important indication that the system is quantum enabled, one larger goal is to determine the most appropriate role of cavity quantum mechanics within a quantum network [34]. We suspect cavity mechanics can play a unique role in quantum information via the ability to facilitate wavelength conversion. In particular, one can envision the mechanical oscillator as a mediator for quantum-state transfer between photons in two different cavities to which the mechanical oscillator is simultaneously coupled [35]. Such controlled quantum protocols require reversible energy exchange to take place faster than loss of energy to the environment. This regime of strong coupling is routinely achieved for various atomic and solid state qubits and photons. For cavity mechanical systems, the single-photon coupling rate is typically in the range of $gx_{\rm zp}/2\pi \sim 10$ Hz, which is much smaller than other rates in micromechanical cavity systems, which are typically kHz–MHz. However, it is possible to achieve a regime of strong many-photon coupling [15, 36], as pioneered in the work in Ref. [7]. Here they realized a coupling $\Gamma = gx_{\rm zp}|\alpha|$ greater than both γ and γ_m. While the ability to achieve only many-photon strong coupling precludes protocols that require nonlinearities, there are many things one can imagine doing, including quantum-state transfer, with the linear coupling.

It is possible cavity mechanics could even be useful to convert microwave and optical photons, a wavelength conversion problem that has attracted more and more interest [37, 38]. Superconducting Josephson junction-based qubits and superconducting resonant cavities have emerged as the ideal realization of quantum two-level systems interacting with a single mode of the electromagnetic spectrum. Remarkably, it is now possible to deterministically create arbitrary states of that single mode containing up to 15 photons by controllable interaction with a two-level system [39]. The single mode is a lithographically fabricated electrical circuit resonant at microwave frequencies. The superconducting circuitry provides a highly flexible platform for creating on-demand complex quantum states of a light field. Unfortunately, microwaves can only be distributed efficiently over superconducting cable. Thus, the microwave quantum states are essentially stuck inside of the dilution refrigerator where they are created, and further they cannot be preserved for times longer than $\sim 10\ \mu s$. If, however, it were possible to convert these states to optical wavelengths, they could be transmitted over kilometer distances with negligible loss. They could then be stored for long times in, for instance, atomic ensembles [40].

Explicitly, the position of the mechanical oscillator could tune the resonance frequency of two cavities, one microwave and one optical (Fig. 5). The device is described by the following Hamiltonian

$$H = H_0 + H_E + H_M + \hbar g_{OM} x_{\rm zp} a^\dagger a (d^\dagger + d) - \hbar g_{EM} x_{\rm zp} b^\dagger b (d^\dagger + d) \quad (1)$$

where H_0, H_E, and H_M are the optical, electrical, and mechanical harmonic oscillator Hamiltonians; the two subsequent terms represent the optomechanical (OM) and electromechanical (EM) interaction. The crucial point for the goal of state transfer is that by detuning a drive tone red of the cavity resonance, just as in cavity cooling, a beamsplitter interaction can be induced [41, 42]. With a cavity driven red-detuned at $\Delta = \omega_m$ in the resolved-sideband limit, the coupling in the interaction representation is $\hbar\Gamma_{OM}(\tilde{d}^\dagger \tilde{a} + \tilde{a}^\dagger \tilde{d})$, where $\Gamma_{OM} = g_{OM} x_{\rm zp} |\alpha|$. An equivalent beamsplitter interaction is found for the microwave cavity. Thus, one imagines that by appropriately strongly driving both the optical and microwave cavities red-detuned, we can swap microwave and optical photons that have very different absolute frequencies, but the same frequency difference relative to their respective drives.

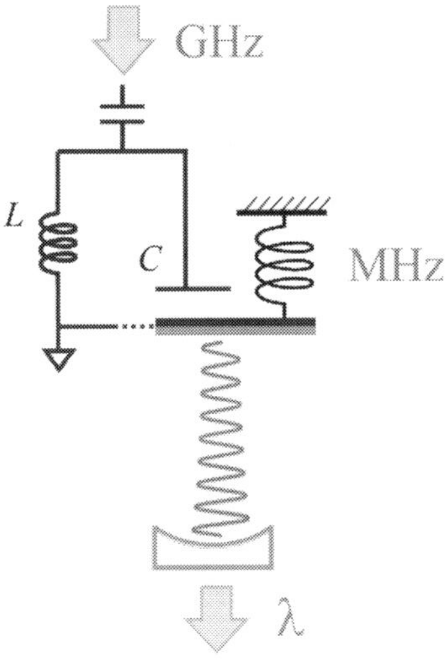

Figure 5. The ability to realize both cavity electromechanics and optomechanics allows one to envision hybrid electro-optomechanical devices. Here the position of a mechanical oscillator changes simultaneously the resonant frequencies of both the microwave and optical resonances. Note no frequency matching is required; the mechanical oscillator in the MHz regime acts as a mediator between the microwave and optical photons.

For coherent exchange of optical and microwave photons, the critical question is whether these swap interactions can be accomplished before decoherence processes destroy the quantum state. Figure 6 demonstrates the various rates that must be considered to answer this question. Very generally, the goal is to achieve strong coupling rates Γ compared to cavity decay γ or mechanical decoherence $\gamma_m n_{\rm bath}$, without exceeding $\omega_m/2$ where bistability occurs. On the practical side, there are a variety of questions to answer about the feasibility of creating the device shown in Fig. 5. While mechanical oscillators can be very versatile elements, it is not a priori clear that a common mechanical oscillator design can work as both a capacitor plate and a mirror. However, as discussed above, there do appear to be materials that could allow for such a hybrid element. In particular, tensioned-drum resonators appear to be good candidates for both electromechanics and optomechanics, and one could envision separating the metallic and dielectric parts of the drum spatially using higher-order mechanical modes.

The quest to control the motion of mesoscopic oscillators is still in its infancy compared to abilities of atomic systems such as trapped ions. Yet, the versatile nature of mechanical oscillators and the inherent coupling to the electromagnetic field that cavity mechanics provides may allow this new system to play an important role in future quantum protocols.

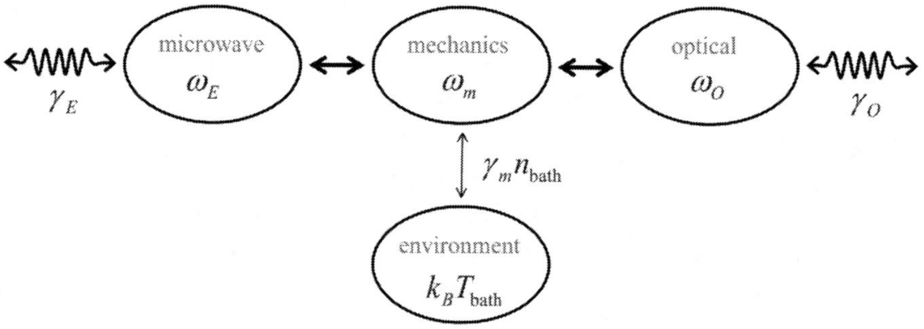

Figure 6. Channels for coupling and loss that are relevant to mechanically mediated quantum-state transfer between photons in two cavities. Here we consider the case of coupling between microwave, E, and optical, O, cavities.

Acknowledgements

We thank J. D. Teufel, R. W. Simmonds, T. Donner, H. J. Kimble, S. B. Papp, D. J. Wilson, A. A. Clerk, and J. Gambetta for helpful discussions. We acknowledge support from the Clare Boothe Luce Foundation, DARPA-QUASAR, NSF, and NIST.

References

[1] Teufel J D, Harlow J W, Regal C A and Lehnert K W 2008 *Phys. Rev. Lett.* **101** 197203
[2] Rocheleau T, Ndukum T, Macklin C, Hertzberg J B, Clerk A A and Schwab K C 2009 *Nature* **463** 72
[3] Gröblacher S, Hertzberg J B, Vanner M R, Cole G D, Gigan S, Schwab K C and Aspelmeyer M 2009 *Nature Phys.* **5** 285
[4] Schliesser A, Arcizet O, Rivière R, Anetsberger G and Kippenberg T J 2009 *Nature Phys.* **5** 509
[5] Park Y S and Wang H 2009 *Nature Phys.* **5** 489
[6] Regal C A, Teufel J D and Lehnert K W 2008 *Nature Phys.* **4** 555
[7] Gröblacher S, Hammerer K, Vanner M R and Aspelmeyer M 2009 *Nature* **460** 724
[8] Wilson D J, Regal C A, Papp S B and Kimble H J 2009 *Phys. Rev. Lett.* **103** 207204
[9] Kippenberg T and Vahala K 2007 *Opt. Exp.* **15** 17172
[10] Wineland D J and Itano W M 1979 *Phys. Rev. A* **20** 1521–1540
[11] Maunz P, Puppe T, Schuster I, Syassen N, Pinkse P W H and Rempe G 2004 *Nature* **50** 428
[12] Leibrandt D R, Labaziewicz J, Vuletic V and Chuang I L 2009 *Phys. Rev. Lett.* **103** 103001
[13] Brennecke F, Ritter S, Donner T and Esslinger T 2008 *Science* **322** 235
[14] Murch K W, Moore K L, Gupta S and Stamper-Kurn D M 2008 *Nature Phys.* **4** 561
[15] Marquardt F, Chen J P, Clerk A A and Girvin S M 2007 *Phys. Rev. Lett.* **99** 093902
[16] Wilson-Rae I, Nooshi N, Zwerger W and Kippenberg T J 2007 *Phys. Rev. Lett.* **99** 093901
[17] Thompson J D, Zwickl B M, Jayich A M, Marquardt F, Girvin S and Harris J G E 2008 *Nature* **452** 72–75
[18] O'Connell A D, Hofheinz M, Ansmann M, Bialczak R C, Lenander M, Lucero E, Neeley M, Sank D, Wang H, Weides M, Wenner J, Martinis J M and Cleland A N 2010 *Nature* **464** 697
[19] Teufel J D, Regal C A and Lehnert K W 2008 *New J. Phys.* **10** 095002
[20] Teufel J D, Donner T, Castesllanos-Beltra M A, Harlow J W and Lehnert K W 2009 *Nature Nanotech* **4** 820
[21] Teufel J D and Simmonds R W 2010 *submitted*
[22] Teufel J D, Donner T, Simmonds R W and Lehnert K W 2010 *in preparation*
[23] Castellanos-Beltran M A, Irwin K D, Hilton G C, Vale L R and Lehnert K W 2008 *Nature Phys.* **4** 929
[24] Corbitt T, Chen Y, Mueller-Ebhardt H, Innerhofer E, Ottaway D, Rehbein H, Sigg D, Whitcomb S, Wipf C and Mavalvala N 2007 *Phys. Rev. Lett.* **98** 150802
[25] Anetsberger G, Gavartin E, Arcizet O, Unterreithmeier Q P, Weig E M, Gorodetsky M L, Kotthaus J P and Kippenberg T J 2010 *arXiv:1003.3752v1*
[26] Ekinci K L and Roukes M L 2005 *Rev. Sci. Instrum.* **76** 061101

[27] Hertzberg J B, Rocheleau T, Ndukum T, Savva M, Clerk A A and Schwab K C 2010 *Nature Phys.* **6** 213
[28] Verbridge S S, Parpia J M, Reichenbach R B, Bellan L M and Craighead H G 2006 *J. Appl. Phys.* **99** 124304
[29] Verbridge S S, Shapiro D F, Craighead H G and Parpia J M 2007 *Nano Lett.* **7** 1728
[30] Southworth D R, Barton R A, Verbridge S S, Ilic B, Fefferman A D, Craighead H G and Parpia J M 2009 *Phys. Rev. Lett.* **102** 225503
[31] Unterreithmeier Q P, Faust T and Kotthaus J P 2010 *Phys. Rev. Lett.* **105** 027205
[32] Zwickl B M, Shanks W E, Jayich A M, Yang C, Bleszynski A C, Thompson J D and Harris J G E 2007 *Appl. Phys. Lett.* **92** 103125
[33] Bokaian A 1990 *J. Sound and Vib.* **142** 481
[34] Kimble H J 2008 *Nature* **453** 1023
[35] Tian L and Wang H 2010 *arXiv:1007.1687v1*
[36] Dobrindt J M, Wilson-Rae I and Kippenberg T J 2008 *Phys. Rev. Lett.* **101** 263602
[37] André A, DeMille D, Doyle J M, Lukin M D, Maxwell S E, Rabl P, Schoelkopf R J and Zoller P 2006 *Nature Phys.* **2** 636
[38] Stannigel K, Rabl P, Sorensen A S, Zoller P and Lukin M D 2010 *arXiv:1006.4361v1*
[39] Hofheinz M, and Ansmann H W, Bialczak R C, Lucero E, Neeley M, OConnell A D, Sank D, Wenner J, Martinis J M and Cleland A N 2009 *Nature* **459** 546
[40] Eisaman M D, André A, Massou F, Fleischhauer M, Zibrov A S and Lukin M D 2005 *Nature* **438** 837
[41] Zhang J, Peng K and Braunstein S L 2003 *Phys. Rev. A* **68** 013808
[42] Akram U, Kiesel N, Aspelmeyer M and Milburn G J 2010 *New J. Phys.* **12** 083030

AUTHOR INDEX

AUTHOR INDEX

Institute of Physics
Dirac House, Temple Back
Bristol BS1 6BE UK

CURRAN ASSOCIATES INC
proceedings
.com